MATTER, SPACE AND MOTION

MATTER, SPACE AND MOTION

Theories in Antiquity and Their Sequel

Richard Sorabji

Cornell University Press

Ithaca, New York

ᒪᑕ

First published 1988 by Cornell University Press

Library of Congress Cataloging-in-Publication Data

Sorabji, Richard
  Matter, space, and motion.

  Bibliography: p.
  Includes index.
  1. Physics—Greece—History.   2. Philosophy, Ancient
3. Greece—Antiquities. I. Title.
QC9.G8S67   1988   530'.0938   87–47984
ISBN 0-8014-2194-2

Printed in Great Britain

5-24-90

Contents

To Dick, Cornelia and Tahmina

Introduction

Some few philosophers and others continued to be interested in physical theory after the third century B.C., but for most of the rest of Antiquity, a period of some eight hundred years, there were few new ideas, little increase in knowledge and only sporadic activity ... The Epicureans put their faith in a reworked Atomism ... The Stoics reverted to a vitalistic four-element theory ... Among some neo-Platonists of the fourth and fifth centuries A.D. there flowered a new enthusiasm for physical theory, mixed up with attempts at astrology, alchemy and magic ... Unfortunately, the neo-Platonists, true to their Platonic allegiance, did not believe in the systematic empirical investigation which alone could have made their ideas truly fertile.[1]

This is an excellent statement of a view that has been widespread about the relation of the first two hundred years of Ancient Greek Philosophy to the next nine hundred, and the onus is on those who wish to change the picture. But a re-evaluation is already under way, and indeed is far advanced for the so-called Hellenistic period, which runs from 323 to 30 B.C. and beyond. I have written before about Democritus, the Epicureans and atomist theories of matter in antiquity,[2] so I shall not repeat my suggestions about them. But I will address myself to the Stoics, who belong to the Hellenistic period, and to the Neoplatonists, who flourished up to the late sixth century A.D.

This is in no way to belittle the importance of the early period down to Aristotle's death in 322 B.C. I shall on the contrary discuss the innovative ideas of Alcmaeon, Archytas and the Pythagoreans on space and time, the account of chemical mixture in Anaxagoras, the treatment of space and matter in Plato's *Timaeus* and above all Aristotle and his influence. It has recently been argued that the Stoics had no idea what Aristotle said, even though their school was founded only a generation after his death, and only a thousand metres away from his own school in Athens.[3] But I do not believe that Aristotle's

---

[1] Edward Hussey, 'Matter theory in Ancient Greece', ch. 2 in Rom Harré, ed., *The Physical Sciences Since Antiquity*, London and Sydney 1986, 24, commenting on ancient theories of matter.

[2] Richard Sorabji, *Time, Creation and the Continuum*, London and Ithaca N.Y. 1983, chs 2, 5 and 22-5.

[3] F.H. Sandbach, *Aristotle and the Stoics*, Cambridge Philological Society, supp. vol. 10, 1985.

influence disappeared so completely in the Hellenistic period. I have argued in the past for the close attention paid to Aristotle in the next generation by Diodorus Cronus and Epicurus in the reformulation of atomist theory.[4] Similarly, I shall now argue that the Stoics paid exact, if selective, attention to him in formulating their theories of chemical mixture and of vacuum. Admittedly, Aristotle came finally into his own only when Porphyry made it Neoplatonist orthodoxy in the third century A.D. that Aristotle and Plato were in agreement. The reverberations of that view will become apparent at intervals through the book.

The book is divided into three interconnecting parts. In the first, on matter, I argue that modern theories of matter have ancient analogues not only at the atomic level, but also in connexion with field theory, which is applied in modern physics at the sub-atomic level. According to Einstein, we often do best to think of matter at that level not in terms of particles, but as space endowed with properties. I find the ancient counterpart of that, not only in Plato's treatment of space, which was ambiguous, and swiftly robbed of its impact by being taken metaphorically, but also in late Greek treatments of what is called prime matter. Prime matter came to be seen as a kind of mobile extension (not space) and that extension, endowed with properties, was ordinary matter or 'body'.[5]

Not all Greeks agreed that matter was solid, if that means that stuffs and bodies cannot be in exactly the same place as each other. The Stoic theory of chemical mixture, which allowed such inter-penetration of stuffs, was at least as reasonable, I believe, as Aristotle's, which denied it. It was subsequently exploited by Neoplatonists and Christians for religious and metaphysical reasons, and that exploitation in turn had its effect on literature, as well as on religion and science. But the Stoic theory needs careful examination. For besides their belief that stuffs could co-exist in the same place, they also embraced a theory of categories. This theory had the anti-Aristotelian effect of reducing things, for example, a man's qualities, to body. But this time their intention was not to treat qualities as so many *extra* bodies packed into a man's body; it was rather a reductionist attempt to represent his qualities as one *single* body, his *pneuma* or spirit disposed in various different ways.[6] It does not therefore illustrate their other anti-Aristotelian belief that there can be more than one body in the same place.

Another familiar thesis about bodies, that they are just bundles of qualities, had its origins in various writings of Plato and Aristotle and its culmination in Neoplatonism.[7]

Turning to space and place, we find some of the most important

---

[4] Richard Sorabji, op. cit., chs 2, 22 and 24.
[5] Chapters 1 to 3.
[6] Chapters 5 to 7.
[7] Chapter 4.

developments in the Hellenistic period. For Aristotle's immediate successors questioned his definition of place as something two-dimensional, his denial of its inertness and infinitude and his rejection of motion in a vacuum. This counter-attack enabled later Greeks to take a more Newtonian view of space than their mediaeval successors, who were much more captivated by Aristotle. The battle on the finitude or infinitude of space was ingeniously fought on both sides, but no Greeks attained to the modern idea of closed space, that is, of space which is finite without having a boundary. What we do find in antiquity, I believe, is the corresponding idea of closed time, but only in one text, where its presence seems to have been overlooked.[8]

The revolution in dynamics represented by impetus theory was an invention of late Neoplatonism, and arose out of reflection on Aristotle, in particular, on his account of natural and of forced motion, and his attack on motion in a vacuum. The late Neoplatonists also converted Aristotle's purely dynamical argument for an infinite deity to keep the heavens *moving* for ever into an argument that such a deity is needed to keep them in perpetual *being*. In postulating a divine sustainer, their aim was to harmonise Aristotle with Plato. But, by an irony, their ideas were transmitted through the Islamic world, and finished by making Aristotle that much safer for Christians in the thirteenth century who read their Aristotle through Neoplatonist spectacles.[9]

When these late Neoplatonist arguments are re-examined, the resulting interpretation sometimes has implications for the succeeding Islamic discussions, and in the course of Chapter 15 I shall venture an interpretation or reinterpretation of some of the Islamic texts.

Following a number of suggestions that an index locorum would have been useful in my earlier books, I shall finish with an index locorum covering my last two books, *Necessity, Cause and Blame* and *Time, Creation and the Continuum*, as well as this volume.

Those whose interest is philosophical, rather than historical, will find the meatiest discussions in the defence of the idea of closed time (Chapter 10) of that of bodies in the same place (Chapters 5 and 6) and of various types of field theory (Chapters 1 to 3). They may also find interest in the ancient arguments for and against infinite space (Chapter 8), and in the various attempts to rescue Aristotle's definition of place from collapse (Chapter 11).

I have been very fortunate in the number of places at which I have been able to try out versions of these chapters since 1984. Rather than thank my many hosts, I have tried to acknowledge helpful conversations with individuals chapter by chapter. I should, however, express my gratitude to one or two institutions in particular. One is the Council of the Humanities of Princeton University, at whose invitation I tried the ideas out in a twelve-week seminar, as a guest of the

[8] Chapters 8-12.
[9] Chapters 13-15. Many of the relevant texts are to be translated in the series *The Ancient Commentators on Aristotle*, ed. Richard Sorabji, London and Ithaca N.Y., of which the first volume was published in 1987.

Princeton Classics Department and in particular of David Furley, in 1985. Another is Bristol University, which invited me to be its first Read-Tuckwell lecturer, and organised a seminar on later theories of space to fit with my own seminar on ancient theories, throughout the Spring term of 1986. My two Read-Tuckwell lectures are here presented as Chapter 10, and the Read-Tuckwell Foundation has helped to meet the costs of typing and indexing. So also has the British Academy and I wish to thank them both. I am grateful to Eric Lewis, Harry Ide and John Ellis for the thoughtful way in which they have planned and executed the index locorum. The typing was meticulously done, as ever, by Dee Woods.

Some of the ideas in this book have appeared or are appearing more or less simultaneously in other publications, but often in a different form, as follows:

'Analyses of matter, ancient and modern', Presidential Address, *Proceedings of the Aristotelian Society* 86, 1985-6, 1-22.

'Theophrastus' doubts on place and natural place', in W.W. Fortenbaugh and R.W. Sharples, eds, *Theophrastus as Natural Scientist*, Rutgers Studies in Classical Humanities 3, 1988.

'Proclus and his predecessors on place and the interpenetration of bodies', in J. Pépin and H.D. Saffrey, eds, *Proclus – lecteur et interprète des Anciens*, Actes du Colloque Proclus, C.N.R.S. Paris 1987.

'Simplicius: prime matter as extension', in I. Hadot, ed., *Simplicius – sa vie, son oeuvre, sa survie, Peripatoi* vol. 15, Berlin 1987.

'Closed space and closed time', *Oxford Studies in Ancient Philosophy* 4, 1986, 215-31.

'John Philoponus', in Richard Sorabji, ed., *Philoponus and the Rejection of Aristotelian Science*, London and Ithaca N.Y. 1987.

'The Greek origins of the idea of chemical combination: can two bodies be in the same place?', in John Cleary, ed., *Proceedings of the Boston Area Colloquium in Ancient Philosophy* 4, 1987-8.

'Infinite power impressed: the transformation of Aristotle's *Physics* and *Theology*', in Sarah Hutton and John Henry, eds, *Science, Education and Philosophy: Studies in Intellectual History in Honour of Charles Schmitt*, forthcoming.

Part I
Matter

CHAPTER ONE

Body as extension endowed with properties: Simplicius on Aristotle

I want to draw attention to a recurrent theme in the analysis of matter, according to which matter is *extension endowed with properties*. Or since the word 'matter' has a special use in connexion with Ancient Greek thought, it may be less confusing to say that *body* has repeatedly been analysed as extension endowed with properties. Under 'body' I include billiard balls, the bodies of persons, stuffs like bronze, what microphysicists investigate and the physical world in general, in other words, matter in the modern, as opposed to the Ancient Greek, sense.

I shall start with a famous text in Aristotle's *Metaphysics*, book 7, ch. 3, where he is discussing matter in the ancient sense, a sense which he himself introduced, employing the word *hulê*, which originally meant timber. By matter here he does not mean body. He is indeed discussing bodies (*sômata*, 1029a13), but matter is the *subject* of properties in a body. The table in front of me may be made of wood. From one point of view, the wood might be thought of as a subject which carries the properties of the table, its rectilinearity, its hardness, its brownness. But according to one persuasive interpretation, Aristotle is looking for the most fundamental subject of properties in a body. He calls it the first subject (*hupokeimenon prôton*, 1029a1-2). The wood of the table is made up of the four elements earth, air, fire and water, and these might be thought of as a more fundamental subject carrying the properties of the wood. But the most fundamental subject would be one which carried the properties of the four elements: hot, cold, fluid and dry. This first subject is sometimes called by Aristotle the last subject, and is known by commentators as first or prime matter. Aristotle finishes by rejecting the idea that matter deserves the honorific title of substance (*ousia*, 1029a27), but this in no way qualifies his conception of it as subject (*hupokeimenon*).

The idea of a fundamental subject has proved persistent but embarrassing. John Locke expressed the embarrassment when, discussing what he called material substance, he described it as 'something, I know not what'. For whereas wood, or bronze, which is what Aristotle starts with in 1029a4, is something perfectly familiar, and even the elements earth, air, fire and water seem familiar enough,

3

it is hard to imagine what the more ultimate subject would be like. Locke's material substance plays more roles than Aristotle's first or prime matter,[1] but it is a descendant, and I assume it plays the role of ultimate subject among others. Locke describes the embarrassment with the following analogy:

> If anyone should be asked, what is the subject wherein colour and weight inheres, he would have nothing to say but, The solid extended parts; and if he were demanded, What is it that solidity and extension adhere in, he would not be in a much better case than the Indian before mentioned who, saying that the world was supported by a great elephant, was asked what the elephant rested on; to which his answer was, A great tortoise; but being again pressed to know what gave support to the broadbacked tortoise, replied, something, he knew not what.[2]

On visiting Ripon Cathedral in Yorkshire some years ago, I was pleased to notice a carving in the fifteenth-century choir stalls which

Unidentified wood carving in fifteenth-century choir stalls, Ripon Cathedral, Yorkshire, drawn by Kate Sorabji.

---

[1] It unites the properties of a body. In more recent thinkers, the ultimate subject or substratum bestows particularity. For these and other roles, see Martha Bolton, 'Substance, substrata and names of substances in Locke's Essay', *Philosophical Review* 85, 1976, 488-513.
[2] John Locke, *An Essay Concerning Human Understanding*, 1690, 2.23.2.

the writers of the guide book could not identify. It appeared to represent the people of the world in a howdah on the back of a great elephant, who in turn rested on something which might have been a tortoise. I mention this not as a serious contribution to the History of Art, but because the picture may provide a mnemonic for some of the other points I want to make.

One thing to notice is that there are several layers of properties superimposed on the ultimate subject, and this is something that has puzzled Locke scholars. Why, they have asked, does he treat extension and solidity as a more fundamental layer than colour and weight? Jonathan Bennett has suggested that extension and solidity have been imported in confusion from the different topic of primary qualities, while John Mackie has expressed agnosticism.[3] In fact, the distinction of the layers of properties seems to be in the tradition of the original passage in Aristotle's *Metaphysics*, where extension, or length, breadth and depth, represent the most fundamental layer of properties to be imposed on the first subject.

Aristotle describes how to get at the idea of the first subject by means of a thought experiment. One is to take a particular body, say, this table, and in one's thoughts strip away its properties in layers. One strips away first the more superficial properties, then length, breadth and depth, in order to find the first subject beneath. (The layered properties do not correspond to those of my earlier example involving the four elements.) Aristotle would insist that the separation here is a separation only in thought. There is no suggestion that the first subject could ever *exist* without having properties. The idea is only that one can *think of* the first subject, without thinking of the properties that it undoubtedly has. The passage runs as follows:

> The subject (*hupokeimenon*) is that of which everything else is predicated while it itself is not predicated of anything else. And that is why we must first determine its character, for the first subject (*to hupokeimenon prôton*) is most of all considered to be substance ... [omitting 1029a2-7].
>
> It has now been stated in outline what substance is: it is not predicated of a subject, but everything else is predicated of it. But we must not merely put it like that, for that is not enough. The statement itself is unclear, and further <sc. on this view> matter (*hulê*) becomes substance. For if this is not substance, it escapes us what else is.
>
> For when everything else is stripped off, evidently nothing remains. For while the rest are active or passive processes or capabilities of bodies (*sômata*), length, breadth and depth are quantities. They are not substances, for quantity is not substance; rather substance is that to which first of all these belong. But when length, breadth and depth are taken away, we see nothing left, unless there is something made definite (*horizomenon*) by these. So to those who look at it in this way, matter alone must seem to be substance.

[3] Jonathan Bennett, *Locke, Berkeley, Hume: central themes*, Oxford 1971, ch. 3; J.L. Mackie, *Problems from Locke*, Oxford 1976, ch.3.

By matter I mean that which is not *in itself* said to be a given anything, nor of a given quantity, nor characterised by any of the other categories which define being. For there is something of which each of these is predicated, and its being is different from that of the predicates. For the rest are predicated of substance, and substance of matter, so that the last thing is *in itself* neither a given anything, nor of a given quantity, nor anything else. Nor is it the negations of these, for negations too will belong to it as accidental attributes.

So for those who think of things from this point of view, it turns out to be matter that is substance (*ousia*). But this is impossible, for separability and being a 'this' are thought to be special characteristics of substance.[4]

What can it be that is left, when in imagination even length, breadth and depth are stripped away? At 1029a17-18, Aristotle says:

We see nothing left, unless there is something made definite (*horizomenon*) by these.

On one interpretation, these words admit that we see *nothing* left, whether this is Aristotle's own view, or a view he reports, in order to correct it. But such an interpretation is hard put to it to make sense of the immediately following words, 'unless there is something made definite by these' (i.e. presumably, by length, breadth and depth). Moreover, once it was decided that there was nothing left, it would be strange to go on, as the argument does, and conclude that this nothing deserved the dignified title of substance (1029a19). The following words, 'unless there is something made definite by these', must then be taken seriously, and not extruded from the text as an inept gloss.[5] But what can they mean? Some commentators, concerned to protect Aristotle from postulating an utterly mysterious first subject, have sought in these words a reference to a more familiar kind of subject, such as bronze (mentioned earlier at 1029a4), or a statue. But how would bronze or a statue fall under the description, 'nothing left, unless there is something *horizomenon* by length, breadth and depth'? And why should Aristotle strip away all other properties, if his aim is to direct attention to bronze or a statue?

In denying that Aristotle is referring to bronze or a statue, I am not relying on an argument which is sometimes used,[6] that if he has asked us to remove the statue's properties from our minds, he cannot still expect us to be thinking of the statue. Much less would I rely on the idea put by Socrates to Theaetetus in Plato's *Theaetetus* that we would need to have the statue's unique distinguishing properties in mind, in

[4] Aristotle *Metaph*. 7.3, 1028b36-1029a28.

[5] One suggestion in an article containing otherwise much of value: Malcolm Schofield, 'Metaph. Z.3: Some suggestions', *Phronesis* 17, 1972, 97-101, at 97.

[6] If I am not mistaken, this is H.M. Robinson's concern in 'Prime matter in Aristotle', *Phronesis* 19, 1974, 187.

order to be thinking of it.[7] Such arguments would be mistaken, for the questions what we are thinking of and what descriptions of it we have in mind may get very divergent answers. How divergent may depend on which of several theories of reference recently canvassed is closest to the truth.[8] On the simplest causal theory, we would still be thinking of the statue, so long as the statue was what *caused* us in the appropriate way to be now entertaining certain descriptions, whether true of the statue or not. But whatever our theory, it should allow for the fact that when we think of the statue, the description of it we have in mind might be as uninformative as 'the subject (whatever it is) of those properties I was thinking of a little while ago (but have now forgotten)'. And it should allow that we might even be incorrect about that description (was it a little while ago?). So much by way of disclaimer. But the disclaimer does not make it any more likely that a statue is what Aristotle is referring to with the words, 'unless there is something made definite by these'.

Our perplexity about what is meant would be resolved, if only we could believe an interpretation of Aristotle which was offered by the Neoplatonist commentator Simplicius in the sixth century A.D., but which has been virtually ignored.[9] Simplicius' commentary on the *Metaphysics* is lost, but elsewhere he suggests that what Aristotle means by his first subject or prime matter is *extension* (*diastêma*).[10] A first advantage of this suggestion is that it would make perfect sense of the description that has proved baffling, 'nothing left, unless there is something made definite by length, depth and breadth'. The expression should be taken together as a unity: nothing *except* what is made definite by these. The length, depth and breadth in question will be *definite* lengths, *definite* depths and *definite* breadths, such as six feet by four feet by two feet. What is left in our thoughts is the *extension* of the table, but with its particular feet and inches ignored. It now becomes entirely appropriate to say that that extension gets its definiteness from (is *horizomenon* by) the particular lengths, breadths and depths which we are ignoring. Precisely such a process of ignoring

---

[7] Plato *Theaetetus* 209B-D. Aristotle himself may believe that in order to think of an *incomposite* entity, we have to have its definition or form in mind (it will have no matter), judging from his view that if we come up with the wrong definition we have not made an error about it, but have rather failed to make contact with it in our thoughts at all. See Richard Sorabji, *Time, Creation and the Continuum*, London and Ithaca N.Y. 1983, ch. 10.

[8] For a recent major treatment of 'thinking of', see Gareth Evans, *The Varieties of Reference*, ed. J. McDowell, Oxford 1982. For causal theories of reference, see Saul Kripke, *Naming and Necessity* (1st ed. 1972), 2nd ed., Oxford 1980; Keith Donnellan, 'Reference and definite descriptions', *Philosophical Review* 75, 1966, 281-304.

[9] An honourable exception is N. Tsouyopoulos, 'Die Entstehung physikalischer Terminologie aus der neoplatonischen Metaphysik', *Archiv für Begriffsgeschichte* 13, 1969, 7-33, esp. 7-20, although I shall be disagreeing with her interpretation. Another briefer mention is found in H.A. Wolfson, *Crescas' Critique of Aristotle*, 1929, 582, which will also be discussed below.

[10] Simplicius *in Phys.* 229,6; 230,19-20; 26-7; 31; 232,21-2; 232,24; 537,13; 623,18-19.

particular inches, while attending to extension, is described by Aristotle in a different context in the *de Memoria*.[11] Simplicius' description of the extension as *indefinite* would mislead, if it suggested that the extension could ever *exist* without having definite measurements superimposed on it. What is true is only that we can think of it, without thinking of its definite measurements.

A second advantage of the suggestion from our point of view, is that it entirely avoids the embarrassment of Locke's something, I know not what. For extension is something perfectly familiar. This, however, is far from being Simplicius' reason for proposing it. On the contrary, he thinks that prime matter ought to be unfamiliar, for it ought to fit the description given by Plato of his 'Receptacle' in the *Timaeus*, that it 'can be apprehended [only] by bastard reasoning' (*nothôi logismôi lêptê*),[12] not by sense perception. In addition, Simplicius thinks of this extension as having a diffuseness that puts it at the opposite extreme from the unity of his supreme God, the One. And so it is close to nothingness, and slides away from being (*apopheugein apo tou ontos*).[13]

Simplicius would not, then, agree about the second advantage, but a third advantage is that the idea of extension as the ultimate subject of properties in a body is, I believe, perfectly viable. It is the idea that a body is some kind of extension endowed with properties. This is not an idea I happen to share at the level of everyday objects: if asked for the subject of properties in a table, I would rather say, 'the table'. But it will be seen in Chapter 3 that a version of the idea has been widely accepted in modern physics at the sub-atomic level, where physical matter is indeed described as an extension endowed with properties, even if the type of extension is a different one.

I should say what kind of extension is relevant to Aristotle's text. It is in a certain sense the *volume* (*onkos*) of the table, but the word 'volume' is ambiguous. I do not mean the cubic size of the statue in exact units of measurement, nor yet its cubic size treated as a determinable with units of measurement left unspecified. In either of these senses, the volume of the table could be the *same* as the volume of the sofa. But I am thinking of the volume of the table in the sense of a certain three-dimensional expanse which moves with the table when it moves, and which is distinct, whatever its measurements, from the volume of the sofa. The volume in this sense *has* size, and it has a particular size, but it is wrong to say that it *is* size. A startling feature of Simplicius' interpretation is that we might have thought of the table's volume as a *property* of it. That would certainly be true of its volume in the (irrelevant) sense of its size. But here we are being invited instead to think of its volume as the *subject* of properties.

The volume is a subject in the sense that we could say with complete metaphysical propriety that the three-dimensional extension is hard,

---

[11] Aristotle *Mem.* 1, 450a1-17.

[12] Simplicius *in Phys.* 229,3, referring to Plato *Timaeus* 52B.

[13] Simplicius *in Phys.* 230,29.

brown and moveable. Although properties like 'is a Chippendale' cannot be applied with equal propriety to the extension, some ancient commentators had a useful device of distinguishing between first and second subjects.[14] In the present example, the extension would be the first subject, and if the extension was a table, the table might be a second subject, suited to having such properties as being a Chippendale. Alternatively, wood might be the second subject, and in that case the wood could be said to be (be the matter of) a table. Of course, the 'is' of 'is hard', 'is (an example of) a Chippendale' and 'is (the matter of) a table' are not the same 'is'. And that means that the various subjects are subjects in correspondingly different ways. But at least all the ways are familar ones.

Simplicius does not say whether the extension of the table is a particular or a universal. His emphasis on its diffuseness and lack of unity allows no hint of particularity. But we shall need to impose the view that the extension is a particular, if it is to play the role of ultimate subject, rather than predicate. On the other hand, we should not follow those modern writers who have reintroduced the idea of an ultimate subject and substratum, in the belief that it would provide a *source* of particularity.[15] On the contrary, the table's extension would have to *derive* its particularity from that of the table. That it should be a particular does not clash with Aristotle's denial at 1029a28 that matter is a *tode ti* (this something), for he does not deny that matter can be a *kath' hekaston* (his expression for a particular).

Because Simplicius' *Metaphysics* commentary is lost, I have only been guessing what he may have said about *Metaphysics* 7.3. It is in the *Physics* commentary that he offers his account of prime matter, and he does so on the basis of Aristotle *Phys.* 4.2, 209b6-11.[16] Aristotle is there describing a tendency to equate matter and *place*, in the mistaken belief that place is extension. The equation presupposes that matter too is extension. Simplicius takes Aristotle to agree, although in fact all Aristotle explicitly commits himself to is that matter is *like* extension, at least in the respect that it needs to be made definite (*hôrismenon*) by properties. To make clear that extension (*diastêma*) needs to be made definite, Aristotle distinguishes it from something else which is already in itself definite, namely, a magnitude (*megethos*), by which he means either a definite quantity, or a thing considered as having a definite quantity:

> In so far as place is thought to be the extension (*diastêma*) of a magnitude (*megethos*), it is matter. For the extension is distinct from the

[14] In Philoponus *in Cat.* 83,13-19; *in Phys.* 578,32-579,18; and *Summikta Theôrêmata* judging from *in Phys.* 156,16, the first subject is prime matter conventionally conceived (i.e. not as extension) and the second subject is prime matter endowed with three-dimensional extension. In Porphyry ap. Simplicium *in Cat.* 48,11-33, the second subject comes at a higher level corresponding to wood or table.

[15] e.g. Edwin B. Allaire, 'Bare particulars', *Philosophical Studies* 16, 1963, reprinted in Michael J. Loux, ed., *Universals and Particulars*, Notre Dame 1976.

[16] Simplicius *in Phys.* 229,5-7; 232,24; 537,9-538,14.

magnitude; it is embraced and made definite (*hôrismenon*) by form, for example by a surface (*epipedon*) and contour (*peras*). And that is what matter and indefinite things are like. For when the contour and passive attributes (*pathê*) of a sphere are stripped off, nothing is left except (*para*) matter.

Simplicius rejects Alexander's view that extension implies form rather than matter and that consequently Aristotle's remarks are inexact, and designed only to make a case for an alien point of view. Alexander's mistake lies in not following through his own remarks distinguishing *extension* from definite *magnitude*, a distinction adequately recognised by other commentators.[17] Extension is connected rather with indefinite diffusion (*aoristos khusis, ekkhusis*) and slackening (reading *paresis*, 538,13). And that fits matter, which is the source of stretching, diffusion and indefiniteness (*diaspasmos, ekkhusis, aoristia*).[18]

The elusiveness of prime matter, as others had conceived it, is brought out by the fact that magnitude had traditionally been taken as a property, not a subject, and had therefore been excluded from the concept of prime matter, which was accordingly described as without magnitude (*amegethes*).[19] On the conventional view, prime matter would also be thought of as unextended. But Simplicius exploits his distinction between magnitude and extension to prise these claims apart: prime matter can lack definite magnitude (*megethos*), and yet be an indefinite extension (*diastêma*, or, as he prefers to say, *diastasis*); it need not be some more elusive subject *underlying* extension. Simplicius' arch-enemy Philoponus, by contrast, takes the conventional view of what Aristotle means. Accordingly, when the quoted passage of Aristotle (*Phys.* 4.2, 209b6-11) associates matter with extension, he has to construe it as concerned not with *prime* matter, but with envolumed matter (*onkôtheisa hulê*).[20] Envolumed matter for Philoponus is not prime matter, but a compound of prime matter with form that gives it volume.[21]

The special character of Simplicius' interpretation has been missed by the few modern commentators who have discussed it. According to one account,[22] Simplicius takes the conventional view of prime matter as a subject that is to be contrasted with extension as a property.

---

[17] Extension here is indefinite (*aoriston*): Themistius *in Phys.* 106,12-13; Philoponus *in Phys.* 515,15-16; 18-19; 520,2-3; 6; 8-9. It is different from magnitude: Themistius *in Phys.* 106,11-12; Philoponus *in Phys.* 520,8-9.

[18] Simplicius *in Phys.* 537,22-538,14.

[19] *amegethes*: Plotinus 2.4.8(11); 2.4.11(4); Philoponus *contra Proclum* 430,16; 25; 431,28; 432,25; 436,17; 22; 24; 439,14.

[20] Philoponus *in Phys.* 515,19.

[21] ibid. 687,31-4.

[22] H.A. Wolfson, *Crescas' Critique of Aristotle*, Cambridge Mass. 1929, 582, interpreting Simplicius *in Phys.* 230,23-33 (translated at the end of this chapter), followed by E. Grant, *Much Ado About Nothing*, Cambridge 1981, p. 272, n.40.

According to a variant of this interpretation,[23] although prime matter is said to *be* dimension,[24] this is to be understood as meaning only that dimension is the first form to be imposed on prime matter.[25] At the same time, a startling interpretation has been offered of what is meant by dimension. It has been taken to be the *weight*, which remains constant when water changes into air. I shall explain the basis of this interpretation in Chapter 3.[26]

The passages we have looked at, *Metaphysics* 7.3 and *Physics* 4.2, are two of the Aristotelian passages in which the idea of prime matter has been detected by some – not all – commentators. The work where it is most commonly, though still not unanimously, found is *On Generation and Corruption*. There are no cross-references among the three contexts, and the context in *GC* is a quite different one. Aristotle thinks that his four elements, earth, air, fire and water, can change into each other. Water changes into air when you boil a kettle. According to his main account of change, something persists (*hupomenei*)[27] through any change, and is the subject (this is the connecting link) first of one set of properties and then of another. This is the role of matter, and there are strong indications that some matter is supposed to persist even when one of the four elements changes into another.[28] The difficulty in the case of the four elements is that it is hard to see what this matter is. When a sword is beaten into a ploughshare, the persisting metal is steel. But just because the elements are elementary (they are not, for example, made up of atoms, in Aristotle's view), it is not obvious what more fundamental thing there is that could be first water and then air. The persisting matter in this case sounds once again like a shadowy entity, or non-entity. And some have taken advantage of certain disclaimers in Aristotle to deny that he believes in prime matter at all.[29]

---

[23] N. Tsouyopoulos, op. cit., followed by Herbert Hunger, *Die hochspruchliche profane Literatur der Byzantiner*, vol. 1 (=*Byzantinisches Handbuch*, part 5, vol. 1), Munich 1978, 30. Wolfson takes much the same view: the first form imposed on prime matter yields a secondary matter, and it is this secondary matter, on Wolfson's view, that is extended and directly underlies the four elements. He is deterred from viewing prime matter itself as an extension by his taking the three dimensions which it is said to lack (230,14; 24-5) as indefinite dimensions, rather than as exact measurements.

[24] N. Tsouyopoulos, op cit., p. 15.

[25] ibid. pp. 16, 17, 19, 20.

[26] ibid. p. 19, on the basis of Simplicius *in Phys.* 232, 21-3 (translated at the end of this chapter).

[27] Aristotle *Phys.* 1.7, 190a11; 1.9, 192a13; cf. 192a25-34; *Metaph.* 12.2, 1069b7-9.

[28] Two important passages are Aristotle *GC* 1.3, 319a29-b4; 2.1, 329a24-35.

[29] There are denials in Hugh R. King, 'Aristotle without prima materia', *Journal of the History of Ideas* 17, 1956, 370-89; Willie Charlton, *Aristotle's Physics, Books I and II*, Oxford 1970, 129-45, and 'Prime matter: a rejoinder', *Phronesis* 28, 1983, 197-211. The orthodox view is defended by F. Solmsen, 'Aristotle and prime matter', *Journal of the History of Ideas* 19, 1958, 243-52; Alan Lacey, 'The Eleatics and Aristotle on some problems of change', *Journal of the History of Ideas* 26, 1965, 451-68; H.M. Robinson, 'Prime matter in Aristotle', *Phronesis* 19, 1974, 168-88; R.M. Dancy, 'Aristotle's second thoughts on substance', *Philosophical Review* 87, 1978, 372-413; C.J.F. Williams, *Aristotle's De Generatione et Corruptione*, Oxford 1982, 211-19. Charlton's view that fluid, dry, etc., take over the role of matter is certainly true for the (non-prime) matter of

Although we nowadays believe that water and steam are made up of atoms, we should face the same problem as Aristotle, if we believed, for example, that electrons and positrons were elementary, that they could change into each other, and that this change required a persisting subject that was first an electron and then a positron.

I know of no explicit hint in this new context of elemental change that extension might be able to play the role of prime matter. Even in the other context in *Metaph*. 7.3, I claimed no more than that extension would perfectly fit an otherwise baffling expression, that it would free us from Locke's embarrassment and that it would provide a conceptually satisfactory view. I had better say now that I do not think that Aristotle reached the point of consciously thinking that extension would play the roles required by these passages. For one thing, he would have said so. For another, he says that prime matter is imperceptible,[30] and I am not clear that that is true of indefinite extension. But even if he did not consciously recognise the claims of extension, I believe that it is profitable to consider how well extension would, or would not, fulfil the various roles of prime matter.

The two roles of prime matter so far discussed are related. One is to be the fundamental subject of properties in a body; the other is to be the persisting subject now of one set of properties and now of another in elemental change, and, I believe, extension could play these two roles very well. The same indefinite extension, altered in its definite measurements and in distribution, might possess now the properties of water and now the properties of air. Prime matter is in addition thought of as a potentiality, and extension too can be regarded that way, because it has a capacity for receiving a very wide range of properties. Prime matter is further viewed as something simple and not further analysable into form and matter, and once again extension might well be thought of that way too. At the same time, we have seen that prime matter is not expected by Aristotle to bestow particularity on a body, nor to unite its properties together, so the inadequacy of extension to these tasks does not disqualify it.

It may be wondered why Aristotle should not instead take the view I mentioned before, that the subject of properties in a table is the table itself. There is also the venerable view, which goes back at least as far as ancient interpretations of Plato's *Theaetetus*, and which will be examined in Chapters 3 and 4, that properties need no subject: that a table is just a bundle (*hathroisma, sundromê*) of properties. (They are then sometimes given another name, for example, *features*). But Aristotle does believe that properties need a subject, and a table would not meet his other requirements for prime matter: it is not something simple and unanalysable, it is not something merely potential, and it will not always persist through change, not, for example, if we burn the

---

compounds in *Meteor*. 4 and some biological works, but it is harder to see if it is true of the prime matter of *elements*.

[30] Aristotle *GC* 2.5, 332a35.

table to ashes. Extension, by contrast, looks as if it would meet his requirements rather well, but it is time to consider some difficulties.

(i) The first and greatest difficulty concerns persistence through change. When I boil a kettle and convert water into steam, or, as Aristotle would see it, into a different element, air, we can still give sense to the idea that the same extension is there, although its particular measurements in feet or inches will have increased, and its other properties will have changed. (The idea of same extension, like the idea of a particular extension, is a derivative one: extension owes its sameness to something else.) Even when a cloud of steam is converted into a multiplicity of water droplets, we can give sense to the idea that the same extension is there, but redistributed. We can do this through any one change, or at least we can, if we ignore the possibility, of which Aristotle was unaware, of converting matter without remainder into energy. But I doubt if we can make sense of the idea that it is the same particular extension that persists through a whole series of such radical changes. This means that Aristotle would need to *modify* his theory of prime matter, if extension was to play the role, so as to allow for the fact that particular extensions are liable to stop existing eventually. The modification ought not to be too damaging to his analysis of change, for it would remain true that an extension could be found persisting as subject through any *one* change, even though different extensions might be invoked for different changes in a long series. Moreover, the modification would have a parallel in what Aristotle seems repeatedly to allow, without even articulating the point very fully, about *form*, that particular forms are liable to start and stop existing.[31] If there is any threat at all to Aristotle's philosophy in such admissions, it is the threat which was already derived by Philoponus from the admission about *form*. If Aristotle allows the creation *ex nihilo* of forms, can he not be forced to allow the subsequent Christian belief in the creation *ex nihilo* of the *world*?[32] But, as Philoponus recognises, this would involve a much more radical departure, for every other case of something's beginning to exist, e.g. the case already mentioned of water droplets forming, is analysed by Aristotle as involving the persistence of some matter both before and after.

(ii) A second obstacle to treating prime matter as extension can be overcome with less difficulty. It arises from the suspicion that prime matter and extension do not belong in the same category. To speak very roughly, Aristotle's first category, substance, contains physical objects, and his incorporeal God. The other categories, quantity, quality, relation and so on, contain the attributes, or, as he would say,

---

[31] Aristotle holds that what starts existing must also stop, and he describes various forms as liable to do one or the other at *Metaph.* 7.9, 1034b16-19; 7.15, 1039b20-7; 8.3, 1043b14-23; 8.5, 1044b21-4; 11.2, 1060a21-3; 1060b23-8; 12.3, 1070a13-26; *Phys.* 1.9, 192b1-2; 7.3, 246a16; b14-16; *DA* 2.5, 417b3; *GA* 2.1, 731b31-5; *GC* 1.10, 328a27-8; 2.11, 338b14-17; *Cael.* 3.7, 306a9-11.

[32] Philoponus *contra Proclum* 340,12-26; 365,3; cf. Philoponus ap. Simplicium *in Phys.* 1141,5-30.

the accidents, of bodies, and are sometimes called the accidental categories. Aristotle does not say to what categories prime matter and extension belong. But it would be natural to connect extension with the category of quantity, whereas the matter of a body might be classed either, like the body itself, in the category of substance, or, as was sometimes suggested after Aristotle,[33] outside the scheme of categories altogether. However, a solution can be found, provided it is remembered that extension is not the same thing as size. That makes it less obvious that extension must be assigned to the category of quantity. And indeed Simplicius appears to assign material extension (that is, matter) to the category of *substance*. For he contrasts material extension or, as he says, material dimension (*hulikê diastasis*) with dimension plain and simple (*diastasis haplôs*). He admits that the latter is a mere accident (*sumbebêkos*) of substance, and therefore belongs in the accidental categories. But material dimension is (like space) not an accident, but a substance (*ousia*), and so belongs to the category of substance.

> Different from these [extensions (*diastêmata*)] again is material dimension (*hulikê diastasis*), which is conceived by reference to stretching and indefiniteness. But neither (*oude* [i.e. no more than material dimension]) is place (*topos*) an accident (*sumbebêkos*); it too (*kai* [i.e. like material dimension]) is substance (*ousia*), for it is not dimension plain and simple (*diastasis haplôs*), but extended space (*diastôsa khôra*).[34]

Simplicius' contemporary Philoponus also lifts the extension of bodies out of the category of quantity and places it in the category of substance. But that will be discussed in Chapter 2. His ground is that, if three-dimensional extension *defines* body, it should be assigned to the category of substance along with bodies.[35]

(iii) A further three obstacles to viewing prime matter as extension concern questions of differentiation. If the prime matter of the four elements is extension how will the matter of the celestial fifth element differ, as on one interpretation Aristotle thinks it does?[36] Alexander of Aphrodisias raised this sort of question in the form of a dilemma.[37] According to the first horn, if celestial bodies have the *same* prime matter, all bodies will be equally perishable. (Simplicius, discussing the same issue in a different context, substitutes 'equally imperish-

---

[33] Doubts are already raised by Aristotle's successor Theophrastus as to whether matter is more than a mere possibility of being, *Metaphysics* VI, 17. For its lying outside the categories, see ps-Alexander *in Metaph*. 464,21-6; Asclepius *in Metaph*. 380,31-5.

[34] Simplicius *in Phys*. 623,18-20.

[35] Philoponus *contra Proclum* 405,24; 26; 423,14-424,11; 424,24; 425,5-6.

[36] See with caution Aristotle *Metaph*. 8.1, 1042b5-6; 8.4, 1044b7; 9.8, 1050b21-2; 12.2, 1069b24-6.

[37] Alexander, *Quaestiones* I 15, 26,30-27,4.

able'.)[38] Philoponus, the Christian, accepts the logic of Alexander's reasoning with pleasure. By giving all bodies the same prime matter (three-dimensional extension), he can ensure that they are all equally perishable, in accordance with his Christian views,[39] and Descartes was to agree that the matter of celestial and terrestrial bodies is the same,[40] but Simplicius could not accept the alleged consequences. The other horn of Alexander's dilemma that if the matter of the fifth element is differentiated, it will no longer be prime matter, for that is *simple*, but will be a *compound* involving differentia and qualities, to provide the differentiation.

The easiest response, I think, is to reject the first horn of the dilemma, and indeed it is not clear that Aristotle himself would accept it. For when he says that celestial bodies have a different matter, he need not mean that their *prime* matter is different, but only that their *elemental* matter is different, in other words, that the fifth element is different from earth, air, fire and water. The difference is that the fifth element is capable only of motion, whereas the other four undergo other kinds of change, including creation and destruction. On that view, what would rob the celestial bodies of their imperishability would be having the same *elemental* matter. But they can and do have the same *prime* matter with impunity, despite the views cited from the ancient commentators.

Simplicius, however, does not think he can allow the celestial bodies to have the same prime matter. To find a difference in the matter, he looks to the Neoplatonist view that everything owes its existence, whether directly or through intermediaries, to the supreme God, the One. The second horn of the dilemma, which Philoponus had revived, assumes that the prime matter of the celestial bodies could differ only by incorporating a differentia, which would turn it into a compound. Simplicius responds that things can differ instead by being produced more or less directly by the One. Or, as he puts it, they can differ in subordination (*kath' huphesin*, as opposed to *sunairesin*), and in their relation to the One from which they proceed (*aph' henos*).[41] But it is hard to see how this answer can fend off the conclusion that the celestial bodies do have exactly the same type of prime matter, namely extension, even if the *particular portion* of prime matter involved (however that is differentiated) has a slightly different causal background. It is probably easier, therefore, to attack the first horn of the dilemma, in the way indicated, than to join Simplicius in attacking the second.

(iv) The other questions of differentiation also have an answer. One is the question how prime matter will differ from *mathematical* extension, or mathematical body, as it was sometimes known, that is, from the

---

[38] Simplicius *in Phys*. 232,8-13

[39] Philoponus ap. Simplicium *in Phys*. 1331,20-5; ap. Simplicium *in Cael*. 89,22-5; 134,16-19.

[40] Descartes *Principles* 2,22.

[41] Simplicius *in Cael*. 135,21-6; cf. *in Phys*. 1133,9ff and Proclus *in Tim*. 2.8.23 (Diehl).

extension of the squares and cubes considered by geometry. An analogous problem is put by Plotinus to the Stoics.[42] The Stoics say that matter is body, but (a) if it is merely three-dimensional, it will be merely *mathematical* body, (b) if it is three-dimensional with resistance (*antitupia*) it will not be simple, but will be a compound of matter with quality.

Alexander and Themistius, followed by the Neoplatonists Syrianus, Proclus, Ammonius and Simplicius, and the Christian Neoplatonist Philoponus, would have an easy answer, because they insist that the place of mathematical entities is in the thought and imagination of mathematicians.[43] Mathematical extension should therefore be easy to distinguish from prime matter. But we can hardly agree with the interpretation according to which Aristotle himself thinks of mathematical extension in this way. He has a doctrine of abstraction, as Simplicius recognises,[44] but whatever that means, Simplicius is wrong to think that it comes to the same as his own view.

Personally, I think that what Aristotle's geometer studies may lie both in the inner (mental) world and in the external world, if I may put it in a way that is not Aristotle's own. It seems to lie in the external world, because Aristotle holds that what the geometer studies is the volume, surfaces and edges of physical bodies, only with the non-geometrical characteristics of the bodies ignored.[45] But there is a passage in which Aristotle describes geometrical thinking as being able to ignore in this way not only the non-geometrical features of a diagram drawn in the external world, but also and equally the non-geometrical features of an inner mental image (*phantasma*),[46] and he says that all thinking involves such an image.[47] There may in addition be a sense in which geometrical features are mind-dependent, even when they belong to the external world. For Aristotle appears to describe how a mere act of thought can bring into actuality in an external diagram a line that has not actually been drawn there.[48] This suggests the view that by attending to some features of an object, while ignoring others, the geometer can bring into actuality there

---

[42] Plotinus 6.1.26(17-23).

[43] Alexander *in Metaph.* 52,13-19; *in Sens.* 111,17-19; ap. Philoponum *in GC* 77,15; ap. Simplicium *in Phys* 526, 16-31; Themistius *in Phys.* 97,17-98,29; 103,18-24; Syrianus *in Metaph.* 12,28-13,3; 84,27-85,15; 91,29-34; 95,29ff; Proclus *in Eucl.* 54,8-12; Ammonius *in Isag.* 11,31-12,6; Philoponus *in Phys.* 500,22; 503,17; *in DA* 58,6-13. Simplicius *in DA* 233,7-17; 277,6-278,6; *in Phys.* 512,19-26; 621,33; 623,15-16. See Ian Mueller, 'Aristotle's doctrine of abstraction in some Aristotelian commentators and Neoplatonists', forthcoming in Richard Sorabji, ed., *The Transformation of Aristotle*, London 1989.

[44] Simplicius *in Phys.* 293,25-6; 512,19-26; *in DA* 233,7-17; 277,1-278,6.

[45] See esp. Aristotle *Phys.* 2.2, 193b22-194a12.

[46] Aristotle *Mem.* 1, 449b31-450a7. For the rendering 'mental image', see the treatment of *phantasma* in Aristotle, *Mem.* ch. 1, and the analysis of its treatment in Aristotle *Insom.* by Pamela Huby, 'Aristotle, *de Insomniis* 462a18', *Classical Quarterly* n.s. 25, 1975, 151-2.

[47] Aristotle *Mem.* 1, 449b31-450a1; *DA* 3.7, 431a16; 431b2; 3.8, 432a3-14.

[48] Aristotle *Metaph.* 9.9, 1051a21-33.

geometrical figures which existed only potentially. And the object may be internal (an image) or external (a table top or charcoal diagram).

One need not accept this unconventional interpretation[48a] of Aristotle's mathematical objects as mind-dependent yet not wholly in the mind, in order to see the possibility that prime matter and mathematical extension may not be two different extensions after all, in which case the problem of differentiation will not arise. There may be just one relevant type of extension found alike in physical bodies and in mental images. But the geometer ignores more features of it than does the physicist who contemplates prime matter. For the physicist is concerned with the ability of this extension to serve as the subject of properties, and as the subject of successive properties when changes occur, while, as Aristotle says,[49] the geometer abstracts from change.

(v) There is one more question of differentiation. If prime matter is extension or volume, how does it differ from place or space? For Aristotle this would be no problem, because he defines place not as a three-dimensional extension, but as a two-dimensional perimeter. But what about Simplicius? His idiosyncratic concept of proper place has indeed been thought to refer to a body's volume.[50] He also suggests a conflation himself, when he says with evident reference to Plato's *Timaeus*, that the cosmos has matter as its space (*khôra*).[51] How, then, would he distinguish prime matter from place? He has a preliminary discussion and a final answer.

The preliminary discussion comes in a passage already cited,[52] where Simplicius calls *matter* an extremely diffuse material dimension (*hulikê diastasis*) and differentiates *place* as an extended space (*diastôsa khôra*). In addition, he distinguishes matter, which he calls a dimension (*diastasis*) as something different from (*allê par'*) four types of extension (*diastêma*). And one of these four extensions may be *place*, for it is described as enmattered (*enulon*: the description of place as enmattered has just been endorsed),[53] but without qualities (*apoion pantêi*) and incorporeal (*asômaton*). The passage should be quoted more fully:

> It looks as if extension (*diastêma*) is spoken of in four ways. One kind consists merely in an unextended formula (*logos adiastatos*), as does the definition of extension. Another resides in thought about dimension

[48a] Contrast J. Annas in A. Graeser, ed., *Mathematics and Metaphysics in Aristotle*, Bern 1987, and the survey by I. Mueller in Sorabji, ed., *Transformation*.

[49] Aristotle *Phys.* 2.2, 193b34. Admittedly Mechanics cannot afford to ignore the motions of what it studies. But that is all right, because Aristotle says that such sciences must treat lines as *physical*, not as mathematical (*Phys.* 2.2, 194a7-12).

[50] S. Sambursky, *The Concept of Place in Late Neoplatonism*, Jerusalem 1982, pp. 18, 22, 167.

[51] Simplicius *in Phys.* 467,37. Aristotle already identified Plato's space (*khôra*) with his own prime matter, *Phys.* 4.2, 209b11-13; *GC* 2.1, 329a14-24. Thereafter Plotinus calls matter the place of all things, 3.6.18(37-8), and Simplicius ascribes to Plato the belief that matter is the *place* of forms, *in Phys.* 643,5-12.

[52] Simplicius *in Phys.* 623,14-20.

[53] ibid. 623,4.

(*epinoia diastaseôs*), as does mathematical extension. Another is enmattered and endowed with natural qualities and resistance, as is body. Another is enmattered, but altogether without qualities and incorporeal. Different from these again is material dimension (*hulikê diastasis*), which is conceived by reference to stretching and indefiniteness. But neither is place an accident; it too is substance, for it is not dimension plain and simple (*diastasis haplôs*), but extended space (*diastôsa khôra*).

But this preliminary discussion does not reveal Simplicius' special conception of place. On that conception, place is *easy* to distinguish from matter, because it looks much more like *form*. It will be seen in Chapter 12 that everything has a proper place, on Simplicius' view, which is like a mould (*tupos*) into which it should fit exactly, if it is to be properly arranged,[54] and which measures its dimensions.[54] This proper place does not stay still, but moves with the thing whose place it is.[55] It is explicitly said to be like form,[56] and in fact is hard to distinguish (*dusdiakritos*) from form,[57] because it is dynamic, drawing bodies together and arranging them.[58] There is no difficulty here, then, about the difference between place and prime matter.

But what about more conventional views of place? The commonest Greek view, which will be discussed in Chapter 11, followed neither Aristotle's nor Simplicius' conception. It took place to be a three-dimensional extension (*diastêma*). We would call this space, rather than place. How would place so conceived differ from prime matter conceived as a body's extension? Aristotle himself gives a hint: we distinguish place from a body's extension only because place stays still when a body (with its extension) moves out of it. If things never moved we should not distinguish a body's extension from its place.[59] This suggests that what differentiates place from prime matter is that place is incapable, matter capable, of motion.

That may seem to raise two difficulties. First, is not place, conceived in that way, that is, as three-dimensional space, capable of moving after all? For example, the vacuous space in my vacuum flask moves with the flask.[60] But the answer is that we treat this mobile vacuum as part of the flask's volume or extension. What is incapable of motion is

---

[54] ibid. 643,12-13. See for the origin of this idea in Proclus 613,9, and in Damascius 645,7-10.

[55] Simplicius *in Phys.* 629,8-12; 637,25-30. See the excellent account by Philippe Hoffmann, 'Simplicius: Corollarium de loco', in *Astronomie dans l'Antiquité Grecque*, Paris 1979, Actes du colloque tenu à l'Université de Toulouse-le-Mirail, 21-3 Octobre 1977.

[56] Simplicius *in Phys.* 629,13-19.

[57] ibid. 638,26-7.

[58] ibid. 626,3-34; 627,29; 631,38; 636,8-13; 637,8; 638,2; 644,11-645,14.

[59] Aristotle *Phys.* 4.1, 208b1-8; 4.4, 211a12-13.

[60] For moving holes, cf. Chapter 5 on whether the vacuum in my flask can be penetrated by body, and Chapter 11 on Aristotle's treatment of a boat's place as an unmoving hole in a river.

rather the three-dimensional space through which the flask moves, not the extension which belongs to it. A second problem is whether by treating prime matter as an extension capable of motion, we have not fallen into Alexander's trap and turned it into a compound of extension with capacity for motion, rather than leaving it as simple, unanalysable and an ultimate subject, to be contrasted with all the properties it carries, even with the capacity for motion. But if there is a difficulty here, it is a difficulty that confronts any interpretation of prime matter. For any interpretation has to view prime matter as endowed with certain inalienable capacities, such as the capacity to receive form. So the present difficulty is not peculiar to the concept of prime matter as extension. Nor, of course, is it a difficulty that Aristotle needs to confront for present purposes, since his own concept of two-dimensional place is easy to distinguish from that of three-dimensional extension.

(vi) A further apparent difficulty for the identification of prime matter with an extension arises from Aristotle's description of matter as not knowable in itself (*agnôstos hath' hautên*).[61] It is known by analogy from examples,[62] and is indicated by negation (*apophasei dêloutai*),[63] presumably by thinking away forms. But there need be no difficulty here, for all these descriptions are applied to ordinary matter, and could apply at least as easily to prime matter, whether or not that is viewed as an extension. The idea would be that one gets to see what matter is, in the extreme case of prime matter, by thinking away all forms and looking for something analogous to the bronze of a statue. A more severe objection to the identification of prime matter with an extension was mentioned earlier – that prime matter is said to be imperceptible.[64] But that showed at most that Aristotle had not thought of extension as playing the roles he wanted, not that it would not play them.

(vii) Two final problems arise from philosophical considerations, the first of which is quite distant from anything considered by Aristotle. Need bodies be extended at all, or could sense be made of the idea of point-like bodies whose presence was detected by their deflecting or obstructing ordinary bodies, or by devouring them, as black holes are said to do?[65] The subject of properties in an unextended body, if such a body there could be, would not be extension. But in that case, all we need do is allow that the role of prime matter could be played by points as well as by extensions.

(viii) The remaining difficulty is that if matter is extension, then bodies will depend for their existence on extension, whereas it is often

[61] Aristotle *Metaph.* 7.10, 1036a9-10.
[62] Aristotle *Phys.* 1.7, 191a7-11; 2.3, 194b23-6.
[63] Aristotle *Metaph.* 10.8, 1058a23.
[64] Aristotle *GC* 2.5, 332a35.
[65] On point-like bodies, see the controversy between Anthony Quinton, 'Matter and space', *Mind* 73, 1964, 332-52, and David Sanford, 'Volume and solidity', *Australasian Journal of Philosophy* 45, 1967, 329-40.

thought that one extension, namely, space, depends for its existence on bodies. I see no problem here, not least because certain theories of modern physics are prepared to view the physical world as dependent on space, in fact, as space endowed with properties, in ways that will be discussed in Chapter 3.

I must not conceal the fact that Simplicius was led to his interpretation of Aristotle by peculiarly Neoplatonist considerations of his own. The diffuseness of extension will have made it a good candidate in his eyes, because it puts prime matter where it should be, at the opposite extreme from the unity of the One. It may be wondered if Simplicius was further motivated by accepting Aristotle's view that Plato's talk of space in the *Timaeus* is an attempt to formulate the concept of matter.[66] That is not so, because although Simplicius, like almost everyone, accepted the identification of matter with what Plato calls space, he did not think that Plato's talk of space was literally meant.[67] What is true, however, is that Simplicius knew the history of interpretations of Plato. One interpretation which he reports is that of Moderatus of Gades (Cadiz), a Neopythagorean of the early first century A.D. Moderatus took Plato's space not as matter, but only as a model (*paradeigma*) for the matter of bodies. But since the model was a quantity (*posotês*), he inferred that for Plato (and the Pythagoreans) the matter of bodies too was a quantity (*poson*), an indefinite quantity.[68] Simplicius did not accept Moderatus' interpretation of Plato, but it provided him with one of the contexts for saying that Aristotle at least understood matter as indefinite extension.[69]

The very different thought processes of a late Neoplatonist do not in this case reveal Aristotle's conscious doctrine. But they open up new vistas by showing how well a certain idea would fit Aristotle's thought – better, I fancy, than anything else that has been proposed. In considering how it would fit, we have been forced to consider a network of interlocking parts of Aristotle's philosophy. Some of the parts would require modification, if extension were to be openly acknowledged as playing the role of prime matter. Most notably, prime matter would have to be perishable in principle. Simplicius' interpretation would also have to be modified by insisting on the particularity of the extension, and its familiarity, as opposed to its diffuseness, or its proximity to non-being. But the modifications to Aristotle ought to be tolerable ones, and they would leave him with a philosophically viable view. The subsequent popularity of views of the same general sort, right up to the present century, will be the subject of Chapter 3.[70]

I will conclude by translating some of the remaining passages in Simplicius.

[66] Aristotle *Phys.* 4.2, 209b11-13; *GC* 2.1, 329a14-24.
[67] Simplicius *in Phys.* 540,15-541,29.
[68] ibid. 230,34; 231,17-20   [69] ibid. 232,24.
[70] For earlier anticipation, see Chapters 2 and 3, and, on the Aristotelians of Plotinus 2.7.1(8-19), Chapter 5.

Simplicius *in Phys*. 537,9-538,14, on Ar. *Phys*. 209b6-11:

Aristotle says that if place is thought to be the extension (*diastêma*) of a magnitude (*megethos*), it will no longer be form, but matter. Now place is thought to be an extension which receives a magnitude. For a magnitude already endowed with form is said to be 'in a place'. In such a magnitude there is, firstly, an indefinite extension (*diastêma aoriston*) viewed in abstraction from, and as deprived of, a boundary or any other accidental property that it receives. Secondly there is a boundary and a defining limit which makes the indefinite extension definite and embraces it, and this is the form, whereas what is left after that has been abstracted, is nothing else but matter or, equivalently, extension. So if place is the extension of a magnitude, and the extension of a magnitude is matter, the conclusion is clear.

(537,19) It is surprising that Aristotle assimilated matter to place in respect of its being extension, and not in respect of its being receptive. But perhaps that was a concession to Plato, as we shall see, or rather he was thinking of extension as something receptive.

(537,22) But if the extension of a magnitude is a certain quantity (*poson ti*), and has a measurement, and a quantity is a form, how can the extension of a magnitude be called matter? It is because material extension is not measured quantity, nor a form, if it does not partake of form in respect of quantity and magnitude, but is an indefinite diffusion (*khusis aoristos*). Aristotle himself shows that he has this sort of idea about it when he says, 'matter and the indefinite are of this sort'. And so when he says 'insofar as place is thought to be the extension of a magnitude, it is matter', by 'the extension of a magnitude' he means the indefinite subject which he previously called infinite (*apeiron*). He also said of the infinite that it is the matter of a completed magnitude.

(537,32) Let no one think, then, that the corporeal dimension (*diastasis*) which is a magnitude and a quantity, or the definite numerical division of a plurality, comes to bodies from matter. From matter there comes only their stretching, diffusion and indefiniteness (*diaspasmos, ekkhusis, aoristia*), which differentiates enmattered forms from matterless ones, and this seems to me to fit the correct conception of matter very well.

(538,3) But, despite having said much towards distinguishing material dimension from a magnitude, Alexander seems to be worried on the ground that every dimension that has magnitude and quantity belongs to form, which is why he was led to add the following comment: 'we must understand the present remarks on matter as not being made in all accuracy, but as being required for proving the premised proposition'. However, as I said, Aristotle clearly distinguishes material dimension from dimension with magnitude by its being indefinite and lacking a boundary, and he perspicuously states that this extension is different from magnitude and is the matter of a completed magnitude and is embraced by form as by a surface and boundary, while being indefinite in its own nature. As I said, the dimension of matter should be thought of not as a magnitude, but as a slackening (reading: *paresis*), and diffusion (*ekkhusis*) of the indivisible intertwining that belongs to form.

Simplicius *in Phys.* 230,23-33, part of attack on view of Stoics and Pericles that matter is qualityless body:

> So perhaps two notions of body should be postulated, the first existing in terms of form, and made definite (*hôrismenon*), by three dimensions (*diastaseis*), the second in terms of a slackening, a spreading and a removal of definiteness from the incorporeal, indivisible, intelligible reality. The second is not given a definite form by three dimensions, but is everywhere slackened, and spilt, and flows from all sides away from being into non-being. And perhaps we should postulate that *matter* is dimension (*diastasis*) of this sort. The corporeal *form* should not be so conceived. Rather, it gives measure and definiteness to this sort of dimension and halts its flight away from being. For we must understand that matter ought to be what differentiates the enmattered from the matterless, and the enmattered differs in having volume (*onkos*), dimension (*diastasis*), divisibility and similar attributes. That is, not attributes of a definite, measured kind, but ones lacking measure or definiteness, and merely capable of being made definite by the measures which form contributes.

Simplicius *in Phys.* 232,21-3, on water changing into air:

> Even though the material dimension (*hulikê diastasis*) remains the same for both of them. For [water and air] are both equally material, and equally divisible and perceptible, and undifferentiated in respect of their matter.

Simplicius *in Phys.* 232,24-30, on *Phys.* 1.7, 191a7. Application to Aristotle of Moderatus' interpretation of Plato and the Pythagoreans:

> You can see from what is said in the fourth book of this work [the *Physics*] that Aristotle himself has the same sort of idea as the Pythagoreans about the dimension and indefinite quantity (*poson*) of matter. For he says there: (quotation of 209b6-9)

Simplicius *in Phys.* 229,5-7, in attack on view of matter as qualityless body:

> And in the fourth book of this work he intends the matter of a magnitude to be an indefinite dimension (*aoristos diastasis*) which is made definite by the magnitude on the formal side.

Simplicius *in Phys.* 230,17-20:

> What belongs in common to all natural, perceptible things as such must be matter. That is one of the most obvious facts, I think. And what is common is being spread into a volume (*onkos*) and dimension.

For Simplicius *in Phys.* 623,14-20, see above.[71]

[71] I am most grateful for comments to Peter Alexander, John G. Bennett, Martha Bolton, Robert Bolton, Abraham Edel, Gail Fine, Pamela Huby, Terry Irwin, Eric Lewis, Izschak Miller, Alexander Nehamas, Paul Pritchard, Larry Schrenk, Gisela Striker and Christian Wildberg.

CHAPTER TWO

Body as extension endowed with properties: Philoponus against Aristotle

I want to introduce an irony. Simplicius' arch-enemy, John Philoponus, finished up with a view of prime matter very similar to the one Simplicius was to take shortly afterwards. But he travelled by an entirely different route. The pagan Neoplatonist school in Athens, where Simplicius taught Philosophy, was closed in A.D. 529 by the Christian Emperor Justinian. (I do not apologise for speaking of a school, nor of a closure: I would consider my school closed if the government forbade us to teach, even though privately funded.) In that very same year, in the rival city of Alexandria, the Neoplatonist-trained Christian John Philoponus, still free to continue philosophising, published an attack on the former Athenian Proclus, in a major work of pro-Christian philosophy, the *de Aeternitate Mundi contra Proclum*. No wonder Simplicius was bitter. If anybody would like a source book of philosophical invective, I can recommend the remarks of Simplicius on Philoponus.

In two earlier works, Philoponus had taken the conventional view of Aristotle's prime matter. In the early *Categories* commentary, for example, he treats it as something which is given volume and three-dimensionality only by the superimposition of form. It sounds, to our ears, like a something, I know not what.

The context of the discussion is the question why Aristotle makes the category of quantity into the *second* category, before the category of quality, or any of the other categories. It is not always appreciated that Philoponus' answer draws on Aristotle *Metaph*. 7.3, the passage discussed in Chapter 1 above.[1] That passage speaks of matter as a first

---

[1] In this I depart from Michael Wolff, *Fallgesetz und Massebegriff*, Berlin 1971, 112-19, but I agree with Ian Mueller, 'Aristotle on geometrical objects', *Archiv für Geschichte der Philosophie* 52, 1970, 156-71, repr. in J. Barnes, M. Schofield, R. Sorabji, eds, *Articles on Aristotle* 3, London 1979.

subject (*hupokeimenon prôton*),[2] and it distinguishes the three dimensions, length, depth and breadth, as the most fundamental layer of properties superimposed on that first subject.[3] Philoponus thinks of prime matter with these three dimensions superimposed on it as forming a *second* subject (*deuteron hupokeimenon*). It is this second subject which takes on the qualities hot, cold, fluid and dry and so constitutes the four elements, earth, air, fire and water. No wonder then, that Aristotle puts the three dimensions, which are here viewed as *quantities*, ahead of the *qualities*, hot, cold, fluid and dry, when he decides on the order of the categories. The explanation comes straight out of *Metaph.* 7.3, and runs as follows:

> In the nature of things quantity (*to poson*) occupies the second position. For, as has often been said, prime matter, which is without body, form or shape before being given volume (*exonkôtheisa*), receives the three dimensions and becomes three-dimensional. And so this, which Aristotle calls second subject (*deuteron hupokeimenon*), then receives qualities and constitutes the elements, so that quality has the third rank among the things that there are, and relatives the fourth.[4]

In his subsequent *Physics* commentary, much of which can be dated to A.D. 517,[5] Philoponus still views prime matter with the three dimensions superimposed, or corporeal extension (*sômatikon diastêma*), as he calls it, as being a second subject.[6] And he still views three-dimensionality (*to trikhêi diastaton*) as a quantity (*poson*), and as an accidental (*sumbebêkos*), if inseparable, attribute of substance.[7] In this regard, his view has not changed. On the other hand, he now wants to revise Aristotle's view to the extent of upgrading the category of quantity, and denying that it need be *posterior* to substance. He adduces two arguments for this upgrading. First, types of substance depend on particular quantities, for you cannot have a man, a drop of wine, or a piece of flesh indefinitely small.[8] Secondly, corporeal and

---

[2] Aristotle *Metaph.* 7.3, 1029a1-2. I take the expression to mean first subject again, and not, as Michael Wolff suggests (loc. cit.), substance, when it reappears in Philoponus *contra Proclum*.

[3] Aristotle *Metaph.* 7.3, 1029a14-18.

[4] Philoponus *in Cat.* 83,13-19.

[5] By a reference at Philoponus *in Phys.* 703,16-17. I leave on one side the plausible thesis of Koenraad Verrycken that parts were revised later, and his further thesis that they were revised after A.D. 529 (K. Verrycken, *God en Wereld in de Wijsbegeerte van Ioannes Philoponus*, Ph.D. diss., Louvain 1985, and 'The development of Philoponus' thought and its chronology', in Richard Sorabji, ed., *The Transformation of Aristotle*, forthcoming, London 1989).

[6] Philoponus *in Phys.* 579,4.

[7] ibid. 561,11-12; 19-23 (in line 23, *akhôriston* should be read). The expression *to trikhêi diastaton* is ambiguous, since it can also refer to that which *has* three-dimensionality.

[8] This problem of indefinitely small drops of wine will arise again in Chapters 5 and 6, in connexion with theories of mixture. For the influence of Aristotle's doctrine of limits on size for natural kinds, see A.G. van Melsen, *From Atomos to Atom*, Pittsburgh 1952.

spatial extension could, so far as depended on them, exist on their own without substances. Spatial extension on its own would be vacuum.[9] It will be enough to quote Philoponus making this last point:

Then it can be added that none of the categories exists without mutual interweaving. You cannot find one category existing without others being interwoven, not even substance itself which is said to be capable of existing on its very own. And matter and the second subject (*deuteron hupokeimenon*), I mean the three-dimensional (*to trikhêi diastaton*) and qualityless body, can exist on its own, so far as depends on it, but none the less it never does exist without qualities. And so, with spatial extension (*topikon diastêma*), even if it could exist on its own so far as depended on it (for what prevents a space (*khôra*) from being empty of body, as we said, if we think of a jar as containing no body within?), yet it never remains empty of body on its own. Rather it is like the case of matter in which, as one form perishes, another immediately supervenes. So too in this case, the substitution of bodies never leaves the space empty, but as one body departs, another falls into its place. And so one can never find even this sort of quantity devoid of substance, and perhaps the force (*bia*) of vacuum consists in this sort of quantity never being separated from substance. This would preserve an idea which is expressed so continually that habit has made it into something that seems to be agreed: I mean the idea that quantity cannot exist on its own without substance. For a vacuum can never exist separated from body.[10]

In the *contra Proclum* of A.D. 529, Philoponus reaches his final view, which is repeated in later work.[11] He dispenses with prime matter as conventionally conceived, as being something useless and impossible. In its place, he decides to put something which we can see to be the lowest layer of properties in Aristotle's *Metaphysics* 7.3. Philoponus does not explicitly mention *Metaph.* 7.3, but in effect what he is recommending is that the lowest layer of properties there should be regarded not as properties, but, as he now says, as the 'first subject' (*prôton hupokeimenon*).[12] And what is that layer? It is length, depth and breadth, or, as Philoponus says, 'the three-dimensional' (*to trikhêi diastaton*). Only Philoponus understands these not in Simplicius' manner as definite measurements, but as indefinite (*aoristos*)[13] – in other words, definite measurements are ignored. He does not explicitly call the three-dimensional an extension (*diastêma*), but he calls it a

[9] Philoponus *in Phys.* 578,5-579,18.
[10] ibid. 578,32-579,18.
[11] Philoponus *contra Proclum* XI 1-8, pp. 405-45 (Rabe); recalled in his *contra Aristotelem* ap. Simplicium *in Cael.* 134,10-17; 135,26-136,1; in a lost work summarised by Simplicius *in Phys.* 1331,20-5; and in the Chronicle of Michael the Syrian, book 8, ch. 13, p. 108 of Chabot's French translation; cf. *Opif.* 37,18-27.
[12] Philoponus *contra Proclum* 406,10-11; 414,3; 425,11-12; 426,22-3; 428,23-5; 433,4-5; 440,6-8.
[13] ibid. 405,26; 424,10; 16; 24.

volume (*onkos*),[14] and qualityless body,[15] which in his earlier work had been described as an extension.[16] The final convergence with Simplicius comes when he recommends calling this three-dimensional extension prime matter.[17] Of course, the two philosophers disagree entirely about what Aristotle himself meant by prime matter. But on what *ought* to be meant, Simplicius' subsequent theory was, at least in the respect mentioned, the same.

The shift in Philoponus' views is marked by his use of the phrase 'second subject'. In his earlier works, the *Summikta Theôrêmata* (probably)[18] and the *Categories* and *Physics* commentaries, he views Aristotelian prime matter as the first subject, and that matter endowed with three dimensions as the second subject. But in the *contra Proclum*, in promoting three-dimensional extension to first subject, he explicitly draws a contrast with being second subject.[19] I think it is a distraction to draw attention to Porphyry's rather different use of the phrase 'second subject'.[20] That seems to me to be merely a divergent reflection of the same common source: Aristotle, *Metaphysics* 7.3.[21]

It is not every kind of three-dimensional extension that can play the role of first subject for Philoponus – not, for example, space. In his earlier writings, he distinguishes between spatial and corporeal extension (*topikon* and *sômatikon diastêma*).[22] Unfortunately, he robs himself in the *contra Proclum* of the criterion by which he had earlier distinguished these two types of extension. For he had distinguished corporeal extension by its having Aristotelian prime matter as its underlying subject,[23] and Aristotelian prime matter is in the *contra Proclum* abolished. A better mark for distinguishing them was suggested in Chapter 1 above, namely, that spatial extension is incapable of motion.

[14] ibid. 424,10; 16; 428,8; 434,4.

[15] ibid. 405,11; 413,6-7; 414,22; 415,2; 4; 426,21-2; 442,17.

[16] Philoponus *in Phys.* 577,13; 687,30-3; 688,30. The three-dimensional was also said to be qualityless body at *in GC* 73,19. I take this passage not to deny the possibility of qualityless body (N. Tsouyopoulos, 'Die Entstehung physikalischer Terminologie aus neoplatonischen Metaphysik', *Archiv für Begriffsgeschichte* 13, 1969, 7-33), but only to deny that the matter that underlies the process of growth can be a qualityless body that exists separately (*hôs khôriston on*) from any qualities, 74,4.

[17] Philoponus *contra Proclum* 405,12; 414,20; 426,22; 428,2; 428,23-5; 435,20; 442,19-20.

[18] Judging from Philoponus *in Phys.* 156,16-17.

[19] Philoponus *contra Proclum* 426,22-3.

[20] Porphyry ap. Simplicium *in Cat.* 48,11-33, used by Michael Wolff, op. cit., 115.

[21] Porphyry applies the expression 'first subject' to matter, just as Aristotle does. But because Aristotle does not specify how the expression 'second subject' is to be used, Porphyry diverges from Philoponus, applying it not to Aristotle's level of length, breadth and depth (as 7.3, 1029a14-17 would suggest), but to the more complex level of complete physical things, like Socrates or bronze (as could be suggested by 7.3, 1029a23-4).

[22] Philoponus *in Phys.* 549,17-552,7; 560,19-561,11; 563,2-5; 575,14; 577,10-16; 579,6; 687,31-5; 688,30-1. The terminology is in Themistius at *in DA* 22,2 and ap. Philoponum *in Phys.* 550,20; 551,2; 16.

[23] Philoponus *in Phys.* 561,11; 577,10-16; 687,31-5.

It was also explained how Philoponus would distinguish corporeal extension from the extension of *geometrical* figures: the latter exists only in the *mind*.[24] As for the supposed distinction between the prime matter of the four elements and that of the celestial fifth element, Philoponus, we saw, was happy to say that there was no difference. He readily concurred, on behalf of Christianity, with the conclusion which Alexander had posed as a threat, that the celestial region would be as perishable as the region down here.[25]

I must now draw attention to a second innovation in Philoponus' *contra Proclum*. For his promotion of three-dimensional extension to first subject is accompanied by another promotion of it to being the form, differentia, essence, or essential attribute of body.[26] His idea is that it performs two disparate but compatible functions: not only does it serve as the first subject of properties, but three-dimensionality also actually defines body, as he repeatedly says in the *contra Proclum*.[27]

Philoponus' idea that three-dimensional extension could *on its own* define body represents a departure from his earlier views.[28] It also means that the convergence with Simplicius is only partial, for whereas Philoponus thinks in the *contra Proclum* that his indefinite three-dimensional extension can be called *body*,[29] Simplicius denies that prime matter is body,[30] unless perhaps in a secondary sense.[31] On this issue I believe that Simplicius is right, and Philoponus' earlier views were better: three-dimensional extension is not sufficient *on its own* to define body. His new view that it is sufficient means that he would be prepared to define body simply as extension, rather than as extension endowed with properties. On the other hand, he presumably still holds that the extension always is as a matter of fact endowed with properties, and that it can exist without properties only 'so far as depends on it'.[32] This qualification, in a passage already translated, has plausibly been taken to mean that it *cannot but* be endowed with properties, even though we ignore those properties when we want to

---

[24] ibid. 500,22; 503,17.

[25] Philoponus ap. Simplicium *in Phys.* 1331,20-5; ap. Simplicium *in Cael.* 89,22-5; 134,16-19.

[26] Philoponus *contra Proclum* 405,24-7; 423,14-424,11; 424,24; 425,5-6; 427,8; 435,21-2. Michael Wolff helpfully draws attention to this second promotion, op. cit. at 118-19 and *Geschichte der Impetustheorie* 1978, 151-2, but because he understands *hupokeimenon* as substance, not as subject, he omits to draw attention to the first promotion. For a different assessment of Wolff on this point, see Christian Wildberg, *Philoponus' Criticism of Ether*, Ph.D. diss., Cambridge 1984.

[27] Philoponus *contra Proclum* 414,10-17; 418,25-6; 419,3; cf. *Opif.* 37,21.

[28] Three-dimensionality had not been considered an adequate definition earlier, *in Phys.* 505,8-9; 561,6-7; 561,22-3.

[29] Philoponus *contra Proclum* 405,11; 16; 19; 412,28; 413,2; 6-7; 414,16; 22; 415,2; 4; 7; 17-18; 417,22; 26; 418,7; 25; 419,3; 421,11; 20-1; 424,18-19; 426,21-2; 442,17.

[30] Simplicius *in Phys.* 201,25-7; 228,17-230,33; 232,8-13.

[31] ibid. 230,21-7.

[32] Philoponus *in Phys.* 579,5-6.

form a clear conception of it in itself.[33]

Unsatisfactory as the definition of body as three-dimensional extension may be, it has interesting consequences for Aristotle's theory of categories. For in making three-dimensional extension the *essence* of body, he is turning it into *substance*. Indeed, the Greek word he uses for essence, *ousia*, can be equally well translated essence or substance, when he says that three-dimensional extension is the *ousia* of body,[34] and when he says that it is not an ordinary quantity (*poson*), but an essential or substantial quantity (*poson ousiôdes*).[35] This means that it falls under the category of substance (*hupo tên ousian*),[36] and not after all under the category of quantity, where he had placed it earlier.[37] This completes a disruption of Aristotle's treatment of the categories of substance and quantity, which began in the *Physics* commentary, where we saw him challenging the subordination of the one category to the other.

An analogous disruption in his opponent Simplicius was noticed in Chapter 1. For Simplicius elevated both prime matter and space into the category of substance.[38] Others, we saw there, had suggested that prime matter might fall outside the categories altogether.[39] These developments may explain something which has been thought to originate with Patrizi in the sixteenth century.[40] Patrizi translated into Latin what he (wrongly) took to be Philoponus' commentary on Aristotle's *Metaphysics*, and he certainly knew Philoponus' *contra Proclum* and the *Physics* commentaries of Simplicius and Philoponus. His treatise on space, which has been translated into English, shows him steeped in the views of Greek antiquity.[41] He there goes further than Simplicius on the subject of space, and suggests that it belongs outside the Aristotelian categories, on the grounds that it exists independently of the physical world. It is neither a quantity, nor a substance, in the sense of the categories, although there is another sense in which it is a substance. It has been shown by others that Patrizi's idea was influential. It recurs in Gassendi, Charleton and Newton, who applies it to all extension.[42]

---

[33] This is David Sedley's interpretation of passages like *in Phys.* 579,5-6 in his 'Philoponus' conception of space', in Richard Sorabji, ed., *Philoponus and the Rejection of Aristotelian Science*, London and Ithaca N.Y. 1987.

[34] Philoponus *contra Proclum* 405,26; 424,9; 24; 425,5-6.

[35] ibid. 405,24; 424,6.

[36] ibid. 424,5.

[37] Philoponus *in Cat.* 83,13-19; *in Phys.* 561,11-12; 19-23.

[38] Simplicius *in Phys.* 623,18-20.

[39] ps-Alexander *in Metaph.* 464,21-6; Asclepius *in Metaph.* 380,31-5; with preliminary doubts raised in Theophrastus *Metaphysics* VI 17.

[40] Edward Grant, *Much Ado About Nothing*, Cambridge 1981, 187 n. 40, 194, 204-6.

[41] Francesco Patrizi, *de Spacio Physico*, probably 1587, translated into English by Benjamin Brickman, *Journal of the History of Ideas* 4, 1943, 224-45 (see 240-1 for the discussion of Aristotle's categories), and see further John Henry, 'Francesco Patrizi da Cherso's concept of space and its later influence', *Annals of Science* 36, 1979, 549-73.

[42] Details on Gassendi in E. Grant, op. cit., 199, 204-6, 209; John Henry, op. cit., 568; Charles Schmitt, *Gianfrancesco Pico della Mirandola (1469-1533) and his Critique of Aristotle*, The Hague 1967, ch. 5, 143-4. On W. Charleton see: J.E. McGuire, 'Body and

What needs reconsideration is whether the idea arose *de novo* in Patrizi, or whether it was not strongly suggested to him by the cluster of ideas which has just been alluded to in the ancient commentators on Aristotle.

The following are some of the principal passages in Philoponus' *de Aeternitate Mundi contra Proclum*:

(405,23-7) That not all quality or quantity is an accidental attribute; there is substantial quantity (*ousiôdes poson*) and quality. And that the thing in bodies which is independent (*authupostaton*) [sc. of any substratum] and substance *tout court* (*haplôs ousia*) is the indefinite three-dimensional which is the ultimate subject (*eskhaton hupokeimenon*) of everything.

(424,4-11) Just as there is such a thing as substantial quality which is referred not to the category of quality, but to the category of substance (*ousia*), as being a substantial (*ousiôdês*) differentia, so presumably there is also such a thing as substantial quantity (*poson ousiôdes*) and this precisely is the three-dimensional (*to trikhêi diastaton*). For the only thing found in bodies which is independent [sc. of any substratum] and is the actual substance (*ousia*) of body is a sort of three-dimensional volume (*onkos tis trikhêi diastatos*) indefinite as regards magnitude or smallness.

(424,23-425,14) Thus the substance of body is nothing other than the indefinite three-dimensional which is made definite by the differentia of smallness and largeness, and which receives the differentiae which create the species of bodily substances. It produces the particular substances of bodies, I mean the substance of fire, of the sun, of the moon and of everything else. With this shown, it is clear that the three-dimensional is not an accidental quantity, for in that case it could come into, and go out of, existence without a body being destroyed. But as it is, we cannot even think of body without the three-dimensional. So it is the substance of body. If, then, the three-dimensional is actually the substance of body as such, and it alone remains unchanged amidst the changes in bodies, as has been shown, then there is no argument to show that incorporeal matter must underlie it as its subject. It itself is the first subject (*prôton hupokeimenon*) underlying all natural forms, and further it is from it and from the substantial qualities in combination (*suntithemenôn*) that there come into being the bodies which are made real, that is, fire, water and so on.

(428,7-10) The three-dimensional is not a composite of an underlying subject and form, but is a simple volume (*onkos*), and has its being in this

void and Newton's *de Mundi Systemate*: some new sources', *Archive for the History of Exact Sciences* 3, 1966, 233. On Newton see: E. Grant, op. cit., 242, 244; J.E. McGuire, loc. cit., and 'Existence, actuality and necessity: Newton on space and time', *Annals of Science* 35, 1978, 463-509. In Newton (*de Gravitatione*, tr. A. Rupert Hall and Marie Boas Hall, 132), the un-Philoponan denial that extension is substance naturally gets an un-Philoponan justification.

fact, and underlies all other things as their subject.

(428,14-25) If this is impossible, and it has been shown that nothing else underlies the three-dimensional as its subject, but that it itself is the basis of everything, then it is clear that it is the simplest thing and the matter of all things, and is not a composite of anything else. Since this is so, and since the three-dimensional is independent [sc. of any substratum], and no change is found in it, there is no argument capable of establishing that an incorporeal matter underlies bodies as their subject. Natural things must be analysed into that on which they ultimately depend, whether one wants to call it the first subject (*hupokeimenon prôton*), or matter, for we will not dispute about names.

(440,6-8) Hence it is clear therefore that the first subject of all things and their matter is three-dimensional.[43]

---

[43] I have benefited from the comments of David Sedley and Christian Wildberg.

CHAPTER THREE

Body as extension endowed with properties: antecedents and sequel

Philoponus was not the first to think in the way he did. Three centuries earlier, in the third century A.D., Plotinus, the founder of Neoplatonism, put an analogous argument into the mouth of an imaginary opponent. Since prime matter is conventionally thought of in separation from magnitude (misleadingly, it is said to be without magnitude, *amegethes*), how can it contribute, as it is supposed to do, to dimension (*diastasis*) and magnitude (*megethos*)? Moreover, how can it perform its role of receiving properties, for where, if it lacks magnitude, will it receive them? What need, then, of prime matter? Should we not kick it out, and allow that it is the next layer up, magnitude, that receives the various qualities and is their subject? This next layer up corresponds to the length, depth and breadth of Aristotle *Metaph*. 7.3 and to the three-dimensional extension of Philoponus, which he similarly promotes into being the first subject of properties:

'And why is anything else needed for the composition of bodies besides magnitude (*megethos*) and all qualities?'
    There is need of something to receive them all.
    'This, then is the volume (*onkos*). But if volume, then, presumably magnitude (*megethos*). But if it is without magnitude (*amegethes*), it will not even have anywhere to receive them. And if it were without magnitude, what would it contribute, if it contributes neither to form and quality, nor to dimension (*diastasis*) and magnitude, which, wherever it occurs, is thought to come to bodies from their matter. And in general, just as acting and making and times and movements are things that exist without having a foundation of matter in them, so there is no need either for bodies, which come first, to have matter. They can each be what they are as wholes, and be more complex when their structure is produced by the mixture of a larger number of forms. So this "without magnitude" of matter is just an empty name.'[1]

In Chapter 5, it will emerge that certain Aristotelians may have pre-

[1] Plotinus 2.4.11(1-14).

supposed a conception of prime matter as extension by suggesting that in chemical mixture the prime matters of the ingredients were juxtaposed (as if prime matter took up space), and only the qualities mingled.[2]

These are the only genuine anticipations known to me of Philoponus' view. But it will seem that there is an enormous lacuna in my discussion. What about Plato's *Timaeus* and what about the Stoics? Certainly, Aristotle understands Plato to be introducing a first subject of properties and identifying it with *space*. Indeed, he criticises Plato for this: space cannot play the role of prime matter. But is he right to interpret Plato in this way?

In *Timaeus* 48E-53C, Plato describes the state of the physical world before God brought order to it, and he introduces the idea of a receptacle. It turns out that the receptacle is space (*khôra*),[3] and I take it, contrary to some ancient interpretations,[4] that he means space in a literal sense. He is viewing space as a receptacle which receives qualities,[5] and the qualities in turn are viewed as copies[6] of the eternal Platonic Forms. It can further be said that space receives everything perceptible that comes into being[7] including bodies,[8] but the talk of bodies needs interpretation. If we take the four elements, earth, air, fire and water, we should think of them not in the way we ordinarily think of bodies, but in qualitative terms. Because they are for ever changing into one another, and not remaining the same, we cannot use the words 'this' (*tode, touto*) or 'being' (*einai*) of them, words which are evidently taken to imply stability over a period. These words should be reserved for space itself, whereas we should use only the words 'what is such and such' (*to toiouton*) for other things.[9] That applies to the four elements, to what is made of them, and to anything which comes into being.[10] There is a further complication: in the original state of chaos before God had made things orderly, there were only vestiges (*ikhnê*)[11] of earth, air, fire and water. For one thing, fire would not have had all its distinctive properties, its prickingness for example, before God had packaged it into sharp little corpuscular shapes.[12]

Plato's language is so fluid that it has given rise to opposite interpretations. According to Cornford, Plato thinks of fire as a bundle

[2] References in ch. 5, nn. 49, 50, if the matter under discussion includes prime matter.

[3] *khôra*: Plato *Tim*. 52A8; B4; D3.

[4] Simplicius and Alexander ap. Simplicium *in Phys*. 539,8-542,14; Philoponus *in Phys*. 516,5-16; 521,22-5.

[5] Qualities: Plato *Tim*., *ideai* 50D7; *genê* 50E5; *morphai, pathê* 52D6; *dunameis* 52E2.

[6] Copies: Plato *Tim*., *mimêmata* 50C5; *aphoiômata* 51A2; *eikôn* 52C2; *phantasma* 52C3.

[7] All *genesis*, Plato *Tim*. 49A5-6; all that has *genesis* 52B1; all the visible and otherwise perceptible that has come into being 51A4-5.

[8] Bodies (*sômata*): Plato *Tim*. 50B6.

[9] Plato *Tim*. 49D5-50A4; 50B3-4.

[10] ibid. 49E7; 50A3-4.

[11] ibid. 53B2.

[12] ibid. 55D6-56B6.

or 'combination of qualities',[13] without any subject to which they belong. Examples of such qualities are hot and white.[14] On Cornford's view, when we are told to think of fire as what is such and such (*to toiouton*) and ever recurrent (*aei peripheromenon*),[15] it is a recurrent combination of qualities which is so described. According to Aristotle's rival interpretation, the properties which characterise fire *do* have a subject, and that subject is *space*. He identifies Plato's space with his own prime matter.[16] It was not always agreed that Plato intended his talk of space literally. But the identification of his 'receptacle', whether that was space or not, with prime matter was accepted by another of Plato's companions, Hermodorus,[17] and was practically universal in antiquity.[18] I shall ignore the non-literal interpretation of space for the moment, and ask what evidence there is that Plato thinks of his receptacle, or space, as the *subject* of properties.

The idea of a receptacle or receiver does not yet tell us whether space is viewed as a *subject*, or merely, for example, as a *container* of qualities. Nor are we much helped by its being called an *ekmageion*, something onto which smears are wiped off,[19] or by being told that what enters it are imprints, imprinted by the eternal Forms (*ektupôma, tupôthen, ektupoumenon*).[20] But it is further said that what enters it patterns it right through (*diaskhêmatizein*) and makes it appear different at different times (*phainesthai allote alloion*).[21] Again, that part of the receptacle which is made fiery appears as fire (*pepurômenon ... pur phainetai*), that which has been made fluid appears as water (*hugranthen ... hudôr*).[22] It is repeated that it is made fluid and fiery and receives other characters, so that it presents every sort of appearance to the sight (*pantodapên idein phainesthai*).[23] This begins to lend colour to Aristotle's interpretation of space here as the subject of properties.

[13] F.M. Cornford, *Plato's Cosmology*, London 1937, 180-1; A.C. Lloyd, 'Neoplatonic logic and Aristotelian logic I', *Phronesis* 1, 1955-6, 58-79, at 62, describes Plato *Tim*. 49-50 as the *locus classicus* in Middle Platonism for the view that particulars were like bundles of particular qualities. Certainly, Albinus(?) uses fire as an example of a bundle in a passage to be discussed in Chapter 4, but for our purposes it will matter whether it is a bundle only of qualities.

[14] Plato *Tim*. 50A2-3.

[15] ibid. 49E5-7.

[16] Aristotle *Phys*. 4.2, 209b11-13; *GC* 2.1, 329a14-24. As prime matter, it will be coextensive with body and finite.

[17] So Simplicius *in Phys*. 247,30-248,19; 256,35-257,4.

[18] The assumption is too widespread to be worth cataloguing, but for some of the references see J.C.M. van Winden, *Calcidius on Matter*, Leiden 1959; Willie Charlton, *Aristotle's Physics, Books I and II*, Oxford 1970, 141-5.

[19] Plato *Tim*. 50C2.

[20] ibid. 50C5; D4; D6.

[21] ibid. 50C3-4.

[22] ibid. 51B4-6.

[23] ibid. 52D4-E1.

So the nurse of the generative processes was made fluid and fiery
(*hugrainomenê, puroumenê*) and received the forms (*morphai*) of earth
and air and underwent all the other qualifications (*paskhousa pathê*)
that go with these, and presented every sort of appearance to the sight.[24]

Aristotle is further impressed by the fact that Plato compares the
receptacle to gold which somebody moulds into many different
shapes.[25] Gold would have been one of his own examples of matter,
although not of *prime* matter. Plato even uses the same preposition
'out of' (*ek*) as Aristotle associates with his matter, when he imagines a
man moulding shapes *out of* gold.[26] There are also connexions with
Aristotle's descriptions of prime matter, for, if he is speaking of prime
matter in *Metaph*. 7.3, he says that it is a subject which *in itself* has no
particular characteristics.[27] And Plato's account of the receptacle looks
at least superficially similar, when he insists that the receptacle has no
qualities of its own, or it would obtrude them and receive the other
qualities badly.[28]

So far, the receptacle, or space, may sound like a subject of
properties, and like Aristotle's matter. But let us notice the rival
evidence, first about the properties. It is said not only that the
receptacle is *made* fiery and fluid, but also, more *cautiously*, that it
appears (*phainesthai*) different at different times, now as fire now as
water, with every other sort of *appearance* to the sight. Aristotle would
not mind the cautious talk of appearance, insofar as the point is merely
that the receptacle has no characteristics *of its own*. But Plato throws
more doubt on the properties than this. If somebody should point to one
of the shapes that is being moulded out of gold and ask what it is (*ti pot'
esti*), it would be safer not to answer with the name of a shape, but
simply to say 'gold', because the shapes are too fleeting to have being.[29]
One obstacle, then, to the receptacle being a subject of properties is a
lack of being in the properties, although Aristotle thinks that this
passage simply represents a mistake on Plato's part.[30]

There is a further obstacle: although the receptacle sounds like a
subject of properties when it is compared to the gold *out of* which
shapes are moulded, it turns out that Plato attaches no special
importance to this way of talking. For he often uses the more
ambiguous preposition 'in': the qualities come into existence *in* the
receptacle, which merely provides a seat (*hedra*) for them.[31]

But there is something worse: even if Plato's space is a subject of

---

[24] ibid.

[25] ibid. 50A5-B5.

[26] ibid. 50A6 with Aristotle *Phys*. 2.3, 194b24. See the excellent discussion by Friedrich
Solmsen, *Aristotle's System of the Physical World*, Ithaca N.Y. 1960, 118-43, esp. 122.

[27] Aristotle *Metaph*. 7.3, 1029a20-6.

[28] Plato *Tim*. 50C1-2; 50D7-51A7.

[29] ibid. 50A5-B5. You are lucky if they can be described even as 'what is such'.

[30] Aristotle *GC* 2.1, 329a17-21.

[31] In: Plato *Tim*. 49E7; 52A6; C4; seat: 52B1.

properties, it is very doubtful that we are to think of *fire* as diverse regions of space which are momentarily characterised by fire-like properties. For when Plato tells us to call fire 'what is such and such', or 'what is at any time (*dia pantos*) such and such', he adds that fire is: what is such and such, and always alike *as it comes round again* (*aei peripheromenon homoion*).³² This last description of fire as *coming round again* is not at all suited to space, or regions of space:

> We should not use these designations for any of them, but should speak in this way of each and every one: 'what is such and such and always alike as it comes round again'. And indeed what is at any time such and such is fire, and so for anything whatever that is generated.³³

Plato does not, then, keep to a single account. There is partial truth in Aristotle's interpretation, to the extent that Plato's space can be described, with qualifications, as a subject of properties, since it is patterned and made fluid and fiery, and *appears* as water or fire. But as for fire itself, or rather the vestigial type of fire, *that* is not treated as space endowed with properties. This is not to say that Cornford's alternative interpretation of fire is necessarily right. The recurrent 'what is such and such' (*to toiouton*) could be, as he says, a combination of qualities. But Plato could be thinking of it instead as a thing qualified, albeit a thing too fleeting to be described as a being, or a 'this'.

There remains the complication of how vestigial fire in the period of chaos relates to fire as we know it, which has been packaged into shaped corpuscles by the Creator-God. But it is natural to assume that space is neither more nor less a subject of properties in the orderly world produced by God than in the chaotic world that precedes it.

Not only was Plato's account very fluid. It was also less influential on subsequent treatments of prime matter than might have been expected. This was partly because of a major criticism made by Aristotle: place is immobile, whereas prime matter is not, so they cannot be identical. This is what Aristotle means by saying that a thing's place, unlike its matter, is separable from it.³⁴ To overcome this objection, one would have to be prepared to go as far as Newton, whose view will be discussed below, and say that the ultimate subject, or prime matter, in a *moving* body would be now this, now that, portion of immobile space. Aristotle makes an indirect dig at Plato in another passage, the now familiar *Metaph*. 7.3, when he says that matter cannot be called a 'this', and so does not deserve the title of 'substance' or 'being' (*ousia*),³⁵ echoes of the designations which Plato had allowed his receptacle.

³² ibid. 49E5-7.
³³ ibid. 49E4-7.
³⁴ Aristotle *Phys*. 4.2, 209b23; 31; 4.4, 211b36-212a2.
³⁵ Aristotle *Metaph*. 7.3, 1029a28-30.

Aristotle's criticism was, I suspect, instrumental in reducing the influence of Plato's discussion. Space was not subsequently cast in the role of prime matter, perhaps because of Aristotle's point that space is immobile. And there was another development, perhaps designed to protect Plato from Aristotle's charge that he had cast space in that role. It was quite widely suggested that Plato's talk of space was not intended in a literal sense.[36] For both reasons, both because of Aristotle's criticism, and because of the (possibly resultant) non-literal interpretation, Plato's flirtation with space as subject was not taken up. Nor are the subsequent treatments of prime matter as extension derived from him. This is why Plato's *Timaeus* has not figured larger in the opening three chapters. At most we saw Moderatus of Gades taking Plato's receptacle as a model (*paradeigma*) for some kind of extension that serves as prime matter. The influence of Plato's treatment of space has, I believe, been too readily detected. It has been found behind the promotion of magnitude (*megethos*) by Plotinus' opponent, behind Gregory of Nyssa's theory that things are just bundles of ideas, behind Basil of Caesarea's denial that qualities have a substratum and behind the attribution by Simplicius and others to Aristotle of a concept of prime matter.[37]

Another authority that has been misleadingly identified as the source for Philoponus' view of prime matter as extension is the Stoics. His equation of prime matter in the *contra Proclum* with 'the three-dimensional' is at least verbally in accord with part of one Stoic view of body, as he himself acknowledges.[38] He also applies the Stoic term 'qualityless body' to the three-dimensional.[39] And Zabarella, who accepted Philoponus' doctrine in the sixteenth century, took it to be the same as that of the Stoics.[40] None the less, I think that the apparent similarities are misleading.

In writings earlier than the *contra Proclum*, Philoponus is actually opposed to the Stoics. For while they had said that matter was three-dimensional and qualityless body, Philoponus still recognised, beneath the level of three-dimensional qualityless body, an Aristotelian matter, which was not body at all.[41] It is true that the *contra Proclum* removes this point of difference by eliminating Aristotelian

[36] Simplicius *in Phys.* 539,8-542,14, including a report of Alexander; Philoponus *in Phys.* 516,5-16; 22-5.

[37] Plotinus' opponent: E. Bréhier, *Plotin, Ennéades II*, Budé edition, Paris 1956, 51. Basil and Gregory: A.H. Armstrong, 'The origin of the non-materiality of body in Plotinus and the Cappadocians', *Studia Patristica* 5, Berlin 1962, repr. in his *Plotinian and Christian Studies*, London 1979. Attribution of prime matter: Hugh R. King, 'Aristotle without prima materia', *Journal of the History of Ideas* 17, 1956, at 388-9.

[38] Philoponus *contra Proclum* 414,3-5.

[39] Philoponus *in Phys.* 156,10-17; *contra Proclum* 405,11; 413,6-7; 414,22; 415,2; 4; 426,21-2; 442,17.

[40] Giacomo Zabarella, *de Rebus Naturalibus Libri XXX*, Frankfurt 1607 (1st ed. 1590), *de prima rerum materia, liber secundus*, col. 211.

[41] e.g. Philoponus *in Cat.* 83,14-17: 'prime matter which is without body, form or shape before being given volume (*exonkôtheisa*).'

matter. But, first, this convergence with Stoic views results from dissatisfaction with Aristotle rather than from love of the Stoics, and, secondly, it calls into question another point of agreement with the Stoics. For how can we now justify saying that the level of the three-dimensional still deserves to be called '*body*', when it is no longer supported by Aristotelian prime matter? Philoponus' distance from the Stoics becomes clear again when we turn from the justification to the motives for thinking of matter as body. The Stoics believed that matter was something real and something acted on, that acting or being acted on was the criterion for being fully real, and that only body could satisfy this criterion. It is doubtful that Philoponus would accept any of this. He is at great pains to insist elsewhere that light, colour, heat and impetus can act, even though they are incorporeal. Nor is it clear that he would agree that matter can be acted on. Certainly, it can receive qualities, but he is keen to protest that it does not undergo change in the process.[42]

Nor does Philoponus in *contra Proclum* XI 1-8 use any of the three further arguments which Simplicius gives, whether or not they are Stoic arguments, for the Stoic view that matter is body.[43] Instead, he appears to have two reasons for describing matter as body. First, if matter were incorporeal, bodies would be composed wholly of the incorporeal, since their other constituent is incorporeal form.[44] Secondly, the three-dimensional (which is now viewed as matter) constitutes the actual definition of body, and so cannot but be body.[45] This second reason is completely un-Stoic in spirit. For one thing, it is not so much a ground for applying the word 'body' to matter, as a ground for applying it to the three-dimensional which, for reasons unconnected with the Stoics, has come to be viewed as matter. More decisively, the argument trades on a definition of 'body' which the Stoics would reject. For according to one tradition, they insist on bodies having resistance (*antitupia*),[46] while according to alternative evidence, they ignore all other criteria, even three-dimensionality, and think of a body simply as that which acts or is acted on.[47]

A final contrast with the Stoics comes in Philoponus' description of qualityless body as an extension (*diastêma*).[48] The Stoics think of it as essentially extended, but not, so far as I know, as an extension. Consequently, they do not provide an example of the type of view I am

---

[42] Philoponus *contra Proclum* 412,15-28; 413,24-414,5; 414,16-20.

[43] Simplicius *in Phys.* 227,26-228,17.

[44] Philoponus *contra Proclum* 443,6-13; 22-3. Cf. Plotinus 6.3.8 (29-34), to be discussed in Chapter 4, who allows merely perceptible substance to be composed of non-substances.

[45] Philoponus *contra Proclum* 414,10-17; 418,25-6; 419,3.

[46] Stoic body is three-dimensional *with resistance*: Plotinus 6.1.26(20); ps-Galen *de qualitatibus incorporeis* 10 (=*SVF* 2.381).

[47] When qualityless matter is called body, this seems to be solely on the ground that it is acted on. See Chapter 6.

[48] Philoponus *in Phys.* 577,13; 687,30-3; 688,30.

considering, according to which body is extension, or extension
endowed with properties.

I shall turn now from the antecedents of these views to the sequel in
later centuries. I say 'the sequel' because I shall not be claiming actual
influence in more than a few instances. But there has been an
interesting claim of direct influence, which I shall mention first, in
order to get it out of the way. The distinction between definite and
indefinite extension, which we found in Simplicius and Philoponus,
was repeated by Averroes in the twelfth century and by Giles of Rome
(Aegidius Romanus) in the thirteenth. Morever, Giles identified one
kind of indefinite extension with the 'quantity of matter' (*quantitas
materiae*), and thus opened the way, it has been said, for the
introduction of the concept of mass.[49] But one commentator has
allowed Simplicius a much more crucial role in this development than
he could claim on my interpretation. For Simplicius says that his
so-called 'material dimension' (*hulikê diastasis*) remains the same
when water changes into air, and that the water and air are equally
perceptible and do not differ in their matter.[50] It has been inferred that
it is the material dimension which is perceptible, and that the only
thing that could remain the same and be perceptible is the *weight* of
the water and the air.[51] The reference to weight then makes Simplicius
the ultimate forerunner of the idea of mass. But of course I have taken
material dimension to be not weight, but simply the extension of a
body, with its measurements, and anything else that limits its
diffuseness, ignored.

I shall, therefore, turn instead to a development that is found in the
seventeenth century in Descartes and Newton. Superficially, Descartes sounds like Philoponus, when he says that extension constitutes
body.[52] When Descartes further claims that, if we strip away what is
not entailed by the nature of body, we will eventually get down to
length, breadth and depth,[53] he reminds us of Philoponus' view that
three-dimensionality defines body. But there are important differences, and one is that Philoponus distinguished corporeal extension (the
extension of bodies) from spatial extension. Descartes, by contrast,
insists that there is no difference: there is only one extension which we
regard in different ways, viewing it sometimes as space and sometimes
as body.[54] Moreover, Descartes goes further than Philoponus in

[49] A. Maier, 'Das Probleme der quantitas materiae' in her *Die Vorläufer Galileis im 14
Jahrhundert*, Rome 1949, 26-52; N. Tsouyopoulos, 'Die Entstehung physikalischer
Terminologie aus der neoplatonischen Metaphysik', *Archiv fur Begriffsgeschichte* 13,
1969, 7-33, esp. 7-20; M. Wolff, *Fallgesetz und Massebegriff*, Berlin 1971, 146-8.

[50] Simplicius *in Phys.* 232,21-3, translated at the end of Chapter 1.

[51] N. Tsouyopoulos, op. cit., at pp. 18-19, followed by Herbert Hunger, *Die
hochsprachliche profane Literatur der Byzantiner*, vol. 1 (=*Byzantinisches Handbuch*,
part 5, vol. 1), Munich 1978, 30.

[52] Descartes *Principles of Philosophy* (1644), part 2, secs 8-11.

[53] ibid., sec. 11.

[54] ibid., secs 10-12.

downgrading the properties superimposed on extension. We saw in Chapter 2 that Philoponus was prepared to define body simply as extension, rather than as an extension endowed with properties. But we also saw him introducing the qualification that the extension can exist without properties only 'so far as depends on it'.[55] That qualification has plausibly been taken to mean that although we ignore the properties it carries, in order to form a clear conception of it, it does none the less have to carry properties.[56] Descartes, by contrast, probably intends to go further than Philoponus, when he says that no properties enter into the nature of body, other than extension in length, breadth and depth, since all other sensible qualities can be removed from body, while it itself remains in its entirety.[57] And he certainly does go beyond Philoponus with his reductionist view that the properties we do perceive in body are all reducible to divisibility and motion.[58]

The collapse of the distinction between corporeal and spatial extension was not new with Descartes. John Buridan in the fourteenth century had refused to draw a distinction, insisting that space was nothing but the dimension of a body.[59] And Franciscus Toletus in the sixteenth century also went some way to blurring the distinction. In the very act of writing about Philoponus and his corporeal extension, he treated it as a kind of space, albeit an intrinsic space (spatium intrinsecum).[60]

Newton took a different tack. He insisted against Descartes that body was much more than extension. But in his early de Gravitatione, he accepted that it was a kind of extension endowed with properties. Where he differed from Descartes was in insisting that the extension in question was space. Space had existed without properties until, at the time of creation, God bestowed properties on it, and thereby created the physical world. This yields a view about moving bodies like that which I extrapolated above from Plato's Timaeus: a moving body is to be construed as a series of spaces sequentially endowed with properties. Newton explicitly gives it as a reason for his view that the appeal to space avoids an unknowable prime matter:

> Between extension and its impressed form there is almost the same analogy that the Aristotelians postulate between the materia prima and substantial forms, namely when they say that the same matter is

[55] Philoponus in Phys. 579,5.

[56] David Sedley, 'Philoponus' conception of space', in Richard Sorabji, ed., Philoponus and the Rejection of Aristotelian Science, London and Ithaca N.Y. 1987.

[57] Descartes Principles of Philosophy, part 2, secs 4 and 11.

[58] ibid., sec. 23

[59] Buridan Questions on the Physics book 4, qu. 10, fol. 77v, col. 1, and book 4, qu. 2, fol. 68r, cols 1-2, in Johannes Buridanus, Kommentar zur aristotelischen Physik, Frankfurt 1964, facsimile reprint of Paris 1509; 'Spacium non est nisi dimensio corporis'. So E. Grant, 'The principle of the impenetrability of bodies in the history of concepts of separate space from the middle ages to the seventeenth century', Isis 69, 1978, 551-71; Much Ado About Nothing, Cambridge 1981, 15.

[60] Toletus, Commentaria una cum Quaestionibus in Octo Libros Aristotelis de Physica Auscultatione, fols 123r, col. 2 – 123v, col. 1, Venice 1580.

capable of assuming all forms, and borrows the denomination of numerical body from its form. For so I suppose that any form may be transferred through any space, and everywhere denote the same body. They differ, however, in that extension (since it is *what* and *how constituted* and *how much*) has more reality than *materia prima*, and also in that it can be understood, in the same way as the form that I assigned to bodies. For if there is any difficulty in this conception, it is not in the form that God imparts to space, but in the manner by which he imparts it.[61]

It is still more surprising that remote analogues of the ancient treatments of prime matter recur in modern physics, in the development of field theories. The suggestion is that at the sub-atomic level it is often better to think of physical matter not in terms of particles, but in terms of a field endowed with properties. The idea of a field is itself variously interpreted. In the eighteenth-century version of Joseph Priestley, we are invited to conceive of physical matter as extension endowed with powers of attraction and repulsion.[62] In the twentieth century, Einstein was able to express his field theory by saying:

We may therefore regard matter as being constituted by the regions of space in which the field is extremely intense ... There is no place in this new kind of physics both for the field and matter, for the field is the only reality.[63]

'Matter' is here used in the sense of physical material, not, of course, in Aristotle's sense of substratum. The idea is that instead of talking of particles or bodies in microphysics, we should talk of regions of space in which properties are manifested with a certain intensity. Others would prefer to talk of regions of *space-time* in which the properties are manifested. Talk of a region of space-time would replace not so much the talk of a body as the talk of the *career* or *history* of a body. Einstein's version has recently been compared with the theory of Plato's *Timaeus*.[64] But on my account, the *Timaeus* is extremely ambiguous.

[61] Newton *de Gravitatione*, as translated by Rupert A. Hall and Marie Boas Hall, 138-41, extract from 140-1. For discussion, see J.E. McGuire, 'Space, infinity and divisibility: Newton on the creation of matter', in Z. Bechler, ed., *Contemporary Newtonian Research*, Dordrecht 1982, 145-90.

[62] P.M. Heimann and J.E. McGuire, 'Newtonian forces and Lockean powers: concepts of matter in eighteenth-century thought', *Historical Studies in the Physical Sciences*, 3, 1971, 233-306.

[63] This and similar statements are quoted, without the original reference, by many sources, for example, by Milic Čapek, *The Philosophical Impact of Contemporary Physics*, Princeton 1961, 319, and Louis de Broglie, *Nouvelles Perspectives en Microphysique*, translated by A.J. Pomerans as *New Perspectives in Physics*, New York 1962.

[64] Edward Hussey, *Aristotle's Physics, Books III and IV*, Oxford 1983, xxxi; 'Matter theory in Ancient Greece', ch. 2 in Rom Harré, ed., *The Physical Sciences Since Antiquity*, London and Sydney 1986, 20-1.

In some versions of field theory, the role of properties is played down in a manner reminiscent of Descartes' reductionism. According to a theory advocated at one time by John Wheeler, matter is just space-time itself with a certain structure, and there is no need to postulate any properties over and above space-time and its structure, in order to give a full description of the world of material things.[65] Similarly, in 1870, W.K. Clifford imagined mobile regions of curved space as being the ordinary matter of the world.[66]

What is the attraction of such theories? Various reasons have been given for rejecting talk of particles in microphysics, in favour of talk of space, or space-time, with or without properties. One disadvantage of talking of particles is that, given the conditions at that level, one cannot talk of a particle persisting through change. One has rather to talk of it as being annihilated and replaced with a rapidity that rivals the description of vestigial fire in Plato's *Timaeus*. A second disadvantage is that a large group of particles, the so-called bosons, must be admitted to be indistinguishable from each other in many contexts. If one persists in talking in terms of particles in those contexts, one must treat them as distinct, even though there is nothing to distinguish them.[67]

I have picked out some of the more interesting analogues of the ancient 'extension' theories, but there are many others, and their motivations are extraordinarily varied. Thomas Aquinas was puzzled in the thirteenth century as to what the subject can be in the Eucharist which carries such properties as colour. It cannot be the bread with which we start, nor Christ's body with which we finish, because neither of these lasts right through the transformation. He concludes that in this special case it is the extension, or dimensive quantity, of the bread which carries the properties.[68] Ḥasdai Crescas in the late fourteenth

[65] See J.A. Wheeler, *Geometrodynamics*, New York 1962; Charles W. Misner, 'Some topics for philosophical enquiry concerning the theories of Mathematical Geometrodynamics and of Physical Geometrodynamics', in K.F. Schaffner and R.S. Cohen, eds, *Proceedings of the Biennial Meeting of the Philosophy of Science Association 1972*, Dordrecht 1974, 7-29, esp. 26; Lawrence Sklar, *Space, Time and Space-Time*, Berkeley, Los Angeles 1974, 166.

[66] W.K. Clifford, *The Common Sense of the Exact Sciences*, 1870, repr. N.Y. 1946, discussed by J. Graves, *The Conceptual Foundations of Contemporary Relativity Theory*, Cambridge Mass. 1971.

[67] See Michael Redhead, 'Quantum field theory for philosophers', *Proceedings of the Biennial Meeting of the Philosophy of Science Association*, Dordrecht 1984, vol. 2, and Heinz Post, 'Individuality and physics', *The Listener*, 10 October 1963, 534-7, which is summarised by W. Quine, 'Whither physical objects?', in R.S. Cohen, P.K. Feyerabend, M. Wartofsky, eds, *Essays in Memory of Imre Lakatos*, Boston Studies in the Philosophy of Science 39, 1976, 497-504, at 498-9.

[68] Thomas Aquinas *Summa Theologiae* IIIa, q. 77, a. 2; *Commentary on the Sentences of Peter Lombard*, book 4, dist. 12, q. 1, art. 2, solutions 1-3. See E. Sylla, 'Autonomous and handmaiden science: St. Thomas Aquinas and William of Ockham on the physics of the Eucharist', in J.E. Murdoch and E. Sylla, eds, *The Cultural Context of Medieval Learning*, Boston Studies in the Philosophy of Science 26, 1975, 349-96; id., 'Godfrey of Fontaines on motion with respect to quantity of the Eucharist', in A. Maierù and A.

and early fifteenth centuries draws on two cosmological ideas. First, it had been a tradition among some of Aristotle's commentators up to Averroes that the heavens were simple, not a compound of matter and form.[69] Secondly, Crescas draws on the idea of Farabi and Averroes that corporeal form, the form that makes things bodies, is just the indefinite dimensions which were of such interest to Simplicius and Philoponus.[70] He then takes Averroes to mean that the heavens consist of these indefinite dimensions, and he adds his own view that the lower elements, earth, air, fire and water, consist of these indefinite dimensions endowed with the qualities hot, cold, fluid and dry.[71] Such a view has been detected in others as well.[72]

Corresponding ideas are also found in modern philosophy. On the one hand, there is Carnap's view, elaborated by Quine, but resisted by Russell, that instead of saying that a physical object was blue, we might pick out a portion of space-time, giving four numbers as coordinates, and describe the portion as blue.[73] On the other hand, Strawson regards it as feasible, though revealingly cumbersome, to eliminate talk of particular bodies, and describe the physical world by treating places, times, or place-times as subjects endowed with features.[74]

Paravicini Bagliani, eds, *Studi sul XIV secolo in memoria di Anneliese Maier*, Rome 1981.

[69] Aristotle mentions as one option that the celestial substance might have no matter, though the option he pursues is that it has matter which is not generable or perishable like the matter of other things, *Metaph.* 8.4, 1044b7-8. The reference must be to proximate matter (in this case the fifth element), not to prime matter, since that is *never* generable or perishable. Themistius takes up the first option and declares that the rotating body lacks matter, *in Cael.* 14,13-14, while earlier Dexippus maintains that the celestial substance consists either of form or of matter, but not both, referring to Aristotle *Metaph.* 12.1, 1069a30. Earlier again, Alexander (*in Metaph.* 22,2-3; 169,18-19; 375,37-376,1; *Mixt.* 229,8) says that the heavens have some body (the fifth element) as a substratum, but it should not be called matter, presumably because the heavens suffer only motion, not creation or destruction. Alexander does not offer the lack of matter as a solution to the question raised in *Quaest.* 1.15, 26, 29-27,29 (see Chapter One above) whether or not the heavens have the *same* matter as the sublunary elements, because that question concerns *prime* matter. Simplicius refers to the view that the heavens are matterless, but prefers to treat the fifth element as a special kind of matter, *in Cael.* 133,29-134,14.

[70] The passages in Farabi, Joseph ibn Ẓaddiḳ and Averroes are translated by Wolfson, op. cit., pp. 585-6, 588.

[71] Crescas, *The Light of the Lord*, proposition 10, part 2, translated by Wolfson, op. cit., 261-3, with comments.

[72] Wolfson, op. cit., 600, translates a passage concerning a later thinker, Leo Hebraeus. S. Pines cites the Islamic philosopher Suhrawardî as holding that body is indefinite dimension, but without giving the context of this belief: *Beiträge zur islamischen Atomenlehre*, Berlin 1936, 84; id., 'Études sur Awḥad al-Zamân Abu' l-Barakât al-Baghdâdi', *Revue des Études Juives* 4, 1938, 31, repr. in *Collected Works of Shlomo Pines*, vol. 1., Jerusalem and Leiden 1979.

[73] Rudolf Carnap, *Logische Syntax der Sprache*, Vienna 1934, p. 11, translated by Amethe Smeaton as *Logical Syntax of Language*, London 1937, pp. 12-13; Quine, 'Whither physical objects?'; Bertrand Russell, *Human Knowledge, its Scope and Limits*, London 1948, 89-96.

[74] P.F. Strawson, *Individuals*, London 1959, ch. 7, Doubleday Anchor edition 1963, esp. pp. 224-32.

Historians have perhaps been too free in finding 'extension' theories in later thinkers. I doubt if they should be attributed to William of Ockham,[75] and his predecessor Peter Olivi, or to Spinoza.[76]

Discussions later than antiquity have two interesting features which set them apart. First, many of them are concerned not to *analyse* our ordinary talk of particular bodies, but to *replace* it. Secondly, those that refer to space or space-time prevent these from being merely relative to bodies. Talk of particular bodies may even be dropped. Portions of space are distinguished by reference to properties or features, rather than to bodies. And physical matter depends for its existence on space or space-time rather than the other way round.[77]

---

[75] Weisheipl has argued that Ockham is not identifying extension and matter, but eliminating one of the two terms of this identification: extension is not a *res absoluta*; there is only matter, and the fact that its parts are spatially distinct. See James A. Weisheipl, 'The place of John Dumbleton in the Merton School', *Isis*, 50, 1959, 439-54, at 443-4; id., 'The concept of matter in fourteenth-century science', in E. McMullin, ed., *The Concept of Matter in Greek and Medieval Philosophy*, University of Notre Dame 1963, 1965², 147-69, at 157-62.

[76] Jonathan Bennett, *A Study of Spinoza's Ethics*, 88ff, esp. 91, 98, 103-5. Particular things for Spinoza are states or modes dependent on the one substance, God (*Ethics*, part 2, proposition 25, corollary). Bennett suggests that Spinoza may think of particular bodies, though not minds, as regions of space (God here being identified with space), endowed with properties. Others have seen the suggestion that particulars are modes or states more in line with the thesis to be discussed in Chapter 4 below, as saying that a particular thing is a bundle of states.

[77] I am very grateful for discussion to Bas van Fraassen, Ted McGuire, Michael Redhead, Gisela Striker and Christian Wildberg.

CHAPTER FOUR

Bodies as bundles of properties

It was seen in Chapter 3 that Cornford took fire in Plato's *Timaeus* to be treated as just a bundle of qualities. That opens up an entirely different approach to the analysis of body. The idea so far has been that body is extension endowed with properties. The new idea is that the properties in a body, in a statue, for example, do not need a subject at all, whether that subject be an extension, a something I know not what, or even the statue itself. Rather a statue is simply a bundle of properties. In consistency, it should not then be described as *having* properties: it merely *includes* properties, as a bundle of sticks includes, but does not have, sticks. This last implication is seldom acknowledged, and the statue is still described as having properties. But such talk must be reinterpreted, for it treats the statue as a *subject* of properties, instead of as a whole made up of properties as parts. At most, it rejects the need for *prime matter* as subject, but many different theories would be compatible with that.

Some kind of bundle view has been detected very early in Greek thought, but its early manifestations are of less interest, for they are due to a failure to perceive a clear difference between properties and stuffs. Thus the Presocratic Anaximander (born 611 B.C.?) has been understood to say that the hot, the cold, the fluid and the dry separated out and became earth, air, fire and water. It is said that the hot here is conceived like an ingredient stuff.[1] Empedocles (c.495-c.435)[2] and more tentatively Anaxagoras (c.500-c.428)[3] have been credited with similar views according to which some or all bodies consist simply of qualities treated like stuffs, although in the case of Anaxagoras I shall dissent from this interpretation in Chapter 5 below.

Bundle theory becomes more interesting once properties have been distinguished as a different kind of thing from stuffs, as they are, for

---

[1] F.M. Cornford, 'Anaxagoras' theory of matter', *Classical Quarterly* 24, 1930, 14-30, 83-95, repr. in R.E. Allen and D.J. Furley, *Studies in Presocratic Philosophy*, vol. 2, London 1975, 299 and n. 59, citing Aristotle *Phys.* 187a20 and Simplicius *in Phys.* 24,24.

[2] Cornford, loc. cit.

[3] Gregory Vlastos, 'The physical theory of Anaxagoras', *Philosophical Review* 59, 1950, 31-57, repr. in Allen and Furley, op. cit., 323-53; Malcolm Schofield, *An Essay on Anaxagoras*, Cambridge 1980.

example, in Aristotle's scheme of categories, and before that in Plato. It then becomes a startling thesis that bodies might none the less be bundles of properties. It might be expected that Aristotle's introduction of prime matter as an ultimate subject of properties would be an insuperable obstacle to the introduction of bundle theory, but this is not so. We have seen how elusive was Aristotle's conception of prime matter. Moreover, Plotinus objects not to its elusiveness, but to its lack of being and reality. It is close to being a nonentity, a mere 'shadow upon a shadow'.[4] Plotinus' objection was influential, and was followed by Porphyry[5] and by Simplicius.[6] It makes no difference that Simplicius identifies prime matter with something comparatively familiar (extension). Given this background, it is not altogether surprising to find that Neoplatonists who accept Aristotelian prime matter none the less talk as if for Aristotle individuals are just bundles of properties. The Christian Basil of Caesarea goes even further, and denies the existence of an ultimate subject (*hupokeimenon*) outright. He says that if you mentally stripped away a thing's qualities, there would be *no* subject left.[7] His brother Gregory of Nyssa pointedly omits any reference to a subject in two of the three passages that will become relevant below.[8]

I shall start the story from Plato, whose *Timaeus* was discussed in Chapter 3. Cornford based his 'bundle' interpretation of the *Timaeus* on Plato's saying that fire is not a *this*, but 'what is *such and such*, and always alike as it comes round again', and on his illustrating what he means by 'such and such' with the examples of hot and white.[9] I took Cornford's interpretation to be possible, but not mandatory. But I now want to move on to another work of Plato's, the *Theaetetus*. There Plato's Socrates talks to Theaetetus about the special characteristics which Theaetetus shares with no one else, for example, his peculiar type of snubnosedness and, he adds, 'the others (sc. the other unshared characteristics) of which you are composed (*ex hôn ei su)*'.[10] Taken strictly, the expression implies that Theaetetus is just a bundle of unshared characteristics. The phrase is only a passing one, and nothing hangs on pressing the implication. But the insistence that the characteristics of which he is composed should not be ones common to him and others (*koina*, 209A10) may have been influential, as we shall see.

Still more influential was a second passage in the *Theaetetus* which

---

[4] Plotinus 6.3.8(34-7).

[5] Porphyry ap. Simplicium *in Phys*. 230,34-231,7.

[6] Simplicius *in Phys*. 230,29.

[7] Basil in *Hexaemeron* 1.8,21A-B.

[8] Gregory in *Hexaemeron* PG 44, col. 69B-C; *de Anima et Resurrectione* PG 46, cols 124B-D.

[9] Plato *Tim*. 49E-50A.

[10] Plato *Theaetetus* 209C7, stressed by Julius Moravcsik, 'Sumplokê eidôn and the genesis of *logos*', *Archiv für Geschichte der Philosophie* 42, 1960, 120, followed by John Rist, *Plotinus, The Road to Reality*, Cambridge 1967, 105.

actually uses a word for bundle: *hathroisma* (collection).[11] The thing called a collection is not, I believe, the class of all men or stones, but rather each individual man or stone. That Plato is concentrating on individuals is clear from what immediately precedes.[12] He is describing a theory of perception which will make it necessary to resist the temptation to call things 'somebody's', 'mine', 'this', or 'that'. The temptation will surely not be felt about the class of stones, but only about individuals. And the evidence that they are individuals is confirmed, when we hear that we should speak of the things in question as coming into existence, being produced, being destroyed and altering. The treatment of individuals as bundles serves to keep in view the main point of the passage, the lack of stability introduced by the theory of perception into the physical world, whereas a stress on the class of all stones would seem to lack motivation. It is, then, each individual stone which is being called a collection or bundle, and if the kind can be called a bundle, it will be only thanks to the individuals.

If this is correct, it will resolve some of the ambiguities that have been found in the passage, and I would render it as follows:

> We must speak this way both about the [sc. ingredients] taken singly (*kata meros*) and about many of them collected together (*hathroisthentôn*) into a collection (*hathroisma*), to which people give the name of man, or stone, or of each animal, or other kind (*eidos*).

A man or stone is a bundle, then, but a bundle of what? The commonest answer is a bundle of qualities,[13] but this seems impossible, for a few lines earlier a white stone was treated not as consisting of qualities only, but as consisting partly of something called a 'parent', which generated qualities.[14] The parent, then, must be one of the ingredients, along with qualities, in the bundle which constitutes the stone. If so, *Theaetetus* 157B-C does not, like 209C, treat man as a bundle just of properties, but it may have encouraged others to do so. A man remained the commonest example of an individual in the literature, but theories of the individual would apply equally to stones and all bodies.

The word *hathroisma* used by Plato was subsequently associated with some entirely different doctrines in Aristotle, which are not concerned with body at all, but with mind. At the end of the *Posterior Analytics*, in book 2, ch. 19, Aristotle is describing how we form

[11] *Theaetetus* 157B8-C2. This text and many of those that follow have been collected by A.C. Lloyd, 'Neoplatonic logic and Aristotelian logic', *Phronesis* 1, 1955-6, 158-9, and *Form and Universal in Aristotle*, Liverpool 1981, 67-8, and I am very grateful to him for drawing my attention to them.

[12] *Theaetetus* 157B1-8.

[13] F.M. Cornford says 'perhaps', *Plato's Theory of Knowledge*, London 1935, 48, n. 1; John McDowell offers it as one of two main interpretations, *Plato Theaetetus*, Oxford 1973, 143.

[14] *Theaetetus* 156E4-7. I thank Alan Towey for this observation.

universal ideas. It has been debated whether he is talking about forming universal *concepts*, or universal *propositions*, but I believe that he rightly ignores this distinction.[15] To acquire the *concept* of a lunar eclipse is to learn the *proposition* that a lunar eclipse is a so-and-so. And this applies whether we are talking of the pre-scientific concept described in *An. Post.* 2.8 of lunar eclipse as some sort of lunar loss of light, or of the scientific concept, which is what eventually interests Aristotle, and which he describes as a lunar loss of light due to the earth's acting as a screen. The process of concept-formation, as he conceives it, might be described as one of making bundles in the mind, and Themistius does use the description,[16] although Aristotle himself does not. A plurality of sense-images (*aisthêmata*) leaves memory as a residue, many memories constitute experience, from experience or the universal in the mind comes the origin of technology and scientific understanding (*tekhnê*, *epistêmê*). That is how reason (*logos*) is formed.[17] There is no suggestion that the process is one of amalgamating mental images, but in a quite different work Aristotle does suggest that that happens in the use of the imagination for deliberating what to do. In deciding whether to do this or that, in order to get the bigger share, you have to use one measure for the different alternatives, and so must be able to make one image (*phantasma*) out of many.[18] In the light of these passages, many things came to be given the name *hathroisma* or *athroisma* and its cognates by Hellenistic authors: universal conceptions (*ennoiai*) formed by bundling *phantasmata*, technology (*tekhnê*), reason (*logos*), mathematical universals, and such universals as man, or body.[19] Equivalent words, concurrence (*sundromê*, our syndrome) and conglomeration (*sumphorêsis*) were also introduced.[20] But if the universal man is a bundle of animal, mortal and rational, or body is conceived as a bundle of shape, size, resistance and weight, this does not yet imply that *individual* men or bodies are bundles.

The first really clear statement of the theory that an individual is

[15] This complements my argument in *Time, Creation and the Continuum*, ch. 10, that non-discursive thought is the entertaining not of isolated concepts, but of whole propositions, at least for Aristotle. As regards Plotinus, see the reply to me by A.C. Lloyd, 'Non-propositional thought in Plotinus', in *Phronesis* 31, 1986, 258-65. For earlier debate on whether Aristotle is discussing propositions or concepts in *An. Post.* 2.19, see Jonathan Barnes ad loc. in *Aristotle's Posterior Analytics*, Clarendon Aristotle Series, Oxford 1975, and F. Solmsen, *Die Entwicklung der aristotelischen Logik und Rhetorik*, Berlin 1929, 95-101

[16] Themistius *in An. Post.* 63,20 and 29.

[17] Aristotle *An. Post.* 2.19, 99b36-100a9.

[18] Aristotle *DA* 3.11, 434a7-10.

[19] Universal conceptions: Theophrastus ap. Sextum *M* 7.224. Technology, reported by Sextus *M* 7.373. Reason: Chrysippus ap. Galenum *de Hipp. et Plat. plac.* 5.3(160), p. 421 M (=*SVF* 2.841, line 21). Mathematical universals: Stoics ap. Proclum *in Eucl.* 12.6-7; 15.17. Man: The Academy, ap. Sextum *M* 7.276-7. Body: Epicurus ap. Sextum *M* 9.257.

[20] e.g. Sextus *M* 7.278; Epicurus *Letter to Herodotus* in Diogenes Laertius *Lives* 10,68-9.

just a bundle (*athroisma, sumphorêsis*) of properties is to be found in Epicurus, who formulates it only in order to reject it:

> Now as for shapes and colours and sizes and weights and all the other things predicated of a body as permanent properties (*sumbebêkota*), whether belonging to all bodies or to those that are visible and themselves knowable through sense perception, we should not think that these are independent natural substances (*phuseis*): that is not conceivable. Nor on any account should we think that they do not exist, nor that they are some distinct incorporeal entities accruing to the body. Nor should we think that they are parts (*moria*) of the body, but [we should think], that the whole body cannot have its own eternal nature composed wholly of (*katholou ek*) all these things conglomerated (*sumpephorêmenôn* [the word later used in Plotinus]) as when a larger collection (*athroisma*) is composed of actual particles (*onkôn*), either of primary ones [i.e. atoms], or of [other] quantities smaller than the whole in question.[21]

Whatever Epicurus' target may be in this repudiation of bundle theory, it will not be the Stoics. They did not compose individual bodies entirely (*katholou*) of properties as parts. For one thing, they believed in prime matter. Admittedly, Sextus tells us that they thought of virtue as a *part* (*meros*) of the good man.[22] But this connects with the doctrine, to be discussed in Chapter 6, that each quality of a body is to be identified with that body (or its matter, or *pneuma*, or soul, or reason) *disposed* in a certain way. Thus a man's courage and prudence are not two *different* parts of him, but are his *pneuma* (the all-pervasive gas) disposed in two different ways. Again, his weight and height are not two *different* parts of him, but are his body, or matter, or *pneuma*, disposed in two different ways. They are thus the *same* part of him, and we cannot connect a *distinct* part with each quality. At most, a *localised* quality, for example, the colour of an eye, may be identified with a *part* of the *pneuma* disposed in a certain way. This may indeed explain Plutarch's use of the plural, when he reports the Stoics as thinking that qualities are *pneumata* (plural) and tensions of air (*tonoi aerôdeis*) which shape the parts of matter in which they are found.[23] It sounds as if the colour of my eye may be a *part* of my *pneuma* in a certain state of tension. But this still falls short of turning each distinct quality into a *distinct* part of me.

In spite of this, the Stoics did after all make a certain contribution to bundle theory. For Chrysippus brought into prominence the idea of a unique quality or bundle (*sundromê*)[24] of qualities which distinguishes an individual from all others, and which lasts as long as the individual. It has been argued[25] that one reason for his doing so was to meet the

---

[21] Epicurus *Letter to Herodotus* in Diogenes Laertius *Lives* 10,68-9.
[22] Sextus *M* 11.24 (*SVF* 3.75)
[23] Plutarch *Sto. Rep.* 1054A-B (*SVF* 2.449).
[24] Dexippus *in Cat.* 30,20-6; cf. Porphyry *in Cat.* 129,8-10.
[25] David Sedley, 'The Stoic criterion of identity', *Phronesis* 27, 1982, 255-75.

sceptical question put by the Academy, whether the Stoic 'wise man' had a means for discriminating infallibly between individuals and for avoiding deception.[26] But for the next example of a bundle theory applied to individuals, we should turn to a Platonist text, the *Didaskalikos*, a work of disputed authorship which summarises Plato's philosophy. The mss ascribe it to one Alkinoos; another attribution is to Albinus the Middle Platonist of the second century A.D., and another suggestion is that it is an epitome by the Stoic Alcinous of the account of Plato given by the Stoic Arius Didymus at the end of the first century B.C.[27] Certainly, the work combines Platonic, Aristotelian and Stoic doctrine. The relevant point is that the author speaks of fire and honey as each being an *athroisma*, and the context suggests that he is talking of individual instances of fire and honey, because he is discussing them in the context of objects of sense perception, which Aristotle at least declared to be particulars.[28] Unfortunately, however, he does not make clear what they are bundles of. Certainly, the bundle is likely to include qualities, because he has just been talking about those. On a reasonable interpretation, his discussion progresses from qualities like whiteness, to a thing which is white, to a bundle including not just whiteness, but all the manifold qualities of fire and honey. But it is not likely that the bundle consists solely of qualities, because two lines before the relevant piece of text, he reveals himself as a believer in matter (*hulê*), and so we should expect matter to enter the bundle as well. The passage runs as follows:

> And again among perceptibles some are primary, like the qualities (*poiotêtes*) such as colour, white (*khrôma leukotês*); others are incidental (*kata sumbebêkos*) such as what is white what is coloured (*to leukon to kekhrôsmenon*); and after that is the *athroisma* (collection) such as fire, honey. Hence in perception one kind will be perception of the primary and will be called primary, another of the secondary will be called secondary. Perception discriminates the primary perceptibles and the secondary with the aid of (*ouk aneu*) judgmental reason; but the collection is discriminated by judgmental reason with the aid of perception.[29]

If this work does not after all make the individual into a mere bundle of qualities without matter, we should turn instead to three Neoplatonist

---

[26] Cicero *Academica* 2.19-26; 56-8; 85-6 (partly in *SVF*, but very scattered; see Adler's index to *SVF*).

[27] The *Didaskalikos* is printed in C.F. Hermann's edition of Plato, vol. 6, Leipzig 1921-36, and in the Budé edition, ed. P. Louis, Paris 1945. For attributions, see J. Freudenthal, *Der Platoniker Albinos und der falscher Alkinoos*, Berlin 1879; M. Giusta 'Albinou epitomê ho Alkinoou didaskalikos?', *Atti Tor., Classe di Scienze morali, storiche e filologiche* 95, 1960-1, 167ff; John Whittaker, 'Parisinus Graecus 1962 and the writings of Albinus', *Phoenix* 28, 1974, 320-54; 450-6. There is an English translation by George Burges in vol. 6 of *The Works of Plato*, Bohn Classical Library, London 1854.

[28] Albinus (?) *Didaskalikos* in Hermann's *Plato*, vol. 6, p. 155, line 35.

[29] ibid. 156, 2-10. The example of fire suggests the influence of Plato *Tim.* 49-50.

texts which all have some reference to Aristotle, and which have been cited by A.C. Lloyd.[30] The Neoplatonists, like the Stoics before them, accepted a conception of prime matter, in some version or other.[31] But in the context of discussing Aristotle, they appear to omit it from their descriptions of individuals. Individuals are treated as if they were just bundles of properties. First, Porphyry, in his introduction to Aristotle's *Categories*, the *Isagôgê*, says that Socrates is composed of (*ek*) individuating qualities (*idiotêtes*), the *athroisma* of which would never come into being identically in anything else. These individuating qualities remind us of the Stoics and of Plato's *Theaetetus* 209C, discussed above.

> Socrates is called an individual, and so is this white thing and this son of Sophroniscus who is approaching (if Socrates is Sophroniscus' only son). So things like this are called individuals because each is composed of individuating qualities (*ex idiotêtôn sunestêken hekaston*), the conglomeration (*athroisma*) of which would never come into being identically in anything else. For the individuating qualities of Socrates would not come into being identically in another particular.[32]

Then Proclus ascribes a similar theory explicitly to the Aristotelian Peripatos. The individual comes into being out of (*ginesthai ek*) the concurrence (*sundromê*) of accidental properties (*sumbebêkota*), whose *athroisma* would never come into being in anything else.

> The Peripatos was wrong about the individual, for it thought that the individual came into being out of (*ginesthai ek*) the concurrence (*sundromê*) of accidental properties (*sumbebêkota*). This is why it defined the individual as: that whose conglomeration (*athroisma*) would never come into being in anything else. Thus they put together the better thing out of (*epoioun apo*) the worse, that is, out of accidental properties.[33]

Finally, Philoponus, commenting on the text of Aristotle's *Posterior Analytics* that was discussed above, explains that Socrates consists of (*sunistasthai ek*) individuating qualities (*idiotêtes*).

> Sense perception sees Socrates and Alcibiades, and takes an impression not only of the particular individuating qualities (*merika idiômata*) in them. (The particular individuating qualities are one man's being

[30] A.C. Lloyd, loc. cit.

[31] I say 'in some version or other', because Koenraad Verrycken has argued that Philoponus' *in An. Post.* was revised after 529, and so may presuppose his new and unorthodox conception of prime matter as three-dimensional extension (see 'The development of Philoponus' thought and its chronology', in Richard Sorabji, ed., *The Transformation of Aristotle*, forthcoming, London 1989, which summarises part of his Ph.D. diss., Louvain 1985).

[32] Porphyry *Isagoge* 7,19-27. Mario Mignucci, however, offers a different interpretation of the passage in work in preparation.

[33] Proclus ap. Olympiodorum *in Alcibiadem* (Creuzer) 204.

long-haired and pale, the other not), but also of some of the shared
characteristics that it finds in them, that is, their being animals, or
rational, or something of that sort. It then sends this along in the first
instance to the imagination ... For sense perception grasps not only
particulars, that is, the accidental (*sumbebêkota*) and individuating
qualities (*idiotês*) of which the particulars consist (*ex hôn sunestêkasin*),
but also the universal man, that is, some of the things which constitute
the universal man.[34]

How did these Neoplatonists get the idea that an Aristotelian
individual is just a bundle of qualities? Lloyd rightly points to
Plotinus' criticism of Aristotle's treatment of substance in the
perceptible (that is, the physical) world. Plotinus objects that Aristotle
does not deal with true substance in the intelligible world of Platonic
Forms. His sensible substance is a mere conglomeration (*sumphorêsis*)
of matter and qualities – Plotinus here repeats Epicurus' word.
Aristotle, he says, fails to realise that sensible substance gets its being
from true substance in the world of Forms, whereas matter does not
have or bestow being. It is a shadow upon a shadow, a picture and an
appearance. Porphyry and subsequent Neoplatonists tried, in differing
degrees, to reinstate Aristotle's view, and to argue that it was
complementary to Plato's[35] (the theme of the harmony of Plato and
Aristotle will become relevant to more than one issue in this book).[36]
But Plotinus' treatment of Aristotle's matter as a shadow would be
enough to leave them thinking that the only ingredient worthy of
mention in Aristotle's *sumphorêsis* was the qualities. This is what
Plotinus says:

> But then is perceptible substance some conglomeration (*sumphorêsis*) of
> qualities and matter? And are all these substance, when they are fixed
> together upon a single matter? But when each is taken separately, will
> one be quality and another quantity? Or will they be many qualities?
> And is that whose absence does not allow being to be completed a *part* of
> this substance, whereas what supervenes on a substance when it has
> come into being has its own position and is not hidden in the mixture
> that makes the so-called substance? I do not mean that when it is there
> with the others it is substance, completing a single volume of such and
> such a size and kind, but that elsewhere when it is not completing
> anything, it is a quality. I mean that not even in the former case is any of
> them a substance, but the whole composed of all of them is a substance.
> There should be no objection, if we make perceptible substance out of
> non-substances. For not even the whole is true substance; it imitates it,

[34] Philoponus *in An. Post.* 437,21-438,2.

[35] See P. Hadot, 'L'harmonie des philosophies de Plotin et d'Aristote selon Porphyre
dans le commentaire de Dexippe sur les Catégories', in *Plotino e il neoplatonismo in
Oriente e in Occidente*, Rome 1974, 31-47, translated into English in Richard Sorabji,
ed., *The Transformation of Aristotle*.

[36] See Chapters 12 (on Iamblichus' intellective theory of place) and 15 (on God as a
source of infinite power). For the supposed harmony of Plato and Anaxagoras see
Chapter 10 (on Anaxagoras' multiple worlds).

for true substance has being independently of the others which
accompany it and the others which are generated out of it, because it
truly is. But here what underlies does not generate, and is not adequate
to be a being, because the others do not come from it, but it is a shadow,
and a shadow upon a shadow, a picture and an appearance.[37]

The last Greek thinker to be discussed is the Christian Father
Gregory of Nyssa, who introduces an idealised version of bundle
theory. The problem Gregory considers is how an incorporeal God
could have created a corporeal universe, since cause ought to be like
effect. His solution is to say that a body is just a bundle (*sundromê*) of
thoughts and concepts (*ennoiai, noêmata*) that were in God's mind. The
concepts listed include familiar qualities like light, heavy, hot, cold,
colour, shape. They also include extension, but this is given no special
role as it is in the extension theories discussed in Chapters 1 to 3
above. Although Gregory uses the word matter (*hulê*), what he means
by it is only body, not prime matter. An underlying subject
(*hupokeimenon*) for the properties is mentioned only in one of the three
passages, and in omitting it elsewhere, Gregory is probably following
his brother Basil, whose explicit denial of an underlying subject was
referred to above. Insofar as Gregory does mention an underlying
subject, this may seem safe enough to him, after Plotinus and
Porphyry had deprived it of all reality. The passages run as follows:

> This being so, no one should still be cornered by the question of matter,
> and how and whence it arose. You can hear people saying things like
> this: if God is matterless, where does matter come from? How can
> quantity come from non-quantity, the visible from the invisible,
> something with limited bulk and size from what lacks magnitude and
> limits? And so also for the other characteristics seen in matter: how or
> whence were they produced by one who had nothing of the kind in his
> own nature? ... By his wise and powerful will, being capable of
> everything, he established for the creation of things all the things
> through which matter is constituted (*di' hôn sunistatai*): light, heavy,
> dense, rare, soft, resistant, fluid, dry, cold, hot, colour, shape, outline,
> extension. All of these are in themselves thoughts (*ennoiai*) and bare
> concepts (*noêmata*); none is matter on its own. But when they combine
> (*sundramein*), they turn into matter (*hulê ginetai*).[38]

The corporeal creation is thought of in terms of properties which have
nothing in common with the divine. And in particular it produces this
great difficulty for Reason (*logos*). For one cannot see how the visible
comes from the invisible, the solid and resistant from the intangible, the
limited from the unlimited, or what is in every way circumscribed by
quantitatively conceived proportions from what lacks quantity and
magnitude, and so on for everything which we connect with corporeal

---

[37] Plotinus 6.3.8(19-37). I thank Hilary Armstrong for allowing me to see his
translation in advance of publication.
[38] Gregory *in Hexaemeron* PG 44, col. 69B-C.

nature. But we can say this much on the subject: none of the things we connect with body is on its own a body – not shape, not colour, not weight, not extension, not size, nor any other of the things classed as qualities (*poiotês*). Each of these is an idea (*logos*), but their combination (*sundromê*) and union with each other turns into a body (*sôma ginetai*). So, since the qualities which fill out (*sumplêrômatikos*) the body are grasped by the mind and not by sense perception, and the divine is intelligent, what trouble is it for the intelligible (*noêtos*) being to create the concepts (*noêmata*) whose combination (*sundromê*) with each other produces corporeal nature for us?[39]

There is an opinion about matter which seems not irrelevant to what we are investigating. It is that matter arises from (*hupostênai ek*) the intelligible (*noêtos*) and immaterial. For we shall find all matter to be composed of (*sunestanai ek*) qualities (*poiotês*) and if it were stripped bare of these on its own, it could in no way be grasped in idea (*logos*). Yet each type of quality is separated in idea (*logos*) from the subject (*hupokeimenon*), and an idea (*logos*) is an intelligible not a corporeal way of looking (*theôria*) at things. Thus, let an animal or a log be presented for us to consider, or anything else which has a corporeal constitution. By a process of mental division we recognise many things connected with the subject and the idea (*logos*) of each of them is not mixed up with the other things we are considering at the same time. For the ideas (*logoi*) of colour and of weight are different, and so again are those of quantity and of tactile quality. Thus softness and two-cubit length and the other things predicated are not conflated with each other, nor with the body, in our idea of them (*kata ton logon*). For the explanatory formula (*hermêneutikos horos*) envisaged for each of these is quite individual according to what it is, and has nothing in common with any of the other qualities which we connect with the subject. If, then, colour is intelligible and so is resistance and quantity and the other such properties, and if upon each of these being removed from the subject the whole idea (*logos*) of body would be removed: what follows? If we find the absence of these things causes the dissolution of body, we must suppose their combination (*sundromê*) is what generates material nature. For a thing is not a body if it lacks colour, shape, resistance, extension, weight and the other properties, and each of these properties is not body, but is found to be something else, when taken separately. Conversely, then, when these properties combine (*sundramein*) they produce material reality. Now if the conception (*katanoêsis*) of these properties is intelligible (*noêtos*), and the divine is intelligible in its nature, it is not strange that these intellectual (*noeros*) origins for the creation of bodies should arise from an incorporeal nature, with the intelligible nature establishing the intelligible properties, whose combination (*sundromê*) brings material nature to birth.[40]

I am hesitant in translating *logos* as 'idea' and, in calling this version of bundle theory idealistic, I am relying not on that, but on the talk of

[39] Gregory *de Anima et Resurrectione* PG 46, cols 124B-D.
[40] Gregory *de Hominis Opificio* ch. 24, PG 44, cols 212-13.

thoughts and concepts (*ennoiai, noêmata*). The analogy with Berkeley's
theory that physical objects are bundles of ideas is remarkable,[41] and I
have elsewhere argued that the case of Gregory refutes the idea that
idealism had to wait for Berkeley.[42] The analogy is made closer by
what Berkeley says about the creation of the world. A supposed
advantage of his theory is that it saves us from a view of creation which
he considers absurd, that the world, which is conventionally conceived
as a corporeal substance existing outside the minds of spirits, should
be produced out of nothing by the mere will of a spirit. There is, by
contrast, no puzzle, he says, about how a spirit can at will produce
ideas.[43] Gregory sees comparable advantages in his own theory.
Further, Gregory's concession in talking in one passage about an
underlying subject (*hupokeimenon*) is matched by the concession which
Berkeley makes in his later work *Siris*.[44] He there says that the
ancient Greek concept of prime matter differs from the concept of
underlying matter that he is attacking, in being so negative a notion
that he can afford to allow it. I take it that he is not reneging on his
bundle theory, but rather treating the ancient concept of prime matter
as so negative as to be compatible with it.

Of course there are differences between the two thinkers. According
to Berkeley, the Creation takes the form of God making the ideas in his
mind perceptible to others, whereas, according to Gregory, it consists
at least partly in God's bundling the ideas together into a concurrence
(*sundromê*). Moreover, there is a difference to which I did not attend
before. Gregory thinks that God has first to lay down a certain void and
nothingness (*kenôma kai outhen*), which is to be understood as a power
(*dunamis*) for accommodating (*khôrêtikê*) qualities (although it is not
treated as a *subject* of qualities). This is how he understands the claim
in Genesis that at first the earth was without form and void.[45] A third
difference is that Berkeley insists, unlike Gregory, that ideas can exist
only in a mind, and seizes the opportunity, which Gregory misses, of
inferring the 'necessary and immediate dependence of all things' on
God.[46] Berkeley's point is that ideas can exist only so long as they are
entertained by a mind, and that accordingly the physical world would
perish, if God were to stop entertaining its component ideas. This
difference flows naturally in turn from a difference of motivation
between the two theories. Berkeley's aim is to combat *scepticism* by
getting rid of anything which, like an underlying subject, is
unknowable, and by making the physical world into something

[41] George Berkeley, *Principles of Human Knowledge*, 1710; *Three Dialogues Between Hylas and Philonous*, 1713.

[42] See Richard Sorabji, *Time, Creation and the Continuum*, London 1983, 290-4; Myles Burnyeat, 'Idealism in Greek Philosophy: what Descartes saw and Berkeley missed', *Philosophical Review* 91, 1982, 3-40.

[43] Berkeley, 3rd of *Three Dialogues Between Hylas and Philonous*.

[44] Berkeley, *Siris* 1744, 306, 317-20.

[45] Gregory *in Hex.*, PG vol. 44, col. 80B-C.

[46] Berkeley, 3rd of *Three Dialogues Between Hylas and Philonous*.

thoroughly knowable, namely, ideas. He therefore has a strong interest, which Gregory lacks, in the status of the ideas *after* the initial act of Creation.

The differences of motivation will appear more starkly if I say more than I have said before about the sources of Gregory's view. The ideas in God's mind are descendants of Plato's Ideas, although Plato did not put his Ideas into a mind at all.[47] They are also descendants of the so-called seminal reasons (*spermatikoi logoi*)[48] which the Stoics and later the Neoplatonists postulated as mechanisms of creation. Indeed, in the quoted passages, Gregory refers to the ideas in God's mind as *logoi*. Two things had happened, at latest by the first century A.D., to bring seminal reasons and Platonic Ideas into the story. First, we find Seneca and Philo in that century representing Plato's Ideas as thoughts in the mind of God.[49] Secondly, we find Philo equating Plato's Ideas with seminal reasons.[50] We can infer that seminal reasons are thoughts in the mind of God, and Gregory seems to have been responding to this tradition.

But there is another more immediate source of Gregory's view. Porphyry, whom we know Gregory to have read, at least in part,[51] raises the very same question as Gregory, and gives a different answer, but one appealing to seminal reasons. For Porphyry asks, in relation to God as cause of the physical world, 'How can the immaterial produce matter?'[52] He replies that it is perfectly possible, because we see the seminal reasons (*logoi*) in a drop of sperm producing all the parts of the body, although the drop of sperm is tiny, and the seminal reasons have no volume at all. So it should be all the more possible for the Creator's Reason (*logos*) to produce everything. Gregory takes up the Creator's seminal reasons and recalls that (in a certain tradition) they are thoughts in the Creator's mind. His innovation is that, instead of having them *produce* the material world, as does Porphyry, he has them *constitute* the material world.

[47] At most, in *Republic* 597B-C, Plato makes God create the Ideas. Not even this is true in the *Timaeus*.

[48] For the Stoics, see e.g. Diogenes Laertius *Lives* 7.135-6 (*SVF* 1.102; 2.580); Aëtius *Placita* 1.7.33 (*SVF* 2.1027; *Dox. Gr.* 306a3 and b3); Arius Didymus fr. 36 in *Dox. Gr.* 468a25 and b25 (*SVF* 1.107; 512; 2.596); and fr. 38 in *Dox. Gr.* 470,10 (*SVF* 1.497); Philo *Aet.* 9.94-103; Origen *Against Celsus* 1.37 and 4.48 (*SVF* 2.739 and 1074); Seneca *Nat. Quaest.* 3.29.3; Plotinus 3.1.7. For Plotinus' own view, see 2.3.16-18; 3.2.2(18-31); 3.3.1-5; 4.3.10-11 and 15; 4.4.11-12; 36; 39; 4.8.6(7-10); 5.9.6(10-24).

[49] Seneca *Ep.* 65.7; Philo *Opif.* 4.16-5.20; 7.29; *de Providentia* 1.21, translated into Latin from the Armenian by Aucher. See John Dillon, *The Middle Platonists*, London 1977, 95, 138.

[50] Philo *Legatio ad Gaium* 55.

[51] See P. Courcelle, 'Grégoire de Nysse, lecteur de Porphyre', *Revue des Études Grecques* 80, 1967, 402-6. This study concentrates on one work of Porphyry's, but I hope that doubts about the other works may be allayed by some of the evidence assembled here. See David L. Balás, 'Gregory of Nyssa and Neoplatonic ontology', *Studia Patristica* 18, 1985.

[52] Porphyry ap. Proclum *in Tim*. 1, 396,5-24.

Gregory's idealised bundles provide a fitting conclusion to our survey of antiquity, because they draw together the two streams that seemed at first so disparate, the bundles constituting a body in Plato and the bundles of the mind in Aristotle. Gregory's bundles are bodies, like Plato's, but the thoughts (*ennoiai*) of which they are composed have a multiple lineage. They recall both the mental items of Aristotle's *An. Post.*, and the mentalised version of Plato's theory of Ideas.

If we ignore the special motives behind Gregory's idealisation of bundle theory, the main motives for some kind of bundle theory in antiquity were the unreality of Aristotle's prime matter and Plotinus' stress on that, coupled with Stoic concern about the distinguishing qualities of the individual. In the bundle theories that have appeared since then, something like the first motivation has sometimes reappeared, the unsatisfactoriness for various reasons of an underlying subject. But in other respects the differences have been considerable.

Gregory's influence continued, according to the argument of a forth-coming book,[52a] and helped to inspire a version of idealism in the Irishman Eriugena (b. c. 810). In the East, we find another bundle theorist in Job of Edessa. He was a Christian and his encyclopaedic *Book of Treasures* was written in Syriac,[53] around A.D. 817, in Baghdad, in a period when translation from Greek and Syriac into Arabic was actively encouraged there. It covers almost all the natural and philosophical sciences taught in Baghdad at the time. In Discourse 1, Job argues (ch. 2) that prime matter is impossible. The basic elements (ch. 1) are hot, cold, fluid and dry, while the most fundamental bodies, earth, air, fire and water, are compounds made of these. Hot, cold, fluid and dry should be viewed as substances (*ûsiyas*).[54] They are not accidents (*gedsê*), or in modern terms properties belonging to a substance (ch. 3). Viewed in isolation, they are not three-dimensional, and they can be apprehended only by the mind (chs 1; 8). Although they can be thought of in isolation, they never exist in isolation from earth, air, fire and water (chs 1; 3). They are created out of nothing by God, as something external to his own nature (this excludes Gregory's idealism, ch. 6), and at the same time as the compounds which they form (ch. 1). Thereafter they persist contin-uously throughout the decomposition and recomposition of compound bodies (ch. 3). This account clearly treats earth, air, fire and water as bundles of qualities. If those qualities are denied their normal Aristote-lian status as accidents, this is not, as in Anaximander, due to a failure to distinguish between substance and accident. It is based on an argument that they fit the definition of substance, being imperishable

---

[52a] Dermot Moran, *The Philosophy of John Scottus Eriugena*, Cambridge forthcoming.

[53] Translated with Syriac text by A. Mingana, Cambridge 1935.

[54] The translation 'substance' is suggested by the sense, is endorsed by Paul Kraus *Jâbir ibn Hayyân*, Cairo 1943, 175, and is allowed by Mingana, p. xxvii. I do not see, however, that these qualities can be *corporeal* substances, or that the view owes anything to the Stoics, as Kraus suggests. For in themselves they are said not even to be three-dimensional.

and (as Aristotle's *Categories* requires) not resident in some further entity (ch. 3).

A few years earlier, Ḍirâr (died A.D. 815 at latest), a theologian of the Basra school, had produced a bundle theory, which apparently had a different basis. At least, it is reported as an attempt to do away with the distinction between substance and accident, by retaining only accidents.[55] A body consists simply of accidents, and these accidents cannot exist in isolation, but only as constituents in compounds. The fragment is too short to tell anything more, but the theory is said to have influenced other Islamic thinkers.

Ḍirâr b. 'Amr: a body [consists of] accidents so composed and assembled as to make a stable body susceptible of ... being changed from one state (or: moment) to another. These accidents are such that a body cannot but have either the feature in question or its contrary, like life and death (one of which will always be present in a body), colours and tastes (for a body will not be without a feature of either type), and similarly weight (heaviness and lightness), roughness and smoothness, hotness and coldness, moistness and dryness, and similarly compactness (?)

Ḍirâr rules out that these accidents should assemble to form bodies at a time when they already exist themselves. He thinks it absurd that this should happen (lit.: be done) to them at any time except at the beginning of their existence: they cannot come into existence except as an ensemble. Thus according to him they can be both united and existent, but not both separate and existent. For if they were both separate and existent, a colour could exist without belonging to something coloured, and life could exist without belonging to something alive.[56]

In the bundle theories of the twentieth century,[57] dissatisfaction with the idea of an ultimate subject or substratum has once again played a

[55] I am grateful to Fritz Zimmermann for this information.

[56] Ash'arî Maqâlât, pp. 305f.

[57] The main twentieth-century proponents are Russell and Stout: Bertrand Russell, *An Inquiry into Meaning and Truth*, London 1940, ch. 6; *Human Knowledge, its Scope and Limits*, N.Y. 1948, part 2, ch. 3; part 4, ch. 8; *My Philosophical Development*, ch. 9; G.F. Stout 'The nature of universals and propositions', *Proceedings of the British Academy* 10, 1921, repr. in his *Studies in Philosophy and Psychology*, London 1936, 384-403. For a variety of other versions, see F.H. Bradley, *Appearance and Reality* 1893, repr. Oxford 1955, p. 16; B. Blanshard, *The Nature of Thought*, London 1939, ch. 17; *Reason and Analysis*, London 1962, ch. 9; D.C. Williams, 'The elements of being', *Review of Metaphysics* 6, 1953, 3-18; 171-92; A.J. Ayer, 'The identity of indiscernibles' in his *Philosophical Essays*, London 1954, repr. in Michael J. Loux, *Universals and Particulars*, Notre Dame 1976; H. Hochberg, 'Universals, particulars and predication', *Review of Metaphysics* 19, 1965, pp. 87-102; 'Things and descriptions', *American Philosophical Quarterly* 3, 1966, 39-47; 'Moore and Russell on particulars, relations and identity', in E.D. Klemke, ed., *Studies in the Philosophy of G.E. Moore*, Chicago 1969; H. Hiz, 'On the abstractness of individuals', in Milton K. Munitz, ed., *Identity and Individuation*, N.Y. 1971, 251-61; H.N. Castaneda, 'Perception, belief and the structure of physical objects and consciousness', *Synthese* 35, 1977, 285-351, esp. 321-6; Ralph W. Clark, 'The bundle theory of substance', *New Scholasticism* 1976, 490-503.

role. But there are differences, and one that we have already noticed in Berkeley's idealised theory is the concern with scepticism. An underlying substratum is unsatisfactory because it would be unknowable, and twentieth-century discussions stress its imperceptibility to the senses. On the other side, those who have reinstated such a substratum have sometimes done so from a concern with particularity which is quite foreign to Aristotle. They have challenged bundle theorists to say what would make a bundle of properties into a particular, if properties are universals, and what would make one particular bundle differ from another.[58] Their hope that a substratum might solve these problems is not to be found in Aristotle, who did not have prime matter in mind, when he said that Callicles and Socrates were differentiated by their matter.[59]

Before leaving bundle theories, we should ask how they compare with the extension theories of Chapters 1 to 3. Many difficulties have been raised against them.[60] My own view is that they have a lot more explaining to do than extension theories, and it is hard to know in advance whether or not they can do it. To give one example, extension theories rely on a set of relationships which are at least familiar, and whose existence is scarcely in doubt, even though they require further analysis – those between a subject and its properties. But bundle theories must decide from scratch what relationships tie together the properties of one bundle, without tying together properties in different bundles, and whether the requisite relationships are available. Mere contiguity will not do: the colour of the tea I have drunk may be contiguous to, without being part of, the bundle that constitutes me, whereas my thoughts may be part of the bundle without being contiguous. Moreover, diachronic relations must be specified as well as synchronic: since modern theorists do not postulate life-long properties, like the Stoics, they may escape having to say what makes a property the same property from one time to another. But they will need to say what makes two different properties found at different times belong to the same bundle. Nor can bundle theorists claim to escape the necessity that confronts extension theorists of referring to distinct regions. For they must acknowledge that different properties are displayed in different bodies and in different parts of the same body. But if bundle theories are at a disadvantage in comparison with

[58] Edwin B. Allaire, 'Bare particulars', *Philosophical Studies* 16, 1963, reprinted with a series of attacks and defences in Michael J. Loux, ed., op. cit.; G. Bergmann, *Logic and Reality*, Madison Wisconsin 1964.

[59] Aristotle *Metaph.* 7.8, 1034a7-8.

[60] See for example D.C. Long, 'Particulars and their qualities', *Philosophical Quarterly* 18, 1968, reprinted in Michael J. Loux, op. cit., along with Edwin B. Allaire, op. cit., and Michael J. Loux, 'Particulars and their individuation'. See also D.M. Armstrong, *Universals and Scientific Realism* I: *Nominalism and Realism*, Cambridge 1978, 77-101. Many problems are tackled, however, including those mentioned here, by Richard Sikora in unpublished work which he has been kind enough to let me see.

extension theories, this is not to say that we need accept either. For everyday purposes, I see no reason to abandon the view that the properties of a statue have a subject, and that it is the *statue*.[61] For scientific purposes, we must be guided by scientific developments. But so far, there has been some preference in micro-physics for theories which supply a *subject* for properties, whether that subject be space, or space-time, or something whose exact character is more controversial, namely, a field.

---

[61] This preference is urged by D.C. Long, op. cit. For my debt to A.C. Lloyd, see nn.11 and 30.

CHAPTER FIVE

Can two bodies be in the same place? The theory of chemical combination* from Anaxagoras to Aristotle

Can two bodies be in exactly the same place? On first reflexion, I believe, most people will think this absolutely impossible, and they will have leading contemporary philosophers on their side.[1] But among the Greeks it was a matter of controversy. One relevance was to theories of chemical combination: must we avoid saying that the ingredients in a mixture have got into exactly the same place? The next two chapters will consider the history of this question from Anaxagoras to the Stoics. But in the Stoics, and still more in the Neoplatonists and Christians, who will be the subject of Chapter 7, there are metaphysical as well as chemical reasons for concern with the issue.

It links also with the analysis of body which was the subject of Chapters 1 to 4. For body is sometimes defined as solid. That can mean various different things: that bodies cannot be in the same place, that they cannot even contract (as the atomists held of their atoms), that they offer some degree of resistance before contracting or getting into the same place. These different meanings of 'solidity' are in various ways connected. But philosophers have not agreed whether solidity in any of these senses should be put into the definition of body. Epicurus and allegedly the Stoics defined body as having resistance (*antitupia*),[2] but Philoponus conspicuously left this out of an otherwise similar definition,[3] and the Stoics did not always use it.[4] The disagreement was echoed in later centuries, when Descartes defined body merely as extension and Locke and Leibniz responded that body must be solid.[5]

---

* In what follows, I shall use the conventional translations 'mixture' and (for the Stoics) 'blending', although the modern term would be 'combination'.

[1] Anthony Quinton, 'Matter and Space', *Mind* 73, 1964, 332-52; David Wiggins, 'On being in the same place at the same time', *Philosophical Review* 77, 1968, 90-2.

[2] Epicurus ap. Sextum *M* 1.21; Stoics ap. ps-Galenum *de qualitatibus incorporeis* 10 (=*SVF* 2.381), and ap. Plotinum 6.1.26(20).

[3] Philoponus *contra Proclum* 11,1-8, pp. 405-45, as described in Chapter 1.

[4] Not when they describe qualityless matter as a body – see Chapter 6.

[5] Descartes, *Principles of Philosophy*, part 2, secs 8-12; Locke, *Essay Concerning Human Understanding* 2.8.8-9; cf. 2.4; Leibniz, *New Essays Concerning Human Understanding* 2.4, and *Conversation of Philarete and Ariste*.

'Solidity' covers more than one type of property in Locke and Leibniz, but the idea that interests me is the idea that bodies cannot interpenetrate, in other words, that they cannot be, and their parts cannot be, in exactly the same place. This idea was not accepted by all the Greeks: the story goes back to the early stages of Greek Philosophy, and I shall start with the Presocratic philosopher Anaxagoras (c.500-c.428 B.C.).

Anaxagoras

Anaxagoras held that there is a portion of everything in everything. There is a tradition that he was interested in nutrition and growth.[6] Eating honey nourishes my flesh and bones, and this is because there is flesh and bone in the honey. There will be, in tinier quantities, gold and lead in the honey, and some of every such stuff. There will also be every kind of opposite quality, density and rarity, for example:

> And because the portions of the large and of the small are equal in number, for this reason too everything must be in everything, and nothing can be separate (*khôris*), but everything has some portion of everything. Because there cannot be such a thing as the least, things could not be separated (*khôristhênai*), nor come to be on their own, but as in the beginning, so now, all things must be together ... In everything there is a portion of everything except of Mind, and in some things Mind too is present ... All other things have some portion of everything, but Mind is infinite, self-ruling, and mixed with nothing. It alone is by itself.[7]

A further part of Anaxagoras' view is that:

> Each thing is and was most manifestly those things of which it has most in it.[8]

This implies that a thing is most manifestly honey, if there is more honey than any other ingredient in it.

How are we to imagine the ingredients in honey mixed? On one view, Anaxagoras restricts the ingredients of honey to qualities.[9] A variant of this qualitative view is that stuffs are just bundles of qualities – the idea which was the subject of Chapter 4 – so that bone would be in

[6] Aristotle *GA* 1.18, 723a10-11; Schol. Greg. Naz. PG 36,911 (=Anaxagoras fr.10 Diels-Kranz); Aëtius 1.3.5 (=Anaxagoras A46 Diels-Kranz); Simplicius *in Phys.* 460,6-22.

[7] From Anaxagoras, frs 6, 11, 12 Diels-Kranz.

[8] Anaxagoras fr. 12 Diels-Kranz, ad fin.

[9] Paul Tannery, *Pour l'histoire de la Science Hellène*[2], Paris 1930, 296ff; John Burnet, *Early Greek Philosophy*[4], London 1930, 263; F.M. Cornford, 'Anaxagoras' theory of matter', *Classical Quarterly* 24, 1930, 14-30 and 83-95; Gregory Vlastos, 'The physical theory of Anaxagoras', *Philosophical Review* 59, 1950, 31-57; Malcolm Schofield, *An Essay on Anaxagoras*, Cambridge 1980, 107-21. Cornford and Vlastos are reprinted together in R.E. Allen and David J. Furley, *Studies in Presocratic Philosophy*, vol. 2, London 1975, see esp. pp. 305-13 (Cornford); 335-41 (Vlastos).

honey only in the roundabout sense that the qualities needed to constitute bone are in honey.[10] But it requires a lot of skill to explain away the references to stuffs alongside qualities in the texts.[11] And the compensating advantages nowadays claimed for the qualitative interpretation are only ones of philosophical economy. In addition, I shall argue below that the qualitative interpretation cannot explain the role Anaxagoras gives to infinite smallness and the absence of minimum size, especially in fragment 6.

A second interpretation envisages that in honey there are patches which largely resemble bone, but also a predominance of patches which largely resemble honey. Only largely, because in the bone patches there are smaller patches which largely resemble honey, but again only largely, because they too contain still smaller patches which largely resemble bone, and so *ad infinitum*. The infinite regress in this patchwork model has rightly been declared benign.[12] Nor should it be objected that if honey is contaminated only in patches, it should be possible to scoop out from around the patches pure honey, and that pure honey would fail to contain a portion of everything, as Anaxagoras wants it to. No such pure honey could be scooped out, because we are to imagine that any portion of honey, however small, contains contaminating patches. The model is not subject to these objections, but as an interpretation of Anaxagoras I think it must fail for a well documented reason.[13] Anaxagoras is said to have believed that many stuffs are homogeneous, having parts and whole all like each other.[14] Flesh, for example, differs from a face in that parts and whole not only contain the same ingredients, but contain them in the same ratio. This excludes flesh being an irregular patchwork containing areas that largely resemble bone. Rather bone needs to be distributed uniformly throughout flesh and, for that matter, throughout honey.

A third interpretation would achieve this uniform distribution by appealing to the blending of liquids.[15] The result is like what one would have if flesh, bone, gold, lead and honey were homogeneous

[10] Vlastos (338 in Allen and Furley); Schofield 116.

[11] Certainly Aristotle thought that there were stuffs alongside qualities in everything, *Phys*. 1.4, 187b1-188a5; 3.4, 203a25-6; *GA* 1.18, 723a10-11; *Cael*. 3.3, 302a32; *GC* 1.1, 314a19-20; a27-8; and his view was followed, as regards stuffs, by Lucretius 1,859ff; Simplicius *in Phys*. 27,8-9; 460,15-28; Schol. in Greg. Naz., PG 36,911. This is also the easiest reading of Anaxagoras frs 4 and 10 Diels-Kranz, the second of which comes from the Schol. in Greg. Naz.

[12] Vlastos in Allen and Furley, 350-1.

[13] Cornford, endorsed by Vlastos in Allen and Furley, 335-6.

[14] Aristotle *Phys*. 1.4, 187a23ff; *Cael*. 3.3, 302a28ff; *Metaph*. 1.3, 984a11; Lucretius 1,830ff; Simplicius *in Phys*. 460,4ff.

[15] Aristotle reports a description of such a blend for different purposes in *GA* 4.3, 769a28. It is explained by George Kerferd, 'Anaxagoras and the concept of matter before Aristotle', *Bulletin of the John Rylands Library* 52, 1969, 129-43 (repr. in Alexander P.D. Mourelatos, Garden City N.Y. 1974, see p. 499), and by Jonathan Barnes, *The Presocratic Philosophers*, London 1979, vol. 2, 23-4. It is also presupposed by Colin Strang in the work cited below.

rather than being, as we should think, composed of molecules or atoms, had at first been isolated from each other (which they were not), and had then been liquidised, stirred together into a perfect blend and re-solidified, to make ordinary impure honey. This would not leave us with isolated bone and isolated honey juxtaposed in particles. (Such particles would fail to contain a portion of everything.) Rather, the blend would be perfect in the sense that wherever there was honey, there was also bone. Alternatively, since the bone in our honey would never in fact have been isolated, we may prefer to think of it as elemental bone.[16] Elemental bone differs from ordinary bone in not being a compound, but differs from isolated bone in that it exists *only* as an ingredient in compounds. Though never isolated, it can become increasingly concentrated, as it does when digestion extracts bone from honey. Bypassing the fiction of isolated bone, the blending model can describe honey as a blend of elemental (i.e. non-compound) bone and honey.

Although the blending model is, I believe, philosophically the most elegant, it shares a difficulty with the qualitative model from which we started. First these models leave us seeking a motive for Anaxagoras' insistence (fragment 1) that in the beginning all things were infinitely small (*panta khrêmata apeira smikrotêta*). Worse, I believe they invalidate the reasoning of fragment 6. Anaxagoras there gives it as a reason (*hote*) why you do not get nuggets of bone or honey in isolation that a minimum size is impossible:

> Because (*hote*) there cannot be such a thing as the least (*toulakhiston*), things could not be separated, nor come to be on their own, but as in the beginning, so now, all things must be together.

But why, on these models, should minimum size give us stuffs in isolation? It cannot be argued on the blending model (to concentratrate on it) that a minimum volume of honey would not have *room* for elemental bone, but only for elemental honey. On the contrary, if there were a minimum volume of compound honey, elemental bone would be accommodated within it not by occupying a smaller *volume*, but by being spread throughout the mixture at a lower *density* than that of the elemental honey.[17] It is only the *density*, not the *volume*, that would be reduced, so there would be no need for the bone to fall below the minimum permissible volume.

---

[16] The distinction is proposed, though not as explicit in Anaxagoras, by Colin Strang, 'The physical theory of Anaxagoras', *Archiv für Geschichte der Philosophie* 45, 1963, 101-18, repr. in Allen and Furley, vol. 2.

[17] The lower density guarantees that the whole is not bone, but honey. In my interpretation I take Anaxagoras to mean that if there *were* a least volume, honey and bone *would* 'come to be on their own', not the converse (that if honey and bone 'came to be on their own', there would be a least volume), since it is not obvious why anyone should think the latter. Jim Lennox (loc. cit. n.38 below) has offered a reason, but *infinite divisibility* ought not to yield a least volume, while *simplicity* does not seem relevant to volume at all.

If none of these models, the qualitative, the infinite patchwork, or the blending model, fits what Anaxagoras says, I would propose a fourth model. Anaxagoras may have thought in terms of an infinitesimal powder. The mixture is like what one would have, if bone and honey had (counterfactually) started in isolation, and had been ground to powder and stirred together, only the grinding would have gone on until the particles were 'infinitely small'. Alternatively, if the fiction of isolated bone is unwelcome, the powder model can be expressed in terms of the idea that elementary (i.e. non-compound) bone and honey exist in ordinary honey in infinitely small units. The units will need to be sizeless, to avoid nuggets of isolated bone or honey. For in a positively sized nugget of bone, there would be inner parts not surrounded by honey; whereas in an infinitesimal powder every bit of bone could be together with some honey.

The meaning of fragment 6 now becomes clear. For if there were a minimum positive size of grind, isolated nuggets are exactly what we would get. When Anaxagoras says that there is no least, he has in mind that there is no least *nugget*. He is not thinking of his infinitesimal units, which would in any case be better described as lacking a size than as having a least size. The doctrine of something of everything in everything will mean that there are infinitesimal units of each kind *either* at every point, *or* at least in every positively sized region however small. Anaxagoras will not necessarily have asked himself which of these two alternatives to choose, or worked out the details of his theory. But at least in broad outline, I think the powder model fits most closely with what he says. And it has the further advantage that it looks like the theory of mixture which Aristotle envisages and attacks, though without naming anyone in *GC* 1.10, 328a15-17. In a mere juxtaposition, he says, such as that of grains of barley and wheat, there will be parts of each barley grain that are not alongside (*para*) any part of a wheat grain. The type of division needed to get them alongside would be impossible. I take it he means that it would have to grind the grains down into infinitesimal units, for then every unit of barley could be alongside some unit of wheat.

It must be admitted that Anaxagoras does not make it clear what his theory is. Probably for this reason, Aristotle tries out more than one interpretation on him. It must also be admitted that the powder model is not an attractive one for Anaxagoras to hold. For it would be a rich source of philosophical difficulties. Although points are sizeless it is less clear how units of matter can be. Aristotle goes further and objects against Anaxagoras that flesh and bone cannot fall below a certain *positive* size and still retain the defining characteristics of flesh and bone,[18] which is not to deny that their matter might be further divided,

[18] Aristotle *Phys.* 1.4, 187b13-21; cf. 187b28-34; *Phys* 6.10, 241a32-b3; *GC* 1.10, 328a23-8; *Sens.* 6,446a7-10. In Chapter 2, we saw Philoponus appealing to this sort of point, to prove that substances depend on quantities. Aristotle has a similar view about perceptible qualities like colour (*Sens.* 6, 446a16-20). There is a limit to the size of colour

but only that the resultant stuff would still be flesh or bone. Again, not only size, but also arrangement would create a problem: the sizeless units of bone, gold and honey would need to be arranged in relation to each other with all the necessary ratios of, say, three honey to two bone and one gold. Could these diverse units coincide at the same point of space? Or are they all to be included in every region however small? And how would they combine to form a larger whole? In other contexts, Aristotle gives it as his view that points can never add up to anything,[19] and that there is no suitable way of arranging them in relation to each other.[20] The optional fiction of bone first isolated, and then infinitely ground, would raise additional puzzles, since the infinite grinding would need to be completed, and the number of steps would be more than finite. In Aristotle's terms, it would form an actual, not just a potential, infinity, which he at least thinks impossible,[21] and he may be making this complaint in the passage cited from *GC* 1.10.[22] He would also object to the number of units in anything being more than finite.

The purpose of considering Anaxagoras' theory was to see if it involved there being two or more bodies, at least stuffs, in the same place. That depends partly on the model chosen. On the qualitative model, there are only *qualities* in the same place, but I have tried to rule that out as a model for Anaxagoras. On the blending model, there will be stuffs in the same place. And the same goes for the powder model, if we take the variant according to which different infinitesimal powder grains collect at the same *point*. On the other variant, according to which all types of powder grain are found in every *region* however small, we shall have a correspondingly reduced sense in which stuffs are in the same place: stuffs of each kind will be found in any *region* however small, but not at every point. The same applies to the patchwork model. So either in a stronger, or in a weaker sense, Anaxagoras does seem to have put stuffs in the same place. He must even allow that stuffs can *move* into the same place, as happens when bone with a little honey is recruited by digestion to join the bone with a little honey in my bones.

Anaxagoras had no inkling of the idea later proposed by Aristotle that ingredients exist only *potentially* in a mixture. If he had had, he would, I think, have been very doubtful about it. For if bone does not exist in *actuality* in the honey, he would have wondered, how can it *explain* the fact that honey nourishes my bones? Such explanation was

---

patches which are perceptible actually (*energeiâi*) and on their own (*khôris*). But this time he concedes there is no limit to the size of patches perceptible in a weak sense simply by virtue of being in the whole (*hoti en tôi holôi*).

[19] Aristotle *GC* 1.2, 316a25-34.

[20] Aristotle *Phys.* 6.1, 231a21-b18.

[21] Infinity is never actual, Aristotle *Phys.* 3.6, 206a14-23; never completely traversed, *Phys.* 3.5, 204b9; 6.2, 233a22; 6.7, 238a33; 8.8, 263a6; b4; b9; 8.9, 265a20; *Cael.* 1.5, 272a3; a29; 3.2, 300b5; *Metaph.* 2.2, 994b30; *An. Post.* 1.3, 72b11; 1.22, 82b39; 83b6.

[22] Aristotle *GC* 1.10, 328a15-17.

part of his aim, when he proposed 'a portion of everything in everthing' in the first place.[23]

Juxtaposition theories

It may have been partly to avoid stuffs in the same place that Anaxagoras' contemporary Empedocles (c.495-35 B.C.) opted for juxtaposition. He described elemental mixture in terms of undivided (albeit divisible) nuggets of earth, air, fire and water in contact with each other.[24] His theory was followed by the full-blown atomism of Leucippus (fl. after 440) and Democritus (fl. after 430). Atomic theory involves not only the juxtaposition of particles, but also their indivisibility in some sense, and further their possession of a limited range of primary qualities which explain the many qualities of ordinary bodies at the perceptible level. All this is found in the ancient atomists, although they did not add an intermediate level of molecules,[25] and this will become relevant later. An atomic theory of juxtaposed particles is what we still favour at one level of analysis. At a sub-atomic level, as remarked in Chapter 3, some kind of field theory is commonly preferred. The closest analogues of modern theory, then, are found on the one hand in the ancient atomists, and on the other in the treatment of prime matter as extension, which was the subject of Chapters 1 to 3.

Aristotle on chemical mixture

Aristotle, however, rejected juxtaposition (*sunthesis*)[26] as not producing genuine mixture (*mixis*). For in juxtaposition, he thought, the ingredients remain unaltered.[27] (Modern theories of molecular bonding would make this untrue, but it was true of ancient atomism.) Aristotle's lead is followed by the Stoics: they too reject the idea of juxtaposition which seems so obvious to us. For an adequate theory of mixture, Aristotle makes at least three requirements. We must avoid mere juxtaposition, which is not mixture. But we must also avoid the destruction of one or both of the ingredients, such as happens when a

[23] It is only on the qualitative model, which I have sought to exclude, that Anaxagoras could be viewed – implausibly – as anticipating Aristotle's idea, and making bone exist in my honey only potentially, in the form of qualities.

[24] For the evidence, see Aristotle *Cael.* 3.6, 305a1ff; *GC* 1.8, 325a6-8; b5-10; b19-22; 1.9, 327a10-11; Aëtius 1.13.1; 1.14.1; 1.16.1; 1.17.1 (Diels *Dox.* 312, 315).

[25] Epicurus does not allow his minimal parts to separate or recombine; so they do not resemble the atoms in a molecule. I have discussed ancient atomism in *Time, Creation and the Continuum*, chs 2, 5, and 22-5. The nearest analogue to our distinction between atoms and molecules is perhaps Plato's distinction between triangles and corpuscles in *Timaeus* 53C-57D.

[26] Aristotle *GC* 1.10, 328a6.

[27] ibid. 327a35-b2. Though the complaint is one made by others, Aristotle appears to respect it.

drop of wine loses its form or defining characteristics on being dissolved in a mass of water.[28] In that case, the ingredients do not survive long enough to mix. Both requirements are laid down at the very beginning of his discussion.[29] Thirdly, Aristotle assumes, as we know from other contexts (see below), that at all costs we must avoid supposing the ingredient bodies are in the same place. It is on this last requirement that Anaxagoras at least in a sense, and the Stoics unequivocally, differ from him. The laying down of requirements is Aristotle's method also in reaching a definition of *place* (see Chapter 11).

Aristotle meets his three requirements as follows. Instead of juxtaposition, we have a homogeneous *tertium quid*. But the original ingredients have not been destroyed, because they still exist *potentially*. On the other hand, we do not have bodies in the same place, because the original ingredients are not *in actuality* existent. This is a very nice balancing act. Aristotle expresses his solution as follows:

> Since some things are potentially (*dunamei*), and others actually (*energeiâi*), existent, ingredients can be in a way and yet not be what they were before being mixed, and need not be destroyed. What comes into being out of them differs from them and exists actually, yet each still exists potentially. This [destruction] was the problem raised by the earlier argument, but the ingredients formerly came together from being separated, and can be separated again (*khôrizethai palin*). So they neither persist in actuality like body or whiteness, nor are they destroyed – neither one nor both – because their power is preserved (*sôzetai hê dunamis*).[30]

Although this account of mixture is ingenious, there are problems. First, what does *potential* existence mean? It is distinguished by Aristotle not only from actual existence but also from the destruction which occurs when a drop of wine loses its form in a mass of water and is obliterated. To show that the ingredients are not destroyed, Aristotle says that they can be re-separated (*khôrizesthai palin*),[31] and re-separation no doubt differs from reconstitution in implying that in some way the ingredients are there all along. Indeed, if earth, air, fire and water are to serve as matter when flesh is formed, it is most important that they should in some sense persist throughout its formation. For, as seen in Chapter 1, Aristotle's whole analysis of change involves the persistence of matter. Change consists in the same matter persisting (*hupomenein*), but acquiring a new form. In the present instance, if earth, air, fire and water are the matter, they should persist through the change in which they acquire the form of flesh.

But how can fire and earth persist, when they are mixed to make

---

[28] ibid. 328a23-8.   [29] ibid. 327a34-b6.
[30] ibid. 327b22-31.   [31] ibid. 327b26-9.

flesh, especially if the defining heat of fire and coldness of water no longer exist in actuality (*entelekheiâi*) in the mixture, but are replaced by a single intermediate temperature characteristic of flesh? This suggests that when we recover the original ingredients, we need after all to reconstitute them, not merely to re-separate them. The loss of the original temperatures is described in the following passage.

> Is it like this, then? There can be more or less heat and cold. When either exists without qualification (*haplôs*) and in actuality (*entelekheiâi*), the other will exist potentially. But when they do not exist completely (*pantelôs*), but one is cold for a hot, and the other hot for a cold, because ingredients destroy each other's extremes on being mixed, then there will exist neither their matter, nor either of those contraries without qualification in actuality (*haplôs, entelekheiâi*), but something inter-mediate (*metaxu*) ... In this way the elements are transformed first, and out of them there come to be flesh and bone and such-like. The hot becomes cold and the cold hot, when they approach the middle, for there neither of them is found (*entautha oudeteron*), but the middle is wide and not an indivisible point. In the same way dryness and fluidity and such-like produce flesh and bone, etc., in accordance with a middle range.[32]

A solution might be found, and the ingredients be saved from temporary destruction, if it was right to translate the closing words of the quotation before last, 'nor are they destroyed ..., because their *power* (*dunamis*) is preserved (*sôzetai*)'. Another translation of the word *dunamis* would tell us merely that their *potentiality* is preserved, which is the point at issue.[33] But if Aristotle is talking about the *power* of the ingredients, he could be telling us more about *how* their potentiality is preserved. In the sequel, he refers more than once to powers, for example to the powers of heating and cooling,[34] and his point could be that the earth, air, fire and water which went to make up my flesh are not destroyed, because their powers of heating and cooling survive, and still go to work. For example, if I step out into the winter frost, the power of heating contributed by fire, even if somewhat modified, will act to counter the new influx of cold.

The powers of heating and cooling must, on this view, be *distinguished* from the heat and cold of the original ingredients. In the *Parts of Animals*,[35] Aristotle explains how different criteria of heat can come apart and exist in separation from each other. Perhaps, then, fire's power of re-heating me, when I step into the frost, could survive, when the other manifestations of its heat have been replaced. There is

---

[32] ibid. 2.7, 334b8-13; 24-30.

[33] ibid. 1.10, 327b30-1. 'Power' is preferred by H.H. Joachim in the Oxford translation, ed. W.D. Ross (Oxford 1931); 'potentiality' by C.J.F. Williams in the Clarendon Oxford Series, ed. J.L. Ackrill (Oxford 1982).

[34] Aristotle *GC* 1.10, 328a29; 2.6, 333a24-7.

[35] Aristotle *PA* 2.2, 648b11-649b8.

now only a *single* intermediate temperature, and this prevents the
ingredients fire and earth from existing in full actuality. But there are
two *distinct* powers, one of heating and one of cooling, and this saves
the ingredients fire and earth from being destroyed. We thus get the
intermediate status between actual existence and destruction that
Aristotle seeks when he talks of potential existence.

Unfortunately, this solution has to face some quite severe difficulties. For one thing, the powers of heating and cooling, which fit
various of Aristotle's definitions of opposites, would have to belong to
one and the same mixture. But he repeatedly denies that opposites can
belong to the same thing.[36] It can hardly be urged that the opposite
powers, which from time to time go to work, still belong to the original
ingredients, rather than to the mixture, for those ingredients now exist
only potentially. If Aristotle wants the advantage of an appeal to
surviving powers, his best course might be to revise his views on
opposites and allow the opposite powers to belong to the mixture after
all.

But there are still other difficulties. Alexander does not agree that
the original powers of fire and earth survive.[37] And the more elaborate
account of chemical processes in *Meteorology* book 4, which I am
inclined to accept as Aristotle's, speaks as if the original powers of
heating and cooling are superseded. If they are superseded, they will
not provide a satisfactory sense in which the ingredient earth, air, fire
and water escape destruction. We cannot allow, and Aristotle does not
want to allow, that the ingredients exist potentially merely in the
sense that they can be reconstituted. For that would be tantamount to
temporary, or revocable, destruction, and the ingredients could not
then be said to *persist* through the formation of flesh, as they are
required to do, if they are to serve as matter.

Aristotle's account of chemistry here poses a threat to his central
metaphysical belief in the persistence of matter. It is a different threat
from the one normally noticed, because it concerns the transformation
of the four elements into flesh, not the transformation of one element
into another. But Jim Lennox has now suggested that Aristotle's
solution may lie at *Meteor.* 4, 385a11ff. There Aristotle cites not the
powers of heating and cooling, but eighteen other potentialities, such
as the powers of being melted by heat or solidified by cold. These
enable us to diagnose (389a5) whether a mixture is, say, of water,
because some of the characteristics (solidifiability and re-meltability)
are *shared* with water. Others are shared with earth, or with a
mud-like mixture of the two. And this gives content to saying that

---

[36] Aristotle *Metaph.* 4.3, 1005b26-32; 4.6, 1011b17-21; 5.10, 1018a25-38; 9.5,
1048a9-10; *Cat.* 6a4. Aristotle allows that there can be a single power for producing
opposite effects: the doctor can use his knowledge to kill or cure (*Metaph.* 9.5). But he
does not redescribe this as a case of two opposite powers inhering in the doctor.

[37] Alexander *Mixt.* 231,9.

water or earth, or both, have persisted potentially. I am less sure, however, that it can give content to Aristotle's further view that *all four* elements persist potentially in every compound.[38]

I shall now turn to a second problem. If Aristotle's ingredients exist only potentially, how can they be *explanatory*? If earth, air, fire and water do not exist in actuality, how can they explain the temperature of my flesh? It is a general principle of Aristotle's that whatever is acted on is acted on by something that is in a state of actuality (*energeiâi*).[39] And it is in any case independently plausible that what does not exist actually will not act.

Aristotle explicitly raises the question how earth, air, fire and water can produce something so different from themselves as flesh, and his answer trades on their existing in actuality when they first enter the mixture.[40] As we have seen, the extreme heat which characterises fire and the extreme cold which characterises earth temper each other and produce an intermediate temperature characteristic of flesh.[41] But what keeps my flesh at the right temperature thereafter? The earth, air, fire and water cannot do so, since they no longer have actual existence. And that is why it was tempting to appeal to the survival of *powers* of heating and cooling to do the job. But if these powers are superseded, it may be better to appeal to an idea found in the *de Anima*, that what is at an intermediate temperature can act as a contrary towards extremes of heat and cold,[42] in which case flesh can presumably temper influxes of these extremes.

A third problem is whether Aristotle can account for mixtures in which the ingredients have *not*, like earth and fire, lost their original qualities. In the cocktail called a Saint Clement's, one can distinctly taste both the lime juice and the orange juice. And in a mixture of wine and water, one can detect the coldness, fluidity and transparency of the water and the flavour of the wine. As the Stoics were later to say of their mixtures, the original qualities '*show forth together*' (*sunekphainesthai*) in the mixture.[43] So why are the original ingredients not present in *actuality*? Aristotle recognises that inessential qualities can persist, as tin leaves its colour in bronze and food its flavour in water.[44] But essential qualities can persist too. Fluidity and coldness *define* water. Nor need the fluidity and coldness be offset by further

---

[38] Aristotle *GC* 2.8. I am very grateful to James G. Lennox for insisting on *Meteorology* 4 in his reply to me, of which a version is to be published with my paper, 'The Greek origins of the idea of chemical combination', in John J. Cleary, ed., *Proceedings of the Boston Area Colloquium in Ancient Philosophy* 4, 1989.

[39] Aristotle *DA* 2.5, 417a17-18.

[40] We can ignore the complication that pure elemental earth, air, fire and water are different from the impure surrogates with which we are most familiar, *GC* 2.3, 330b21-30 and 2.8. At least *impure* earth, air, fire and water exist in actuality when they enter into the mixture.

[41] Aristotle *GC* 2.7, 334a21-3; 334b8-30, quoted in part above.

[42] Aristotle *DA* 2.11, 424a6-7.

[43] Arius Didymus fr. 28 ap. Stobaeum *Eclogae* 1.17, *Dox. Gr.* 464, 1-6 (*SVF* 2.471).

[44] Aristotle *GC* 1.10, 328b13; *DA* 2.10, 422a11-14.

characteristics which count against saying that water persists actually in the mixture. Admittedly, the *degree* of fluidity and coldness will be modified when the water is mixed with wine. But it was widely understood, and much discussed in medical literature, that water differs in different places, and a range of different degrees must have been allowed for at least when ordinary water was under discussion. Even elemental water may admit variations of degree.[45]

One course for Aristotle would be to deny that wine and water is a genuine mixture. This is not the normal interpretation. For he tells us that a single drop of wine in ten thousand gallons of water could not preserve its form or essential characteristics, but would be destroyed and simply add to the volume of the water, whereas by contrast, if '*they*' are equal in their powers, they *will* form a mixture.[46] It is natural to suppose that 'they' includes such ingredients as wine and water. But the alternative would be to assume that Aristotle intends to restrict examples of genuine mixture far more severely. All bodies made up of earth, air, fire and water (and that will be most bodies beneath the heavens) will be mixtures of those elements. But in wine and water, or lime juice and orange juice, the constituent fluids will be in juxtaposed globules, and will not form a genuine mixture. One commentator has taken Aristotle that way, but others assume that wine and water is accepted as a genuine example of mixture.[47]

Aristotle may have better grounds for appealing to juxtaposition in the case where two flavours, odours, or sounds, or two sets of colour stripes, show forth together, than in the case of wine and water. For in the latter case, the qualities which show forth together are flavour on the one hand and coldness, fluidity and transparency on the other, and these qualities are not taken from the same range. But where the qualities which show forth together are two flavours, or two other qualities taken from the same range, Aristotle might well appeal to his theory of the physical basis of such qualities. It might turn out in some cases that the showing forth of two qualities from the same range was incompatible, on his physical theory, with a genuine mixture of the physical ingredients. But his physical theories of colour, odour, flavour and sound would need to be examined case by case.[48]

There remains a final problem with which Aristotle does not deal, and I suspect that certain later Aristotelians formulated a different theory, precisely in order to deal with it. If the receiving body does not have particles juxtaposed with those of the incoming body, why does it

---

[45] Proclus and Simplicius had a theory that superior grades of the four elements go to make up the heavens – see Chapter 7. Aristotle himself allows that the region of elemental fire and air can be made more or less hot, depending on friction, *Meteor.* 1.3, 341a17-32, *Cael.* 2.7, 289a20-35

[46] Aristotle *GC* 1.10, 328a28-31.

[47] H.H. Joachim in his commentary on *GC* (Oxford 1922), at p. 180. Contrast the commentary of Philoponus, which he cites ad loc., and that of C.J.F. Williams in the Aristotle Clarendon Series (Oxford 1982).

[48] For colour, odour, flavour, see Arist. *Sens.*, chs 3-5; for sound *DA* 2.8; for voice *GA* 5.7.

swell until the mixture has a volume equal to the sum of the two
original volumes? Juxtaposition of particles would explain this
perfectly. But Aristotle must treat the increase of volume as a *brute*
fact, not further to be explained. It is, of course, a very familiar fact,
but that should not blind us to the advantage of having an explanation
of it, if we can.

Galen and Arius Didymus describe certain Aristotelians – they do
not include Alexander – as saying that only the qualities, not the
substances or bodies, of the four elements mix.[49] Plotinus reports the
same theory more fully without explicit attribution: the matters of the
two bodies are *juxtaposed* (*paratithenai, parakeisthai*), only the
qualities mix.[50] This was held to be superior to the Stoic belief in total
interpenetration (*di' holôn krasis*), partly on the grounds that the
Stoics could not explain increase of volume and its equality to the sum
of the original volumes.[51] To explain swelling, the neo-Aristotelian
theory reimports the idea of juxtaposition, which Aristotle had
removed, in a way which is thought to retain the spirit of his ideas.[52]
An interesting feature of the new theory is that, if it is applied to prime
matter, it will need to treat it as a sort of *extension*, if the juxtaposition
of matters is to explain the increase in volume. The theory will then
anticipate something of Simplicius' interpretation of prime matter,
discussed in Chapter 1.

Although Aristotle rejects mere juxtaposition as an account of
genuine mixture, Eric Lewis has suggested to me[52] that he might after
all endorse the neo-Aristotelian explanation of swelling, for he thinks
that genuine mixture is preceded by a stage at which small globules of
distinct ingredients are juxtaposed, and therefore need plenty of room.
The qualities of the globules then temper each other, until the whole
mass, requiring as much room as before, has turned into the new kind
of stuff.

Aristotle on bodies in the same place

Aristotle's full-scale opposition to the idea of bodies in the same place,
though presupposed in his theory of chemical mixture, is more clearly
brought out elsewhere. Indeed, he founds a tradition of opposing the
idea. Four times, he uses it as a *reductio ad absurdum* of other people's

[49] Galen *in Hippocr. de Nat. Hom.* (vol. 15, p. 32 Kühn)=*SVF* 2.463; Arius Didymus
Epitome, fr. phys. 4, *Dox. Gr.* 449,1-3 from Stobaeus *Eclogae* 1.17.
[50] Plotinus 2.7.1(8-9); 2.7.2(10-12).
[51] id. 2.7.1(15-19).
[52] In discussion.

views that they would lead to bodies being in the same place,[53] and twice he considers the allegation that certain other ideas would lead to that absurdity.[54] In addition, he reports an objection, without claiming to originate it, which later commentators illustrated by the example of the sea in a cup. It is directed against the idea that one body could be put in the same place as another, without the two taking up more space than either did singly. It would therefore be ineffective against Anaxagoras, since his ingredients, bone and honey, do add to the volume of the whole. The objection is that if you could fit a cupful of sea water into a cupful of wine once, and obtain only one single cupful, then you could repeat the operation indefinitely, until you had fitted the whole sea into a cup, and that is obviously absurd.[55] I shall be returning to the long history of this objection in Chapter 6.

Bodies in the same place: modern controversy

It is time to ask if Aristotle is right that bodies cannot occupy exactly the same place. The issue has been debated in the context of modern Physics. At an earlier date, James Clerk Maxwell wondered whether atoms might pass through each other.[56] And in more recent sub-atomic Physics, it has been suggested, there is a large class of particles, the bosons, which can be in exactly the same place at the same time.[57] For at least two reasons, however, this is not the case. First, the contexts in which one might be tempted to say this are precisely contexts in which, as explained in Chapter 1, it is better not to think in terms of particles, but of fields, or of space-time endowed with properties. Secondly, even if one were to say that particles shared all their states, those states would not include a position, because the contexts in question are contexts in which a precise position cannot be assigned.[58]

[53] *Phys.* 4.1, 209a4-7 against place as body; *DA* 1.5, 409b3 against soul as body; *DA* 2.7, 418b13-18 against light as body; *Cael.* 3.6, 305a19-20 against the four elements or the heavens being generated from body.

[54] The allegation that growth, or motion, without void would lead to two bodies in the same place, *Phys.* 4.6, 213b7; *GC* 1.5, 321a5-10.

[55] *Phys.* 4.6, 213b5-12.

[56] James Clerk Maxwell, 'Atom', *Encyclopaedia Britannica*, 9th ed., 1875-9, cited by Sanford (1983), as below.

[57] See the series in *Philosophy of Science*: Alberto Cortes, 'Leibniz's principle of the identity of indiscernibles: a false principle', 43, 1976, 491-505; R.L. Barnette, 'Does quantum mechanics disprove the principle of the identity of indiscernibles?' 45, 1978, 466-70; Allen Ginsberg, 'Quantum theory and the identity of indiscernibles revisited', 48, 1981, 487-91; Paul Teller, 'Quantum physics, the identity of indiscernibles and some unanswered questions', 50, 1983, 309-19; cf. Bas van Fraassen, 'Probabilities and the problem of individuation', in S.A. Luckenbach, ed., *Probabilities, Problems and Paradoxes*, Encino California 1972; R.N. Hanson, 'The dematerialisation of matter', in E. McMullin, ed., *The Concept of Matter*, Notre Dame Indiana 1963, 549-61.

[58] I am indebted to Teller's article and to conversation with Michael Redhead and Bas van Fraassen.

If the favoured models nowadays are juxtaposition at the level of molecules or atoms and fields at the sub-atomic level, it looks as if the idea of bodies in the same place is bereft of scientific support. But if scientists were to tell us tomorrow that they had after all discovered interpenetrating bodies, could we tell them in advance that they had made a conceptual mistake? I doubt if our concepts would entitle us to do that.

The conceptual possibility, though not the reality, of bodies in the same place has been defended by the philosopher David Sanford.[59] He argues that at least the world could have been arranged, even if it is not in fact arranged, in such a way that bodies were in the same place. And he applies his argument both to individuals, like billiard balls, and to their stuffs. Why should not the world have been so arranged, for example, that two billiard balls could pass right through each other, momentarily occupying exactly the same place? The first trick here is to exploit motion so as to make the coincidence of place easier to imagine. We are not to think that the atoms of one billiard ball are merely juxtaposed with the atoms of the other. Either we should think of the atoms as passing through each other, or we should think of bodies as not having an atomic structure, but as being homogeneous and continuous[60] in the manner of Anaxagoras, Aristotle and the Stoics. Sanford envisages that one billiard ball could have 'Mummy' scratched on it and the other 'Daddy', so that 'Daddy' and 'Mummy' could be reidentified as two billiard balls emerged from the collision. In his later paper, he envisages how such behaviour could have a regular and predictable basis. Things made of Moon rock and things made of Mars rock might interpenetrate with each other, while no other types of rock interpenetrated. (For this regularity there might be no explanation at a lower level.) The wise maker of billiard tables will make them of Earth rock, and players will not mix balls of Moon rock and of Mars rock, except for a joke. After sculpting a bust out of Moon rock, you will not place it on a pedestal of Mars rock, without placing a sliver of Earth rock in between. We may imagine that on passing through each other, the billiard balls double in weight and density, rather than expanding like fluids. Would this be a case of two bodies in the same place?

We have encountered two main challenges to purported examples of bodies in the same place. One question, raised in connexion with Anaxagoras, was: would they still be *bodies*? Sanford has an answer to this, that his billiard balls would be sufficiently like ordinary bodies, in that they would still be *selectively* exclusive, since they would

---

[59] David Sanford, 'Locke, Leibniz and Wiggins on being in the same place at the same time', *Philosophical Review* 79, 1970, 75-82; 'The perception of shape', in S. Shoemaker and C. Ginet, eds, *Knowledge and Mind*, Oxford 1983.

[60] The work of Kripke has suggested that the atomic structure of matter may on the contrary be a necessary truth. Some relevant doubts are expressed in my *Necessity, Cause and Blame*, 219-22.

interpenetrate only with *some* things. The other main challenge to bodies in the same place came from Aristotle: would there still be *two*? When the original billiard balls met, would we not have a new amalgam, a *tertium quid*, rather than *two* bodies? This challenge might be pressed by referring to Leibniz's principle of the identity of indiscernibles: if there are really two bodies, then, according to this principle, there ought to be features to differentiate them. In reply to this challenge I think more care is needed than has sometimes been taken in the handling of what might be called historical properties.[61] It would be no good saying, in the context of the present challenge, that in the amalgam before your eyes two entities can be distinguished by their having the distinct historical properties of *having* been to the East and *having* been to the West. For the challenger will say that, in his view, the two entities which had these properties have by now perished and been replaced by a *tertium quid*, which, being newborn, has no such historical properties. Fortunately, Sanford supplies not only historical properties by which the two billiard balls might be distinguished, but also contemporaneously perceptible ones. Perhaps one billiard ball was striped pink on white, the other green on white, and in the amalgam those colours of stripe can still be seen cross-hatched. Perhaps they were musical billiard balls, one emitting Middle C and one D Sharp: correspondingly from the direction of the amalgam before you the same two notes can still be heard. If two persons were to pass through each other, one might say 'Excuse me' and the other 'You're welcome' in their distinctive tones of voice.

It should also be noticed, as a distinct point, how uneconomical is the suggestion of a *tertium quid*. The two original billiard balls would perish by shrinking in a peculiar pattern, while the *tertium quid* would grow, until it reached a full spherical shape. It would then in turn start shrinking, while two other entities (a fourth and a fifth?) began to grow. The multiplication of entities and of shrinkings and growings seems needlessly elaborate beside the simple description according to which two billiard balls have passed right through each other.

These two types of consideration, diseconomy and differentiation of colour stripe, should be differently treated. The diseconomy should remove the initial appeal of the description according to which we have a *tertium quid*, and should return the burden of argument to the proponents of that description. The differentiation of stripe by no means guarantees that we must have two different entities still before us, but it disarms the charge that it would violate Leibniz's principle to talk of two entities; for we now have differences of feature to refer to, which make it possible to comply with Leibniz's principle. Indeed, with the burden of argument returned to the challenging party, we can at this stage reintroduce merely historical properties, and so increase the range of differentiating features.

[61] For the exploitation, in some contexts legitimate, of historical and dispositional properties, see Sanford, Barnette and van Fraassen, as above.

My agreement with Sanford that in the imagined situation we could speak of two billiard balls in the same place does not exclude our describing the two billiard balls as momentarily forming a single mass. That would no more count against their being two than our speaking of two billiard balls wired together as forming a single mass. Admittedly, the story as so far told could be elaborated in different ways, and some of them would disincline us after all to talk of two bodies in the same place. But to show the conceptual possibility of such a thing, it is enough if there is always a way of elaborating the story which makes that description the best one. I believe that stories could be told quite different from Sanford's in which again we should have a motive for talking of bodies in the same place.

Can other types of extension be in the same place?

Before I turn to the Stoic response to Aristotle, there is a final issue to be considered. Commentators on Aristotle understood him to take his opposition much further, and to deny that any two extensions (*diastêmata*) could be in the same place, whether they were bodies or not. A similar account of vacuum as impenetrable has also been ascribed to Epicurus.[62] Can these interpretations be justified? I believe not. In one of the relevant passages, Aristotle is attacking the idea of a vacuous space. It would have to penetrate the body (say, a wooden cube) which filled it, but then how would it differ from the volume (*megethos*, 216b3; *onkos*, b6; *sôma*, b10) of that wooden cube? And if two such things could be in the same place, why not any number?[63] So far there is no general objection to extensions being in the same place. This is supplied by the commentator Themistius, who constructs and offers to Aristotle a supplementary argument: it is only because bodies are extensions (*diastêmata*), not because they have qualities, that they occupy places. Hence, if you once allowed extensions of any kind to occupy the same place, you would have to allow bodies to do so. Philoponus and Simplicius follow Themistius' interpretation, with variations, though without approving the reasoning.[64] But the whole argument, with its conclusion that extensions cannot be in exactly the same place, is a concoction not to be found in Aristotle himself. And in fact Aristotle does nothing to deny that in a sense the wooden cube coincides with its volume.

In a second passage, Aristotle is attacking the idea that a body's place is an extension (*diastêma*) stretching between its surroundings, rather than being, as he thinks, merely the inner surface of those surroundings. The 'extension' view would result in many places being

---

[62] Brad Inwood, 'The origin of Epicurus' concept of void', *Classical Philology* 76, 1981, 273-85.

[63] Aristotle *Phys.* 4.8, 216a26-b12.

[64] Themistius *in Phys.* 134,1-135,8; Philoponus *in Phys.* 687,7-12; Simplicius *in Phys.* 622,19-21; 682,5-7.

together, and indeed in an infinity of places being in the same place.[65] Themistius takes up Aristotle's illustration of water in a jug and explains that the following extensions would be superimposed on each other: the (immobile) extensions occupied by the whole jug, by a part of it, namely, the water, and by parts of the water; again, the (mobile) extensions between the inner walls of the jug, extensions which are occupied by the water and by the parts of the water; again, the volumes of the jug, of the water and of its parts. Further, if you carry the jug to a new place, you will superimpose the extension between its walls on the new place.[66] But not even Themistius claims this time that the absurdity lies in extensions coinciding. He may rather be thinking that since extensions clearly *could* coincide, so (absurdly) would places, if they were extensions.

There is a third idea that might encourage the present over-interpretation of Aristotle. He twice objects that if the Platonists' geometrical solids were in ordinary physical bodies, you would get two solids (*sterea*) in the same place.[67] Alexander and pseudo-Alexander trade on the fact that geometrical solids were called bodies (*sômata*), and supply the objection that body would go through body.[68] But this is not in Aristotle. Nor is there any sign that he would object to saying that geometrical solids, as he himself conceived them, coincide in a sense with physical bodies. For him geometrical solids just *are* physical bodies with their non-geometrical properties ignored. What he objects to in the Platonist view is that their geometrical solids are independent substances, and *substantial* solids, he thinks, cannot coincide.

In a final passage, Aristotle considers a preliminary puzzle about place, without in any way endorsing its ideas. Place, on one view, has three dimensions, and to have these is the defining characteristic of body. Yet place cannot be body, or two bodies would interpenetrate.[69] If taken seriously, the argument would turn every three-dimensional extension, for example a body's volume, into a body, and this would indeed prevent any three-dimensional extensions coinciding. But despite the occasional use of the word 'body',[70] Aristotle would deny that a body's volume was an ordinary body (it belongs to a different category, the category of quantity), and he denies too that vacuum is a body in the course of one of the anti-vacuum arguments cited above.[71] So he need not extend his ban on interpenetration from ordinary bodies to three-dimensional extensions of *every* kind.

None the less, in certain authors, the ban on interpenetration is extended in an unexpected way to *vacuum*. It is not so extended by

[65] Aristotle *Phys*. 4.4, 211b20-1; 24-5.
[66] Themistius *in Phys*. 116,21-30; 117,4-6.
[67] Aristotle *Metaph*. 3.2, 998a13-14; 13.2, 1076b1.
[68] Alexander *in Metaph*. 201,23; ps-Alexander *in Metaph*. 725,18.
[69] Aristotle *Phys*. 4.1, 209a4-7.
[70] ibid. 4.8, 216b10.
[71] ibid. 216a34.

Aristotle, for he says in a passage already cited,[72] that if there were a void, it or the equivalent extension *would* interpenetrate with bodies, with absurd results. But it has recently been argued that *Epicurus* thought of void quite differently, as an ideal fluid that *yielded* on the approach of bodies rather than being able to interpenetrate with them.[73] Although I am not persuaded by this as an interpretation of Epicurus, the conception of vacuum as impenetrable by body *is* found in the thirteenth and fourteenth centuries. Edward Grant has shown that Duns Scotus, and subsequently others reported by Burley, claimed that a vacuum could yield, instead of penetrating a body. It was thus immune to the objection that extensions would coincide, which some considered impossible. It was also immune to the objection that speed in a vacuum would become infinite. For the vacuum might yield only slowly, bit by bit, and so reduce the speed of an advancing body. Or again, if it could yield, it could also contract and expand, and so (it was claimed) exert resistance.[74]

The idea that vacuum cannot interpenetrate with body must be confronted, because it bears on the definition of body. One might have supposed that a body could penetrate a vacuum, and that two moving bubbles of a vacuum might pass through each other, but that bodies were distinguished from vacuum by the fact that they could not interpenetrate with each other. The present suggestion threatens this distinction, not by allowing bodies to interpenetrate with each other, but by taking the other tack and forbidding them to interpenetrate with vacuum. I think that the prohibition should be resisted. Admittedly, what is penetrated by body can no longer be described as vacuum. But the same thing, under another name, will have been penetrated. The appropriate name in cases where it is immobile will be 'space', and in cases where it is mobile (like the vacuum in my thermos flask) will be 'volume'.[75] Space and volumes which start by being vacuous can both be penetrated by bodies. What needs questioning, then, is not the penetrability of the vacuous, but the supposed impenetrability of body.[76]

---

[72] ibid. 216a26-b2; cf. 4.6, 213a15-19.

[73] Brad Inwood, op. cit.

[74] Edward Grant, 'The principle of the impenetrability of bodies in the history of concepts of separate space from the middle ages to the seventeenth century', *Isis* 69, 1978, 551-71 esp. 564-9, and *Much Ado About Nothing*, Cambridge 1981, 31-8. For the denial of vacuum retreating, see Aristotle *Phys.* 4.8, 216a26-b2; Sextus *M* 10,24-36 and David Sedley, 'Two conceptions of vacuum', *Phronesis* 27, 1982, 175-93.

[75] Moving holes are discussed also in Chapter 1, in connexion with the immobility of space, and in Chapter 11, in relation to Aristotle's treatment of a boat's place.

[76] I am grateful for discussion to John Ackrill, Peter Alexander, David Barlow, Robert Bolton, Wolfgang Detel, Jo de Filippo, Bas van Fraassen, David Furley, Harry Ide, Peter Klein, James Lennox, Eric Lewis, Frank Lewis, Connie Meinwald, Aldo Mignucci and Michael Redhead.

CHAPTER SIX

Can two bodies be in the same place?
Stoic metaphysics and chemistry

Aristotle, we have seen, has a lot of explaining to do, if he is to maintain his theory of mixture. So it is not surprising if the Stoics offer an alternative. Using different terminology from Aristotle, they hold in effect that in a genuine blend (normally called *krasis*)[1] the ingredients exist *actually*. In the words of our sources, the substance, nature and qualities of the ingredients are preserved and remain.[2] But this is not achieved by the juxtaposition (*parathesis*, as they call it) of particles, an idea which they separate off as sharply as Aristotle,[3] and against which they make common cause with him. The actuality of the ingredients combined with the denial of juxtaposition suggests that the Stoics must locate ingredient stuffs in exactly the same place as each other. And this is what our sources repeatedly say of them. The big question, then, will be whether they can meet the storm of objections which they encounter against the idea of bodies in the same place. It makes no difference whether, as various reports allege,[4] the receiving body is not expanded by the influx of a new ingredient. Even if it is expanded – and I shall argue later that they ought to say it is –

---

[1] But according to one of the sources quoted below, the blending of non-fluids was called *mixis*.

[2] Alexander *Mixt.* ch. 3, 216,25-217,2 (*SVF* 2.473); ch. 7, 220,26-34; 221,11-13; Arius Didymus fr. 28, ap. Stobaeum *Eclogae* 1.17, in *Dox. Gr.* 463,24-5; 30 (*SVF* 2.471). The whole of Alexander *Mixt.* is translated, along with other relevant texts, by Robert B. Todd, *Alexander of Aphrodisias on Stoic Physics*, Leiden 1976. I shall avoid the practice of referring to the numbered fragments of von Arnim's *Stoicorum Veterum Fragmenta* (*SVF*) alone. It is important for purposes of evaluation to know who the original authors were, and often to read the surrounding context. Again, *SVF*'s scattered excerpts conceal the drift of consecutive discussions of mixture in Alexander and Plotinus, and (see Chapter 9) of void in Plutarch. Further, von Arnim omits many passages and for these there is no *SVF* number to give. We shall find (Chapters 8 and 9) that excessive reliance on his selection has led to the neglect of such major sources as Cleomedes. Von Arnim is still important as the main collection available, but we can look forward to a new one from Jaap Mansfeld.

[3] Alexander *Mixt.* ch. 3, 216,17-22 (*SVF* 2.473); Arius Didymus fr. 28, ap. Stobaeum *Eclogae* 1.17, in *Dox. Gr.* 463,21-3 (*SVF* 2.471); Philo *de confusione linguarum* 184 (*SVF* 2.472).

[4] Plotinus 2.7.1(17-19); 4.7.8^2(10-11); Alexander *Mixt.* ch. 6, 219,30.

the existence of the ingredient bodies in actuality, without juxtaposition, still suggests that they are in the same place.

The Stoics feel it necessary to *argue* that the qualities (and perhaps that the ingredient stuffs) are preserved in a blend, by saying that the qualities still 'show forth together' (*sunekphainesthai*) in the mixture. And they further argue that the qualities, nature and substance are preserved by pointing to the reseparability (*apokhôrizesthai, khôrizesthai palin*) of the original ingredients.[5] But it has recently been denied that the early Stoics were responding to Aristotle, or even had knowledge of him.[6] I shall oppose this challenging thesis in Chapter 9 as regards the Stoic treatment of vacuum. Jaap Mansfeld has already implied his opposition in connexion with a statement about *mixture* made by the founder of Stoicism, Zeno of Citium. Zeno says that blending occurs through the change of the four elements into each other (*metabolê eis allêla*). The reference to such change had previously been thought unintelligible. Mansfeld argues that it makes sense, as soon as it is taken to refer to Aristotle's view already discussed, that the four elements form tissues by modifying each other's qualities and so changing into each other (*metaballei eis allêla*).[7] In addition, the Stoics are like Aristotle in distinguishing two other cases from genuine mixture. One is identical with his juxtaposition, but the other differs from his remaining case in which the minor ingredient is destroyed. In the Stoics' third type of case, it is not one but *both* ingredients which are destroyed, giving rise to an entirely new kind of stuff. They call this not blending, but fusion (*sunkhusis*).[8]

The divergence over the third type of case may itself be a conscious one. Aristotle illustrates his third type, in which the minor ingredient is destroyed, with the examples of a drop of wine in ten thousand gallons of water and of a drop of flavoured stuff in the sea.[9] His view that the wine will become too tenuous to retain its form, or defining characteristics, is related to the view already discussed, that there is a minimum size for natural substances. Matter is infinitely divisible, but after a certain stage you can only divide it at the cost of losing the type of substance with which you started. What goes for size goes also for

---

[5] See the two passages translated below from Alexander *Mixt.* ch. 3, 216,32-217,2; Arius Didymus fr. 28, *Dox. Gr.* 464,1-6, with footnote on my construction of 'because' (*gar*) at 464,1.

[6] F.H. Sandbach, *Aristotle and the Stoics*, Cambridge Philological Society, supp. vol. 10, 1985.

[7] The Stoic statement is in Arius Didymus, fr. 38, ap. Stobaeum *Eclogae* 1.17, in *Dox. Gr.* p. 470,4 (*SVF* 1.102): *krasin ginesthai têi eis allêla tôn stoikheiôn metabolêi.* Jaap Mansfeld, 'Zeno and Aristotle on mixture', *Mnemosyne* 36, 1983, 306-10, compares Aristotle *GC* 2.7, 334b24-7.

[8] Alexander *Mixt.* ch. 3, 216,22-5 (*SVF* 2.473); 7,220,29-33; 221,11-13; Arius Didymus fr. 28, ap. Stobaeum *Eclogae* 1.17, in *Dox. Gr.* 464,6-8 (*SVF* 2.471); Philo *de confusione linguarum* 184 (*SVF* 2.472).

[9] Aristotle *GC* 1.10, 328a23-8; *Sens.* 6, 446a7-10; cf. *Pol.* 2.4, 1262b17.

density: beyond a certain degree of dilution, the fate of the drop of wine would be merely to add to the total volume of water. On the case of the drop of wine in the sea, the Stoic Chrysippus took a diametrically opposite view. He thought the drop of wine would survive, however thinly spread. At least in one of his works, he is said to have held that it could undergo blending (*krasis*), i.e. could survive, in combination with the whole of the sea and even with the whole of the cosmos.[10] Elsewhere he may make a concession to Aristotle's position, when he says that the drop of wine would perish in combination (*sumphtharêse-tai*) with the sea, after mutually stretching out along with it for a certain distance (*epi poson antiparektathêsetai*).[11] The talk of perishing in combination after doing something else is unexpected on almost any view, but it is hard to agree that the talk of *perishing* refers to a *blend* in which both ingredients *survive*.[12] Chrysippus may rather be conceding to Aristotle this once that, after blending over a *certain* distance, the wine would *thereafter* perish.

Be that as it may, I am not inclined to agree that Chrysippus' example of a drop of wine in the sea has nothing to do with Aristotle's. The argument has been used that Aristotle's drop of wine is placed not in the sea, but in ten thousand gallons of water.[13] But this overlooks Aristotle's second example which places its drop of flavoured stuff in the *sea*. Better evidence for Stoic unawareness of Aristotle could be drawn from Alexander's statement that it was only after Chrysippus that the Stoics were able to learn Aristotle's theory of mixture.[14] But although Alexander can be expected to know about many things, in this case he may merely be drawing an inference, five hundred years after Chrysippus, that no one would have had access to Aristotle's theory until the full edition of Aristotle by Andronicus in the first century B.C.

It is time to quote two of the sources, Alexander and Arius Didymus, who describe the Stoics' three kinds of combination. Alexander is the leading exponent of Aristotelianism and therefore a hostile witness.

> The opinion of Chrysippus on blending (*krasis*) is as follows: He supposes that all substance is unified, and that there goes through it all a certain *pneuma* by means of which the universe is held together (*sunekhetai*) and persists (*summenei*) and is sympathetic with itself.
> He says that for some of the bodies that are mixed (*mignumena*) in this

---

[10] Plutarch *Com. Not.* 1078D-E; cf. Alexander *Mixt.* ch. 4, 217,26-32

[11] Diogenes Laertius *Lives* 7.151.

[12] This is the interpretation of Robert Todd, *Alexander of Aphrodisias on Stoic Physics* 31-2; R.D. Hicks, Loeb edition of Diogenes Laertius ad loc.; H.A. Wolfson, *The Philosophy of the Church Fathers*, Cambridge Mass. 1956, 383. It is also followed by A.A. Long and D.N. Sedley on p. 290 of their invaluable book, *The Hellenistic Philosophers*, Cambridge 1987. A new interpretation is due to be offered by Eric Lewis, different from either of those considered here, in *Bulletin of the Institute of Classical Studies* (forthcoming).

[13] F.H. Sandbach, op. cit.

[14] Alexander *Mixt.* ch. 3, 216,9-12, mentioned by Sandbach in a footnote.

substance the mixtures (*mixeis*) occur by juxtaposition (*parathesis*). Two or more substances are placed together (*suntitheimena*) into the same [volume] and juxtaposed (*paratithemena*) with each other by 'juncture' (*harmê*), as he says. Each substance in such a juxtaposition preserves (*sôzein*) its proper substance (*ousia*) and quality (*poiotês*) in its perimeter (*perigraphê*), as, say, with beans and wheat when these are juxtaposed with each other.

Other mixtures occur by fusion (*sunkhusis*) through and through (*di' holôn*). The substances themselves and the qualities in them perish together with each other (*sumphtheiresthai*), as he says happens with medical drugs, when the mixing ingredients perish together and a different body is generated out of them. He says that other mixtures occur by substances and their qualities being mutually spread out along with each other (*antiparekteinomenai*) through and through (*di' holôn*). This goes with the preservation (*sôzein*) in such a mixture of the original substances and qualities. It is this mixture uniquely which he says is blending (*krasis*). He says that such a mutual spreading out along with each other (*antiparektasis*) of two or more bodies through and through, in such a way that each of them preserves in such a mixture its proper substance and the qualities in it, is the only mixture that is a blending. For it is a unique feature of blended ingredients that they can be separated from each other again (*khôrizesthai palin*), which happens only because the blended ingredients preserve their own natures (*sôzein phuseis*) in the mixture.[15]

Chrysippus used to affirm something of the following sort. What exists is *pneuma* that moves itself towards and away from itself, or *pneuma* that moves itself backwards and forwards. It is taken to be *pneuma*, because it is said to be air in motion. Something analogous occurs with the aether also, so that it and the aether even coincide in the same definition. Such a motion occurs only according to those who think that all substance admits of change and fusion (*sunkhusis*) and composition (*sustasis*) and commixture (*summixis*) and natural association (*sumphusis*) and things like that.

For members of the Stoic school like to distinguish juxtaposition (*parathesis*), mixture (*mixis*), blending (*krasis*) and fusion (*sunkhusis*). They say that juxtaposition is the mutual contact of bodies at their surfaces, such as we see in the case of heaps in which are included wheat and barley and lentils and anything else like them, and in the case of pebbles and sand on the sea-shore.

They say that mixture (*mixis*) is the mutual spreading out along with each other (*antiparektasis*) through and through (*di' holôn*) of two or more bodies, with their naturally associated qualities persisting (*hupomenousai*) as happens with fire and glowing iron. In their case, the mutual spreading out of the bodies [fire and iron] along with each other occurs through and through. And it happens in a similar way too with the souls within us: they are mutually spread out along with our bodies through and through. For the Stoics like body to be mutually stretched along (*antiparêkein*) through body (*sôma dia sômatos*).

They say that blending (*krasis*) is the mutual spreading out along with

[15] ibid. 216,14-217,2 (*SVF* 2.473).

each other through and through of two or more fluid bodies, with their qualities persisting (*hupomenein*). (They say that mixture occurs with non-fluid bodies as well, like fire and iron, or soul and the body that surrounds it, but that blending occurs only with fluid bodies.) [They speak this way of the blending of fluids],[16] because (*gar*) the qualities of each of the blended fluids show forth together (*sunekphainesthai*) out of the blend, for example, the qualities of wine, honey, water, vinegar and the like. That the qualities of the blended ingredients survive (*diamenein*) in such blends is already clear (*prodêlon*) from the fact that the ingredients are often separated from each other (*apokhôrizesthai*) by some device. Certainly, if you dip an oil-drenched sponge into wine blended with water, you will separate the water from the wine, as the water runs back into the sponge.

They say that fusion (*sunkhusis*) is the change (*metabolê*) of two or more qualities in bodies, to produce another quality differing from the earlier ones, as happens with the combination (*sunthesis*) of perfumes and of medical drugs.[17]

Arius Didymus makes it explicit that body goes through body on this theory. Why do the Stoics commit themselves to that? The reason given so far is the difficulty Aristotle has in constructing a different alternative to juxtaposition. But the centre of the Stoics' interest is not in the chemistry of mixture for its own sake, but in some larger metaphysical issues. One such issue is their materialism, according to which only bodies can act or be acted on.[18] Now things that act and are acted on are often not separate from each other. The soul, for example, their view, and it is distributed throughout the body it acts on. Otherwise, as Alexander says in a passage to be quoted shortly,[19] it would alternate with pockets of dead matter. But more important still is the role which the Stoics gave to the gas called *pneuma*. Indeed, the soul simply *is pneuma* in one of its many guises. One of the most important roles of *pneuma* is to hold together the bodies which it permeates. Alexander has just been quoted as saying that *pneuma* permeates and holds together the *whole cosmos*,[20] and elsewhere we are told that it also holds together individual bodies.[21] Or rather, it

---

[16] It is a piece of interpretation that the 'because' refers back to the earlier sentence, and that it explains why the qualities (and perhaps the fluids) are described as all still present, but it is supported by Wachsmuth, who actually bracketed the intervening sentence as a gloss. I have used rounded brackets.

[17] Arius Didymus fr. 28, ap. Stobaeum *Eclogae* 1.17.4, in *Dox. Gr.* 463,14-464,8 (*SVF* 2.471).

[18] Diogenes Laertius *Lives of Eminent Philosophers* 7.55 (*SVF* 2.140); Sextus *M* 8.263 (*SVF* 2.363); Sextus *PH* 3.38; Seneca *Ep. ad Lucilium* 106,2 (*SVF* 3.84, pp. 20-1); Cicero *Academica Priora* 11.39 (*SVF* 1.90); Nemesius *de Natura Hominis* p. 32 Matthaei (*SVF* 1.518); Plutarch *Com. Not.* 1073E (*SVF* 2.525); Aëtius *Placita* 4.20.2 in *Dox. Gr.* 410,6 (*SVF* 2.387); ps-Galen *Hist. Phil.* ch. 23, 'On bodies', *Dox. Gr.* 612,19-613,2.

[19] Alexander *Mixt.* ch. 4, 217,36.

[20] ibid. ch.3, 216,14-17

[21] Galen, *On Cohesive Causes*, ch. 1, translated from the Arabic by M. Lyons, in *Corpus Medicorum Graecorum*, supplementum orientale 2, Berlin 1969 (not in *SVF*).

holds together compound bodies and the two slack (*atonoi*) elements earth and water. The other two elements, air and fire, are taut (*eutonoi*), and so sustain themselves without the need of anything to hold them together.[22] These cohesive functions will be further discussed in Chapters 9 and 14 below. *Pneuma* is the word for *spirit*, and it lies behind such disparate ideas as that of the Holy Spirit and of animal spirits, but I have avoided the translation 'spirit', because not all its uses imply the corporeality which the Stoics intend. *Pneuma* is the *first* example given in the passages translated from Alexander and Arius Didymus, to introduce the Stoic idea of blending. The more everyday examples of blending, by contrast, are cited in three passages by Alexander as designed to make the idea of blending, or of a small body's blending with a large, credible.[23] The impression is created that the Stoics were especially interested in postulating blending in the case of *pneuma*, and used the everyday cases to argue for its feasibility. It has, on the contrary, been argued that the everyday cases are not endorsed by the Stoics, but presented as fictitious aids to the imagination.[24] But this is to go against Alexander's view that the everyday cases were used as 'clear testimony' (*enargê marturia*), to provide persuasion (*pistis*) and establish (*kataskeuê*) the fact of blending. Although I was ready to reject Alexander's opinion on the obscure question of what books the Stoics had available to them up to five hundred years earlier, he is not so likely to be wrong about this very different question. One of his accounts runs as follows:

> Since this is so, they say there is nothing surprising in the further fact that certain bodies, when helped by each other, are so unified (*henousthai*). with each other through and through (*di' holôn*), that they are mutually spread out along with each other (*antiparekteinesthai*) through and through (*di' holôn hola*), while themselves being preserved (*sôzomena*) along with their proper qualities, and that this happens even if some are smaller in volume, and would not be able by themselves to be diffused so far and preserve their proper qualities. For they say that this is the very way in which a cup of wine blends (*kirnasthai*) with a lot of water and is helped by it to spread out that far. They use as clear testimony (*enargê marturia*) that this is the case the fact that the soul, which has its own reality (*hupostasis*) just like the body that receives it, goes through the whole of the body in mixture with it, while preserving its own proper substance (*ousia*). For no part of the ensouled body is without soul. They say that the nature (*phusis*) of plants is similar, and so too is the holding power (*hexis*) in bodies that are held together by a

[22] Alexander *Mixt*. ch. 4, 218,2-6 (*SVF* 2.473), translated below; Plutarch *Com. Not.* 1085C-D (*SVF* 2.444).

[23] Alexander *Mixt*. ch. 4, 217,13 and 33 (*SVF* 2.473): the everyday examples are persuasive evidence (*pisteis*), and used as clear testimony (*enargê marturia*) that a small body can blend with a large. Ch. 12, 227,12-13 (*SVF* 2.475): the idea of body going through body acquires persuasiveness (*pistis*) from the everyday examples, which (ch. 16, 233,14 (*SVF* 2.735)), are used for establishing (*kataskeuê*) the idea.

[24] Robert B. Todd, op. cit., part 2, 45-6.

holding power. They say that fire also goes through and through iron, while each of them preserves its proper substance. And they say that two of the four elements, fire and air, being tenuous, light and taut (*eutonos*) go through and through the other two, earth and water, which are dense, heavy and slack (*atonos*), and that the former and the latter both preserve their proper nature and their continuity. They think that noxious drugs and noxious odours blend with the things that are affected by them and are combined (*paratithesthai* [previously used for juxtaposition]) through and through. And Chrysippus says that light blends with air. That is the view of blending held by Chrysippus and the philosophers who follow him.[25]

Several of these examples of bodies interpenetrating can be reduced to the one case of *pneuma*. For *pneuma* is in one form God and reason, in another soul, in another nature, and in another the holding power (*hexis*) that holds together inanimate things.[26] The identification of soul with something less familiar (*pneuma*) does not prevent it from being offered as clear evidence (*enargê marturia*) for blending. It may be an unfamiliar idea that soul is *pneuma*, but that our soul is a body blended in with our body they think can be made very clear.

Whether or not, as I believe, the Stoics were responding to Aristotle, there is no doubt that the Aristotelians responded to them with a storm of protest against bodies in the same place. But I think the Aristotelians lumped together two entirely different Stoic enterprises. In one enterprise, the Stoics do not put bodies in the same place, but rather *reduce* something to a body. I believe they reduce *pneuma* to fire and to air, reduce a body's qualities to the body itself, and possibly reduce God to matter. These reductions do not involve putting two bodies in the same place, because the reductionist view is that there are not two distinct bodies in question. To understand how the reductions work, we must go to the heart of their metaphysics, and in particular to their theory of *categories*. We can then also understand better the cases in which the Stoics really do put distinct bodies in the same place, which is their second enterprise. But I shall take the reductionist enterprise first.

Enterprise 1. The Stoic theory of categories and their reductionism

(a) Pneuma as air or fire

The first of the cases in which they do *not* put bodies in the same place concerns the inner constitution of *pneuma*. There are many reports that *pneuma* is itself a mixture, being made of the elements air and

[25] Alexander *Mixt.* ch. 4, 217,27-218,10 (*SVF* 2.473).

[26] ps-Galen *Introductio* or *Medicus* ch. 9, Kühn vol. 14 p. 697 (*SVF* 2.716); Plutarch *Sto. Rep.* 1053F (*SVF* 2.449); Diogenes Laertius *Lives* 7.138-9 (*SVF* 2.634); 7.156 (*SVF* 2.774); Themistius *in DA* 35,32-4 (*SVF* 1.158); Galen *Comm. 5 in Hippocratis*

fire.[27] But the suspicion has been put to me[28] that pneuma is not a mixture after all. I should like to support this suspicion, and suggest an entirely different interpretation. First of all, the claim that *pneuma* is a mixture looks like a mere inference. In the passages cited above, Alexander uses the words 'for' (*gar*) and 'I suppose' (*pou*) and makes his conclusion an inference from the fact that *pneuma 'has the substance* (*ousia*) *of* ' air and fire – I shall come back to the meaning of that expression later. He speaks guardedly of *pneuma* being 'compounded in *some* way (*sunkeimenon pôs*) out of fire and air'. Galen calls the ingredients of *pneuma* 'the cold and hot', and merely proposes 'air' and 'fire' as alternative names. If these authors could not find a text actually saying that *pneuma* was a mixture, what then is the relation of *pneuma* to air and fire? I believe the answer is that '*pneuma*' is a designation given sometimes to air and sometimes to fire. Hints that this is the case, although not yet the explanation for it (that will follow), can be found in the way the Stoics interchange, and even equate, '*pneuma*', 'fire', 'air' and other terms.

First, the cohesive function which we have seen assigned to *pneuma* is indifferently assigned to air, fire, aether, pneumatic substance (*ousia*) and to a pneumatic and fiery power imported by air and fire.[29] After assigning it to aether, Diogenes Laertius subsequently *equates* creative fire with a certain fiery *pneuma*,[30] while Arius Didymus, in a passage already quoted, says that air in motion *is pneuma*, and that *pneuma* and aether have the same definition.[31] In the light of these equations, it is not surprising to find Galen interchanging talk of *pneuma* and of air, and again of pneumatic substance and of air and fire in two other passages already cited.[32] What we should conclude from these equations and this interchanging of terms, I believe, is that air or fire can *each* be described as *pneuma* in suitable contexts, depending on the functions which they are thought of as performing: there is, then, no reference to *mixture*. This suggestion can be confirmed by reference to the opening of Galen's treatise *On Cohesive Causes*, where he reports the Stoics as treating air and fire as the two

Epidemiarum 6, Kühn 17B,250 (*SVF* 2.715); Galen *Quod Animi Mores Corporis Temperamenta Sequantur*, Kühn 4.783 (*SVF* 2.787); Galen *Definitiones Medicae* 95, Kühn 19,371 (*SVF* 2.1133).

[27] See for various formulations Plutarch *Com. Not.* 1085D (*SVF* 2.444); Alexander *Mixt.* ch. 10, 224,14a-15 (*SVF* 2.442); ch. 11, 225,6-8 (*SVF* 2.310); id. *Mant.* 26,13 (*SVF* 2.786); Galen *de Hippocratis et Platonis Placitis* 5.3 (*SVF* 2.841); Diogenes Laertius *Lives* 7.156 (*SVF* 2.774); Arius Didymus fr. 28, ap. Stobaeum *Eclogae* 1.17.4, in *Dox. Gr.* 463,16 (*SVF* 2.471); Simplicius *in Cat.* 218,1 (*SVF* 2.389).

[28] I thank Eric Lewis for urging this doubt.

[29] Air, Plutarch *Sto. Rep.* 1053F; pneumatic and fiery power, Plutarch *Com. Not.* 1085C-D (*SVF* 2.444); purer aether, Diogenes Laertius *Lives* 7.139 (*SVF* 2.634); Galen *de Plenitudine*, Kühn 7,525 (*SVF* 2.439).

[30] Diogenes Laertius *Lives* 7.156.

[31] Arius Didymus fr. 28, ap. Stobaeum *Eclogae* 1.17, in *Dox. Gr.* pp. 463-4 (*SVF* 2.471).

[32] Galen *Comm. 5 in Hippocratis Epidemiarum 6*, Kühn 17B,250 (*SVF* 2.715); Galen *de Plenitudine*, Kühn 7,525 (*SVF* 2.439).

tenuous or fine elements, and then says that the Stoics give the name *pneuma* to *all* substance that is fine. The treatise survives in Latin and Arabic versions, and M. Lyons' translation of the Arabic runs as follows (with my additions in square brackets):

> The two active elements [sc. fire and air] have fine parts and the other two thick parts. Every substance with fine parts the Stoics call spirit [i.e. *pneuma*] and they think that the function of this spirit is to produce cohesion in natural and in animal bodies.[33]

Fire and air, then, can *each* be called *pneuma*. But how, it may be wondered, can Stoic terminology be so fluid? I think the answer is given by reference to their so-called theory of categories. It is a matter of controversy whether that theory was thought up only by later Stoics. Even if it was, I believe that at least it articulates the way in which earlier Stoics were already thinking. And if this should be disputed, it is at any rate a theory that can consistently be added on to earlier Stoic formulations, to show how they can be freed from difficulties. According to the theory of categories, every individual can be described in four ways, as mere matter or subject (*hupokeimenon*), as matter endowed with qualities (*poion*), as disposed in certain ways (*pôs ekhon*), or as standing in certain relations (*pros ti pôs ekhon*).[34] What is true, say, of my mere matter need not be true of my qualified, or disposed, or relatively disposed matter,[35] and consequently the four different levels of existence, as they have been called, should not be viewed as the same as each other, although they cannot be described as different either:

> [Posidonius] says that the individually qualified thing (*to poion idiôs*) and the substance (*ousia*) out of which it is [constituted] are not the same, but are not different either, merely not the same, because the substance (reading *ousian*) is part of the individually qualified thing and occupies the same place, whereas things that are said to be different from others must also be separate from them in place, and not seen as part of them. Mnesarchus says it is clear that a thing viewed as individually qualified and a thing viewed as substance are not the same, because things that are the same must have the same attributes.[36]

---

[33] Galen *de Causis Continentibus*, translated from Arabic by M. Lyons as *On Cohesive Causes*, ch. 1, sec. 3, *Corpus Medicorum Graecorum*, supplementum orientale 2, Berlin 1969, p. 53. I owe this confirming evidence to David Sedley. There is a slip of the pen in sec. 2, when Lyons' translation refers to fire and water, instead of fire and air.

[34] Plotinus 6.1.25 (1-33 partly in *SVF* 2.371); Simplicius *in Cat.* 66,32-67,8 (*SVF* 2.369).

[35] Plutarch *Com. Not.* 1083C-D (*SVF* 2.762); Arius Didymus fr. 27, ap. Stobaeum *Eclogae* 1.17 and 20, *Dox. Gr.* 463,6, reporting the Stoic Mnesarchus.

[36] Arius Didymus fr. 27, ap. Stobaeum *Eclogae* 1.17 and 20, *Dox. Gr.* 462,25-463,13, reporting two later Stoics, Posidonius and Mnesarchus. See the excellent account in David Sedley, 'The Stoic criterion of identity', *Phronesis* 27, 1982, 255-75, at 259-62.

If we apply this scheme of categories to the case of *pneuma*, we can see that air or fire could in some contexts be thought of in terms of the *third* category, as *disposed* in certain ways. Their dispositions might include the tautness that we spoke of earlier and perhaps the holding power, unless that belongs in the fourth category of relatives. When air or fire are regarded as disposed in appropriate ways, they can then be described as actually *being pneuma*. They can also be described in those contexts as being a pneumatic and fiery power. We ourselves would naturally think of such a power as more like a quality than a body. But we shall see shortly that the Stoic materialists sometimes thought of qualities as being bodies disposed in a certain way. In that case, the Stoics will simply be referring to different levels of existence, not to entirely distinct entities, when they call the factor that holds bodies together sometimes *pneuma*, sometimes a pneumatic power, sometimes air, fire, or aether.

This reconstruction helps to get rid of a certain vicious circle alleged by Alexander: *pneuma* depends on the existence of air and fire to serve as its ingredients, but air and fire presuppose *pneuma*, presumably to hold them together, and perhaps also because their Creator, God, is *pneuma*.[37] This alleged circle need not trouble the Stoics for several reasons. First, air and fire are already taut (*eutonoi*), we have seen, and need no holding together. The question of God's creating them will be handled below. But finally, it now appears untrue that *pneuma* needs air and fire as *ingredients*: it is not so distinct from them as that.

The reconstruction has the further advantage of helping us to understand something else that has been found mystifying. How is it that all four elements for the Stoics have a tendency to move downwards, that is towards the centre of the cosmos, and yet that two of them, air and fire, are also weightless (*abarês*) and have an upward tendency (*anôpheres*), that is towards the periphery?[38] This double tendency of air and fire can now be seen to fit with a well known and thoroughly familiar doctrine about *pneuma*. *Pneuma* contains two opposite motions, one inwards and one outwards.[39] If *pneuma* just is air and fire disposed in a certain way, it is then quite to be expected that the air and fire in the cosmos should have the same two opposite tendencies.

The reconstruction also explains the fluidity of Stoic language. At

---

[37] Alexander *Mixt*. ch. 10, 224,14a-22 (*SVF* 2.442).

[38] References in Chapter 9. A different answer is offered by Michael Wolff, 'Hipparchus and the Stoic theory of motion', in J. Barnes and M. Mignucci, eds, *The Bounds of Being*, Naples, forthcoming.

[39] The most informative of several references is Nemesius *Nat. Hom*. ch. 2, p. 42 Matth. (*SVF* 2.451). David Furley will have more to say about how the Stoics conceived their two opposite motions (*The Greek Cosmologists* vol. 2, forthcoming, Cambridge). But however they conceive them in the case of cosmic *pneuma*, they must conceive them the same way in the case of cosmic air and fire.

the same time, it throws light on the supposed evidence for the 'mixture' suggestion. Alexander relied partly on the statement that *pneuma* 'has the *substance* (*ousia*) of air and fire'. It can now be seen that that does not mean, as he supposed, that *pneuma* is a mixture. 'Substance' is a name the Stoics give to the first category, matter, and *pneuma* can be said to have the same *substance* as air and fire, because one and the same *matter* can be described as air or fire when viewed as *qualified* in various ways, and as *pneuma* when viewed as both qualified and *disposed*. A further outcome of the reconstruction is that the Stoics postulated fewer cases of bodies interpenetrating than was supposed. *Pneuma* would not itself be a blend, nor would it be blended into air or fire. It *is* air or fire, and as such it would be blended throughout all earth and water, and hence throughout all ordinary compound bodies.

There is a further surprise, if my reconstruction has been correct. For it will be seen in a moment that the Stoics sometimes think of *pneuma* as itself being further disposed in a certain way. When this is put together with my suggestion that *pneuma* is already air or fire suitably disposed, the consequence is that air or fire can be *doubly* disposed. Disposed in one way, they are *pneuma*, but thus disposed, they may be further disposed, to constitute, as we shall see, a quality, for example a character trait.[40] I believe there is nothing untoward about the idea of being doubly disposed. The air (or fire) in a person disposed so as to hold him together can be called his *pneuma*. Disposed so as to hold him together *a little too tightly* (the second disposition), it can be called his nervousness (a quality).

(b) Stoic qualities

To the case of qualities I shall now turn, because it is another locus of difficulties, and provides another instance in which it is not clear that the Stoics really postulated interpenetration. The Stoic view that qualities are *bodies* is frequently recorded. But the further view, that bodies and their qualities are arranged like so many interpenetrating billiard balls, looks again like an *inference* from the first view, an inference designed to embarrass them. Admittedly, Galen reports such interpenetration of bodies as an opinion of Zeno's. But Alexander presents it merely as an inference from Stoic views and Simplicius as an implication they had to agree to, in order to maintain their other beliefs.[41]

If such a view would be embarrassingly strange, the theory of categories may again be invoked, to show that the Stoics need never have held it. They sometimes thought of a quality in terms of the third

[40] I thank Robert Sharples for making this point.
[41] Galen *in Hippocratis de Humoribus* 1, Kühn 16,32, and *de nat. facult.* 1.2 (*SVF* 1.92); Alexander *Mant.* 115,36-116,1 (*SVF* 2.797); Simplicius *in Phys.* 530,9-14 (*SVF* 2.467).

category, that is, as matter, or *pneuma*, or soul, or reason, *disposed* in a certain way (*pôs ekhon*).[42] For the Stoic materialists, each of these, matter, *pneuma*, soul and reason, would be thought of as a *body*. They would not be four *distinct* bodies, for *pneuma is* soul and reason, and all of them are matter variously disposed. The idea about qualities, as I understand it, is strongly reductionist. We are left free to talk about the multiple qualities of a man. But such talk is not perspicuous. For all that actually exists in the world to correspond to it is a single body, the man's matter, *pneuma*, soul and reason. That body is disposed in various ways, but if we talk of dispositions, we should not treat them in the normal way as something distinct from the body they belong to. There is just the body disposed.

Once we see the reductive sense in which a Stoic quality is a body, we can also see why the innumerable qualities that belong to a man are not *distinct* bodies, all interpenetrating in the same place. They are not, for each time we are talking only of *one* body, the *original* one, which is the man's matter, *pneuma*, soul, or reason, but disposed in different ways, as many ways as there are qualities, but ways that are not to be reified. Seneca, for one, tries to get rid of a similar plurality by means of this sort of strategy among others.[43] Admittedly, he warns that he is giving his own independent opinion, but he ascribes to Chrysippus what may be the same view about walking,[44] which the Stoics treated as a quality in a broad sense.[45]

If this reconstruction is correct, the relation to other thinkers' views on qualities is intriguing. The Stoics are by no means saying about qualities what they say about Plato's Forms, that they are not anything at all (*mête tina*).[46] On the other hand, they do not allow a body's qualities to be as distinct from it as Aristotle does, when he separates bodies into a different category from qualities in his own very different scheme of categories. Nor yet is their type of reductionism the same as the Epicurean. Epicurus says some things the Stoics would approve: qualities must be allowed to be existent (*eisin, ontos*), contrary to the earlier atomist, Democritus, who said that sweet, bitter, hot, cold and colour existed only by convention or agreement (*nomôi*), atoms and void in reality (*eteêi*).[47] Further, qualities are not incorporeal (*asômata*), or distinct incorporeal entities accruing to the body (*heter' atta prosuparkhonta asômata*). But Epicurus' reason for denying incorporeality is the different one that incorporeality would imply independent existence (*kath' heauto*), and

---

[42] Seneca *Ep.* 113, esp. 2-4, 7, 11, 14 (partly in *SVF* 3.307, p. 75); Plotinus 6.1.29(11-16) (*SVF* 2.376); Alexander *in Top.* 360,9-13 (partly in *SVF* 2.379); Sextus *M* 11.23 (*SVF* 3.75); id. *PH* 2.81.

[43] Seneca loc. cit.

[44] Seneca *Ep.* 113,23.

[45] Simplicius *in Cat.* 212,17 (*SVF* 2.390).

[46] Arius Didymus fr. 40, ap. Stobaeum *Eclogae* 1.12 (*SVF* 1.65).

[47] Sextus *M* 7.135=Democritus fr. 9 Diels-Kranz.

qualities are not independently existent (*kath' hauta, kath' heautas phuseis*), because they are causally dependent on atoms and void. This is already a kind of reductionism, but the Stoics' reductionism, I think, takes a stronger form:[48] qualities, according to their theory of categories, are not merely causally dependent on bodies; they are not really distinct from (although they are not the same as) the bodies to which they attach. Epicurus never goes as far as that. The relevant passage of his work has already been partly translated in Chapter 4:

> Now here is another point to consider: the name 'incorporeal', according to its commonest usage, is applied to what is thought of as independently existent (*kath' heauto*). But one cannot think of the incorporeal existing independently except in the case of vacuum ...
>
> Now as for shapes and colours and sizes and weights and all the other things predicated of a body as permanent properties, whether belonging to all bodies or to those that are visible and themselves knowable through sense perception, we should not think that these are independent (*kath' heautas*) natural substances: that is not conceivable. Nor on any account should we think that they do not exist (*ouk eisin*), nor that they are some distinct incorporeal entities accruing to the body (*heter' atta prosuparkhonta asōmata*) ...
>
> Now it is often an accidental feature of bodies that they are also accompanied impermanently by things which are neither at the invisible level, nor incorporeal. So by using the name in its most current way, we make it clear that 'accidents' do not have the nature (*phusis*) of the whole which we take together in a bundle (*kata to athroon*)[49] and call a 'body'. Nor do they have the nature of permanently accompanying properties without which a body cannot be conceived ... And we should not banish this self-evident thing from existence (*ontos*) merely because it has neither the nature of the whole it belongs to, which we call by the actual name 'body', nor the nature of permanently accompanying properties. Nor again should we think of them as independent (*kath' hauta*), for that is not thinkable, whether for these or for permanent properties.[50]

The Stoic reduction of a body's qualities to the body itself, variously disposed, in no way impairs, it even enhances, the Stoics' ability to offer explanations. It might previously have seemed that they were handicapped by their view that causal relations hold only between *bodies*: a body causes another body to have an incorporeal predicate true of it.[51] This might appear restrictive, because one wants to be able to say not only that Mr Smith caused the workers to go on strike, but also that Mr Smith's *sliminess* (one of his qualities) caused them to go

---

[48] David Sedley has argued that Epicurus is not consistently reductionist even in a *causal* sense, 'Epicurus' refutation of determinism', in G.P. Carratelli, ed., *Suzetesis, Studi offerti à M. Gigante*, Naples 1983.

[49] The surrounding context, partly translated in Chapter 4, shows that it is only the *concept* of body which is treated as a bundle here, not the concrete body.

[50] Epicurus *Letter to Herodotus*, in Diogenes Laertius *Lives* 10, 67-71.

[51] Sextus *M* 9.211 (*SVF* 2.341).

on strike. The answer, correct so far as it goes, is that for the Stoics all qualities, including sliminess, are bodies, and so can stand in causal relations. But it has often seemed mysterious how qualities could be bodies. We can now see the explanation. Where one wants to say that it was not just Mr Smith, but in particular his sliminess, that caused the strike, the Stoics will have a favoured description of the cause. It is not just the qualified subject, Mr Smith, that caused the strike. It is the subject qualified *and disposed* in a certain way (the way that makes him slimy). They will prefer a description in the third category, to one in the second, when they want to bring out how a *quality* can be efficacious.

There may seem to be a danger that the proposed treatment of qualities would prove too much for the Stoics' own comfort. For if a person's *qualities* are to be regarded as his matter, *pneuma*, soul, or reason disposed in a certain way, what can prevent his location and date and the predications (*lekta*) true of him from being regarded equally as his matter or *pneuma* disposed in a certain way? But then place, time and predications would have as much right as quality to count as bodies, whereas notoriously the Stoics want to treat them as three of the four incorporeals, along with vacuum.[52]

The answer in the case of predications is easy, because our predications are distinguished[53] from qualities. They are not the same as the qualities we predicate because they are mind-dependent. They are also distinguished from another set of bodies, the physical words we use, on the grounds that foreigners hear the words, but do not take in the predications. As regards a person's place and time, it would be hard to think of any disposition of the *pneuma* that would correspond to distinct spatio-temporal locations. His place is in fact sharply distinguished from his body as an extension (*diastêma*) occupied by it.[54] His time might more plausibly be thought of under the category of relation than under the category of disposition, since he arrives at a new time, without this being due to any change in himself. But in fact time is defined only at the cosmic level, and it too is defined as an extension

[52] ibid. 10.218 (*SVF* 2.331). The Neoplatonists Plotinus, Dexippus and Simplicius draw the Aristotelian distinction between places and times, which fit under Aristotle's category of quantity, and *in* a place, *at* a time, which form the separate Aristotelian categories of where and when. About the latter pair, if not about the former, they complain that the Stoic categories fail to accommodate them, and that the Stoics would get into trouble if they tried to accommodate them under the category of the disposed. I do not take this as evidence that the Stoics actually did try to treat a man's location and date as his matter disposed in certain ways: Plotinus 6.1.30(9-15) (*SVF* 2,400); Dexippus *in Cat*.34,19 (*SVF* 2.399); Simplicius *in Cat.* 66,32 (*SVF* 2.369). But John Rist takes it otherwise, 'Categories and their uses', in A.A. Long, ed., *Problems in Stoicism*, London 1971, 38-57, at 51.

[53] Sextus *M* 8,11-12 (*SVF* 2.166).

[54] ibid. 10.3 (SVF 2.505); 10.12 (*SVF* 2.501); *PH* 3.124; Arius Didymus fr. 25, *Dox. Gr.* 460,18-461,3 (*SVF* 2.503); Aëtius 1.20.1, *Dox. Gr.* 317,31-6 (*SVF* 2.504); Themistius *in Phys* 113,7-11 (*SVF* 2.506); Simplicius *in Phys.* 571,24-5 (*SVF* 2.508).

rather than as a disposition or relation. It is said to be the extension of the motion of the cosmos.[55]

(c) God and matter

So far I have resisted the claim that the Stoics are committed to interpenetrating bodies in their account of the internal constitution of *pneuma*, and in their theory of qualities. But Alexander cites as a third example of interpenetrating bodies God and qualityless matter, that is, what Aristotelians would call prime matter,[56] and initially there seems to be evidence enough for his view. For many reports tell us that the Stoic God goes through the whole of matter, as one author says, like honey through the honeycomb.[57] And many reports tell us that God and matter are both bodies.[58] Nor can it be argued that only qualified matter is body. For the sources just cited make it quite clear that it is qualityless matter which they are calling body. And this has to be the case, because other sources tell us that qualityless matter is *acted on* by God,[59] and it is Stoic doctrine, we have seen, that only bodies can act or be acted on. So the isolated reading in the Byzantine lexicon of Suidas must be rejected. Suidas quoted Diogenes Laertius *Lives* 7.134 as making Stoic qualityless matter incorporeal (*asômatous*, instead of

---

[55] Arius Didymus fr. 26, *Dox. Gr.* 461,4-25 (*SVF* 1.93; 2.509; 3, p. 260); Philo *Aet.* 13 (*SVF* 2.509); *Opif.* 26 (*SVF* 2.511); Simplicius *in Cat.* 350,15-18 (*SVF* 2.510); Sextus *M* 10.170 (*SVF* 2.513); Plutarch *Quaest. Plat.* 1007A (*SVF* 2.515).

[56] Alexander *Mixt.* ch. 11, 224,33-225,5.

[57] *God goes through matter*: Proclus *in Tim.* (Diehl) 1,414,4-8 (*SVF* 2.1042); Tertullian *ad Nationes* 2.4 and *adversus Hermogenem* 44 (like honey through the honeycomb, both in *SVF* 1.155); Athenagoras *Legatio* ch. 6 (*SVF* 2.1027); Alexander *Mixt.* ch. 11, 225,1-2 (*SVF* 2.310); 225,19 (*SVF* 2.1044). *God is in matter*: Plutarch *Com. Not.* 1085B (*SVF* 2.313); Diogenes Laertius *Lives* 7.134 (*SVF* 1.493); Alexander *in Metaph.* 78,15 (*SVF* 2.306); *God or pneuma go through the cosmos and all bodies*: Aëtius *Placita* 1.7.33, *Dox. Gr.* 306,5-6 (*SVF* 2.1027); Sextus *PH* 3.218 (*SVF* 2.1037); id. *M* 9.130 (*SVF* 3.370, p. 90); Alexander *Mixt.* ch. 11, 224,15 (*SVF* 2.442); Tertullian *Apologeticum* 21 (*SVF* 1.533).

[58] *God is a body, namely, fire or pneuma*: Alexander *Mixt.* 225,3 (*SVF* 2.310); Hippolytus *Refutatio* 21.1, *Dox. Gr.* 571,9 (*SVF* 1.153); Galen *Hist. Philos.* 16, *Dox. Gr.* 608,19 (*SVF* 1.153); Plutarch *Com. Not.* 1085B (*SVF* 2.313); Eusebius *Praeparatio Evangelica* 3.9.9 (*SVF* 2.1032); Augustine *contra Academicos* 3.17.38 (*SVF* 1.146); Plutarch *de facie in orbe lunae* 926C (*SVF* 2.1045); Aëtius *Placita* 1.6.1, *Dox. Gr.* 292,23-4 and 1.7.33, *Dox. Gr.* 306,1 (*SVF* 2.1009 and 1027); Plotinus 2.4.1(13) (*SVF* 2.320); Servius *in Verg. Aen.* 6,727 (*SVF* 2.1031). *Matter is a body*: Calcidius *in Tim.* §289; Sextus *M* 10.312 (*SVF* 2.309); Simplicius *in Phys.* 227,23-5 (*SVF* 2.326); *God and matter, the two principles (arkhai) are bodies*: Aristocles ap. Eusebium *Praeparatio Evangelica* 15 (*SVF* 1.98); Diogenes Laertius *Lives* 7.134, according to the mss., but Suidas reads incorporeal *asômatous* (*SVF* 2.299); Origen *Against Celsus* 6.7. (*SVF* 2.1051).

[59] God acts on qualityless mattter, (i) to give it quality and shape, Plutarch *Com. Not.* 1076C-D (*SVF* 2.1168); Calcidius *in Tim.* §311 (not in *SVF*); Sextus *M* 9.75-6 (*SVF* 2.311); (ii) to produce earth, air, fire and water, Diogenes Laertius *Lives* 7.134-6 (*SVF* 1.85; 1.102; 2.299); Alexander *Mixt.* ch. 11, 224,32-225,3 (*SVF* 2.310); Aristocles ap. Eusebium *PE* 15 (*SVF* 1.98); (iii) to produce the things in the cosmos, Seneca *Ep.* 65.2 (*SVF* 2.303); Galen *de Qualitatibus Incorporeis*, Kühn vol. 19, p. 476 (*SVF* 2.323); Alexander *Mixt.* ch. 11, 225,18-20 (*SVF* 2.1044).

sômata). But this was already suspect, because it went against all the manuscripts, and it must now be abandoned. It is, of course, surprising that qualityless matter should be a body, seeing that it is thought of as qualityless, and, as Plotinus points out, it must also be thought of in separation from resistance.[60] Perhaps the Stoics are here relying on a definition of body as what can act or be acted on, a definition which ignores qualities and resistance.[61]

The initial appearance, then, is that God and qualityless matter are two bodies in the same place. But our assessment may depend on how God is to be conceived. He has a role as one of the two fundamental causal principles (*arkhai*), for by acting on the other principle, qualityless matter, he produces the four elements, earth, air, fire and water, and everything else in the cosmos. In this role, is he still to be conceived as fire, *pneuma*, or *pneuma* disposed in a certain way? Or should we switch in this context to seeing him in *abstraction* from these, just as qualityless matter is viewed in abstraction from them. Plutarch, Alexander and, following him, Plotinus, we shall see, still think of God in a *concrete* way, in his causal role, that is, as fire, *pneuma* and matter disposed in a certain way. Diogenes Laertius, by contrast, though recognising that God is still a body (*sôma*), seems to view him in his causal role as something more *abstract*. At any rate, Diogenes describes him as formless (*amorphos*), which suggests that he is not to be viewed in this context as fire, *pneuma*, or disposed matter, although he may still depend for his existence on being enmattered in fire or *pneuma*. This is how he puts it:

> They think that there are two principles (*arkhai*) of all things, one active and the other passive. The passive one is qualityless substance, or matter, the active is the reason in it, or God. For God is eternal and fashions everything throughout the whole of matter ... And they say that principles and elements (*stoikheia*) differ. For principles are not generated or destroyed, whereas the elements are destroyed in the conflagration. But the principles too are bodies (*sômata*). They are bodies and formless (*amorphous*), whereas the elements have form (*memorphôsthai*).[62]

If Diogenes is right, I see no way of avoiding the conclusion that God and qualityless matter are two distinct bodies residing in the same place. But by an irony, Plutarch, Alexander and Plotinus, who propound their conception with polemical intent, provide a way of avoiding that conclusion. For if God, in his causal role, is still to be thought of as *pneuma* disposed in a certain way, then Plotinus' description of him is right: he will be, like all *pneuma*, matter disposed

[60] Plotinus 6.1.26(17-23)(*SVF* 2.315); 6.1.28(18-20)(*SVF* 2.319).

[61] Jaap Mansfeld views this as one of two Stoic definitions of body, relying on Sextus *PH* 3.38 and ps-Galen *Hist. Phil.* ch. 23, 'On bodies', *Dox. Gr.* 612,19-613,2 ('Zeno of Citium', *Mnemosyne* 31, 1978, 134-78, at 160-1).

[62] Diogenes Laertius *Lives* 7.134 (*SVF* 2.300).

in a certain way.[63] But matter disposed, the third category, is not distinct from qualityless matter, the first category. For something that falls under the third category is neither distinct from, nor the same as, the qualityless matter that falls under the first category.[64] If God as disposed matter is not *distinct* from qualityless matter, then we do not after all have two bodies in the same place, because we do not have *two* bodies.

In order to assess this latter interpretation, let us consider how effectively it can cope with difficulties. Admittedly, by refusing to see God in his causal role as something abstract, it does prevent one of the two Stoic principles from being simple and unanalysable, in the way that an Aristotelian principle would be. Of the Stoic principles, one, qualityless matter, will be at a *lower* level of complexity than *pneuma*, while the other, God in his causal role, will be at a *higher* level of complexity, because he is *pneuma disposed* in a certain way. The latter principle, then, will on this interpretation, not be simple, but only causally fundamental. But such an outcome need cause us no surprise, for it is the interpretation actually taken by another witness, Plutarch, who makes precisely this point, that God in his role as principle is a compound (*sunthetos*) of matter and intelligence, and is not a simple entity (*haplous*).[65]

Does the interpretation, however, leave the Stoics vulnerable to the objections raised by Alexander and Plotinus? Plotinus doubts that God is sufficiently distinct from qualityless matter to act on it. For if he *is* matter disposed in a certain way, how can he also be what *acts on* matter?[66] Alexander sees a still worse difficulty: God acts on qualityless matter, in order to produce fire and air. Yet he himself is said to be fire or *pneuma*. Even on Diogenes' view, he *presupposes* fire or *pneuma* as the matter he resides in. How then can he also be what produces all the fire and air, and hence all the *pneuma*, that there is?[67] The first looks too like a case of a thing acting on itself, the second too like a case of a thing creating itself, or at least creating what its existence presupposes.

Many suggestions have been made whose intention, or effect, would be to solve these problems. On one view, nothing like self-creation occurs. For we must distinguish between creative fire (*pur tekhnikon*) as what God *is*, and non-creative fire (*pur atekhnon* – I prefer this translation to 'destructive')[68] as what he produces.[69] But unfortunately, I doubt if it can be denied that God produces creative fire, because he

[63] Plotinus 2.4.1(13)(*SVF* 2.320); 6.1.27(6-7)(*SVF* 2.314).

[64] Arius Didymus fr. 27, ap. Stobaeum *Eclogae* 1.17 and 20, *Dox. Gr.* 462,25-463,13.

[65] Plutarch *Com. Not.* 1085B-C.

[66] Plotinus 6.1.27(4-18)(*SVF* 2.314).

[67] Alexander *Mixt.* ch. 11, 225,10-13 (not in *SVF*).

[68] Both fires are destructive, because all fire uses fuel, as Cicero says at *ND* 2.41 (*SVF* 1.504). The Greek means that one of them does not also create.

[69] M. Lapidge, '*Archai* and *stoicheia*: a problem in Stoic cosmology', *Phronesis* 18, 1973, 240-78 at 267-73, and 'Stoic cosmology', in John Rist, ed., *The Stoics*, Berkeley and Los Angeles 1978, 161-85 at 167.

produces everything in the cosmos, and the sun, moon and stars are said to consist of creative fire.[70] On a second view, nothing like self-action occurs, because what God acts on is not qualityless matter as such, but something much more distinct from himself, namely, the passive elements, earth and water.[71] But the texts cited above say that God acts on qualityless matter, and that he does so in order to *produce* earth, water and the other elements. A third way of avoiding anything like self-action would be to say that the qualityless matter underlying God is distinct from the qualityless matter on which he acts. But there are not two distinct matters, for we are told that what God acts on is the whole of matter.[72] On a fourth interpretation, there is nothing like self-action or self-creation, because God in his causal role is an incorporeal entity, and so clearly is neither the matter on which he acts, nor the fire and *pneuma* which he produces.[73] But unfortunately it is in his causal role that God most needs to be a body, according to the Stoic theory of causation. Moreover, the Stoics hold that only bodies have being,[74] and it would be a strange result that God, in his role as a principle, should lack being.

It may be urged that Diogenes Laertius has an advantage over this fourth interpretation, in that he allows that God is a body (*sôma*), while at the same time, he shares the intuition of this interpretation, that God avoids acting on himself, by being something rather abstract, and therefore distinct from the matter on which he acts. But Diogenes still has to confront the other difficulty, that God appears to create the very fire and *pneuma* on which he depends for his existence. I think his abstract interpretation, therefore, has no advantage in this respect over the concrete interpretation of Alexander and Plotinus, but will need a similar solution.

A solution is available, I believe. First, we ought to accept that disposed matter can very well act on qualityless matter. To take a modern analogy, electrified matter might shake itself to pieces. We ourselves could describe this as a case of a thing acting on itself, and indeed cases of self-destruction, of suicide, for example, seem to us perfectly commonplace. So if the Stoics were postulating a case of a

[70] Arius Didymus fr. 33, *Dox. Gr.* p. 467, lines 4-7 (*SVF* 1.120); Cicero *ND* 2.41 (*SVF* 1.504); Achilles Tatius *Isagoge* 11 (*SVF* 2.862); Philo *de visione angeli* Aucher p. 616 (*SVF* 2.422).

[71] Robert Todd, *Alexander of Aphrodisias on Stoic Physics*, Leiden 1976, 34-6.

[72] Diogenes Laertius *Lives* 7.136 (*SVF* 1.102, translated below); cf. 7.134 (*SVF* 2.300, translated above): *pasa ousia, pasa hulê*. Dio Chrysostom *Orations* 36.55 (*SVF* 2.622): *holê ousia*.

[73] F.H. Sandbach, *The Stoics*, London 1975, 73-4; Robert Todd, 'Monism and immanence in Stoic physics', in John Rist, ed., *The Stoics*, Berkeley and Los Angeles 1978, 137-60.

[74] Only bodies have being: Alexander *in Top.* 301,22-3 (*SVF* 2.329); Proclus *in Tim.* (Diehl) 3.95,10-14 (*SVF* 2.521); Plotinus 2.4.1(7)(*SVF* 2.320); Plutarch *Com. Not.* 1073E (*SVF* 2.525). Suidas' false reading which makes God incorporeal (*asômatous*) is favoured also by R.D. Hicks in the Loeb edition and by Andreas Graeser, *Zeno von Kition*, Berlin 1975, 103.

thing's acting on itself, we ought not to complain. But the second point is that they are *not* postulating any such self-action. For their theory of categories denies that *disposed* matter, which is what acts, is the *same* thing as the *qualityless* matter, which is acted upon. It is neither the same nor different.

The remaining problem of self-creation appears difficult, only so long as we think of creation from an absolute beginning. If fire were ever totally non-existent, God would be non-existent, and he would then be in the impossible position of needing to exist, in order to create the fire required for his existence. But some fire is always in existence, and so all God needs to do is to *maintain* it. He appears to do this by acting on his own qualityless matter to ensure at all times that some at least of it is rare, for the rare part constitutes fire.[75] This is perhaps analogous to imagining that electrified matter acts on itself, to maintain its own cohesion. It leaves God perfectly free to increase the total amount of fire available at some stages in the cosmic cycle, and to decrease it at others, although Alexander castigates the latter as self-destruction,[76] just as he castigates the former as self-creation. Diogenes Laertius has left us a description of the cyclical process by which God sometimes turns more of his qualityless matter into fire (by rarifying it) and sometimes more into air, water, or earth (by densification).

> God, Reason, Fate and Zeus are a single thing, and he is called by many other names too. At each beginning, he is on his own, and turns substance as a whole (*pasa ousia*) through air into water. Just as seed is immersed in the case of reproduction, so he too, as the seminal principle of the cosmos, remains behind as something like seed in the fluid, and makes matter easy for himself to work on, in order to generate what comes next. Then he engenders first the four elements, fire, water, air and earth.[77]

I assume that in this process fire remains ever-present, for even when God turns matter into water, fire will remain blended into the water, since water has no cohesion on its own.

If problems of self-action and self-creation can be overcome in this way, we can conclude that God and matter need not be two distinct bodies in the same place, but may rather be different categorial levels of a *single* body. If so, this will provide a third case in which the Stoics are immune to the Aristotelian charge that they put bodies in the same place. They will rather be engaging in their *first* enterprise of reducing things to a single body. But they do, of course, have a second enterprise of putting bodies in the same place. And I will now turn to the cases in which they genuinely do do this, to see why the Aristotelians think it

[75] For the role of density, see David Hahm, 'The Stoic theory of change', in Ronald H. Epp, ed., *Recovering the Stoics, The Southern Journal of Philosophy* 23, supplement, 1985, 39-56.

[76] Alexander *Mixt.* ch. 11, 226,14.

[77] Diogenes Laertius *Lives* 7.135-6.

objectionable. The foregoing discussion should help us to assess how effectively the Stoics can meet the objections.

Enterprise 2. Bodies in the same place

(i) Are they still bodies?

One objection concerns the sense in which things that interpenetrate can still be viewed as *bodies*. There is no problem about many of the everyday examples, because the ingredients are only *selectively* interpenetrative. Wine and water mix with each other, but not with the spoon that stirs them, so there is no doubt about their being bodies. But what about *pneuma*, or God and soul, which are forms of *pneuma*? Or what about light?

It would help if the things alleged to be bodies added to the *volume* or *density* of what they were in. But Alexander complains in the *Mantissa* that many of the Stoics' interpenetrating bodies do not. He takes as examples of interpenetrating bodies light, heat, odour, colour and sound in air, fire in iron and soul in body. As regards light and fire, he complains that they actually thin, rather than thickening, the bodies into which they are mixed.[78] In his *de Mixtione*, Alexander repeats the general objection that what does not add to volume is not a body.[79] Some of Alexander's examples can be discounted, because they are qualities, and I have argued that it is wrong to treat Stoic qualities as so many interpenetrating bodies. The qualities of a thing are a *single* body disposed in different ways. This appears to be true not only of heat, odour, colour and sound, but even of light. At least, Alexander thinks that the Stoics view light as a quality, and hence as a body, namely thinned air, just as darkness too is air, presumably with a different disposition.[80] It is also recorded that sound, or at least voice, is air that has been struck.[81]

As regards *pneuma*, and hence God and soul, the Stoics could make a

[78] Alexander *Mant.* 124,13-14; 16-20; 140,31-141,8.

[79] Alexander *Mixt.* ch. 6, 219,20: by denying that it adds to volume, the Stoics 'destroy the nature of body'. 29-30: such a thing would not even be a body. 32-3: it is clear that it is not a body. 220,5-11: the mixture is not a mixture of bodies. The sentence at *Mixt.* 219,28-32 might be understood as follows: 'Again, as regards those things which are said to blend and which will take up more room in their mixture with each other, it strikes us straight away from these cases that a body which was received into the original space (*en tautôi*) would not even be a body (*mêde sôma einai*). Not at least if [an incoming body] takes up some additional space and is not satisfied with the space of the body that is described as receiving it into itself.' I owe this interpretation of the sentence to David Furley.

[80] Alexander *Mant.* 124,10; 132,37; 139,1 (all in *SVF* 2.432).

[81] Diogenes Laertius *Lives* 7.55 (*SVF* 1.74; 2.140; 3.17, p. 212); Eustathius *in Iliadem* 1158,37 (*SVF* 1.74); Origen *Against Celsus* 2.72 (*SVF* 2.138); Scholia Arati v.1 (*SVF* 2.139); Gellius *Attic Nights* 5.15 (*SVF* 2.141); Galen *de Qualitatibus Incorporeis* 2, Kühn vol. 19, p. 467 (*SVF* 2.384); Simplicius *in Phys.* 426,1-5 (*SVF* 3.19, pp. 212-13).

different reply. If I was right that *pneuma* is fire or air disposed in a certain way, then it can be said that it *does* contribute to volume. Admittedly, earth and water are never found without *pneuma* in them, so there is no question of their volume being increased by some *initial* influx of *pneuma*. But pneuma does make a contribution to the volume they have, and that contribution may vary, as the proportion of *pneuma* is altered.

Besides appealing to the contribution to *volume*, the Stoics might also appeal to the idea of *resistance*. There has been controversy over whether the Stoics assigned resistance to all bodies.[82] But certainly if something can be shown to have resistance, this will count in favour of its being a body. It may be wondered whether interpenetrating bodies can be said to have resistance, but the fact that two bodies inter-penetrate will not of itself prevent them from resisting each other. They may slow each other down, and the interpenetration may be effected only with difficulty. The four elements, fire, air, water and earth, are said to form a series of increasing density.[83] So three of them at least ought to offer resistance. What about fire? It might be answered that it too would create resistance, at least as compared with void, which, in a notable thought experiment of Cleomedes, offers no resistance to the cosmos moving through it,[84] and which offers no resistance either to the expansion of fire during the conflagration.

Returning to water and earth, we may face the objection that these on their own would be too slack (*atonos*) to resist anything.[85] But in fact they are never found without fire or *pneuma* mixed into them,[86] and they end up the densest. If fire and air offer resistance, this has implications in turn for *pneuma*, God and soul. For God and soul are forms of *pneuma*, while *pneuma* is air or fire disposed in certain ways; so we can expect *pneuma*, God and soul to have resistance too. I think, then, that by appeal to resistance, contribution to volume, and selecti-vity of interpenetration, and by denying that qualities interpenetrate in the relevant sense, the Stoics can defend their claim that the interpenetrating items are *bodies*.

[82] This is claimed by ps-Galen *de Qualitatibus Incorporeis* 10, Kühn 19,483 (*SVF* 2.381). Plotinus also offers this view to the Stoics, but only as part of a dilemma: 6.1.26(20) (*SVF* 2.315). Disputed by Margaret Reesor, 'The Stoic concept of quality', *American Journal of Philology* 75, 1954, 40-58, and to be disputed with fuller evidence by Eric Lewis. I agreed with him above, on the grounds that qualityless matter is a body lacking resistance. As noted above, Jaap Mansfield (*Mnemosyne* 1978, 160-1) considers that some Stoics did, and some did not, define body in terms of resistance.

[83] Plutarch *Sto. Rep.* 1053A (*SVF* 2.579); Diogenes Laertius *Lives* 7.142 (*SVF* 2.581); Arius Didymus fr. 21, ap. Stobaeum *Eclogae* 1.10.16, *Dox. Gr.* 458,22-6 (=*SVF* 2.413, lines 20-4); Schol. Apoll. Rhod. 1.498 (*SVF* 1.104); Cleomedes *Kuklikê Theôria* book 1, ch. 1, p. 6, lines 11-15, Ziegler (*SVF* 2.537); Philo *Aet.* 100-103 (*SVF* 2.619); Dio Chrysostom *Orations* 36,55 (*SVF* 2.622).

[84] Cleomedes *Kuklikê Theôria* 1, 1, p. 6, 6-11, Ziegler, discussed in Chapter 8.

[85] Plutarch *Com. Not.* 1085C-D (*SVF* 2.444); Alexander *Mixt.* 218, 5.

[86] Plutarch *Com. Not.* 1085C-D (*SVF* 2.444); Alexander *Mixt.* 218,5; Dio Chryostom *Orations* 36,55 (*SVF* 2.622).

(ii) Is a body enlarged by admixture?

A whole further issue was raised against the Stoics: is a body *enlarged* by admixture? In Chapter 5, I imagined two billiard balls turning into a denser, rather than a larger, mass, as they passed through each other. But I could instead have imagined an increase in volume, and this would be more plausible for *fluids* that passed through each other.

Alexander complains that the Stoics are obliged to deny such enlargement. For part of what is meant, he says, by the claim that one thing receives another in itself, is that the receiver takes up no more room after the reception than before. Surely this is not true of eating food. None the less Alexander concludes that, since the receiver takes up no more room, the Stoics must admit that their receiver and received are not bodies.[87]

Plotinus reports the further complaint, which is presumably also Aristotelian, that reception without enlargement, such as the Stoics allegedly need, never occurs. Such examples may *appear* to arise, as when a sponge is not expanded by an influx of water. But it will be found that what has happened is merely that water has replaced air in the holes of the sponge.[88]

Plotinus is, however, aware that the Stoics might reply they can *allow* enlargement. Why, he asks, should they not say that it is only to be expected? For the admixture of a body will help to produce a new volume just as the admixture of fresh qualities helps to produce a new quality? But here Plotinus sides with the Aristotelians. Just as two qualities dim each other on amalgamation, so two volumes ought to diminish each other, he thinks.[89] Plotinus also suggests a further reply to the Stoics. If they admit enlargement, their opponents, who are again presumably Aristotelians, can reply that this supports their own theory that the matters of the two bodies are juxtaposed, and only the qualities mix.[90]

If we return to Aristotle's original theory, he seems at first sight as vulnerable as the Stoics to Plotinus' attack. If two volumes ought to *diminish* each other upon amalgamation, is he any more able than the Stoics to explain the expansion that occurs? He eschews as much as they do the explanation in terms of mere juxtaposition. But it will be remembered that certain later Aristotelians elaborated his theory so as to incorporate the idea of juxtaposition, and that he might well claim to be in agreement with them. The elaboration does indeed provide an explanation of expansion, which the Stoics may have to treat as a brute fact.[90a] But I think they are still as free as Aristotle to acknowledge that it occurs.

---

[87]Alexander *Mixt.* ch.6, 220,3-11.

[88] Plotinus 2.7.1(17;19;34). The last passage is in *SVF* 2.478.

[89] id. 2.7.1(35-49), partly in *SVF* 2.478; 2.7.2(21-6).

[90] id. 2.7.1(42-4).

[90a] In judging whether they could instead take a leaf out of the Aristotelian book, one should consult Eric Lewis' account, loc. cit., of their concept of *antiparektasis*.

(iii) The sea in a cup

The Aristotelian objection to bodies interpenetrating, that, if you could fit one ladle of sea water into a cup of wine, you could by repeated operations fit the whole sea, was described in Chapter 5. But it should be noted that it works as an objection only on the assumption that the Stoics must deny that the admixture of sea water produces an increase of volume. And that assumption has just been rejected. Despite this, the objection was to be repeated many times, though not always against the Stoics.[91]

(iv) The effect of diffusion on measurement

A further puzzle about measurement and equality envisaged that the Stoics might make exactly the opposite assumption, that admixture *does* produce an increase of volume. The objection was derived from the Stoic insistence that when a litre of wine is blended with a larger volume, say, two litres of water, the wine can spread out thinly, so that the volumes of wine and water are equal. Are we then to say that the water has shrunk, being now equal to one litre of wine, giving us a total of two litres, one of wine and one of water? Or are we to say that the wine has grown to two litres, or rather to three, since there are now three litres of fluid? If so, does that make a total of six litres, three of wine and three of water? And if a litre of new wine is added, and expands to occupy four litres, will there then be twelve?[92]

I think there are two different problems here. Part of the difficulty has been very well brought out by Michael White.[93] It can seem very surprising that an admixture of water can give one's guests a larger volume of wine. But to get rid of surprise, we must distinguish volume from something else, say, mass. It would indeed help the party-giver if a process of mixture could increase the *mass* of wine available. But the *volume* can be increased, for the greater volume is obtained at the price of reducing density. The confusion is not to be laughed at. It has already dogged our footsteps in the interpretation of Anaxagoras. For Anaxagoras is normally taken to think that he cannot allow a minimum volume, since a minimum volume of honey would not have room for smaller amounts of bone in it. It has been overlooked that this

---

[91] Alexander *DA* 20,14-15; id. ap. Simplicium *in Phys.* 530,22-4; Proclus ap. Simplicium *in Phys.* 613,15-20; Philoponus *in Phys.* 505,21-506,2; 559,24-7; *in DA* 328,15; *in DA* 343,33-344,4; *in GC* 90,15-17; *contra Proclum* 281,12-22; Simplicius *in Phys.* 531,8; *in Cael.* 629,25-30. None of these passages is in *SVF*. The original source is Aristotle *Phys.* 4.6, 213b5-12.

[92] Plutarch *Com. Not.* 1078A-B (*SVF* 2.465); Sextus *PH* 3.60-1; 3.96; Alexander *Mant.* 141,9-16; Philoponus *in Phys.* 506,2-14.

[93] I am grateful to Michael White for showing me a draft of his illuminating paper, 'Can unequal quantities of stuffs be totally blended?' *History of Philosophy Quarterly* 3, 1986, 379-89. See also Richard Sharvy, 'Aristotle on mixtures', *Journal of Philosophy* 80, 1983, 441-8.

worry, which may indeed be his, would not arise if he thought of the bone and honey as blended. For in a blend, volumes, unlike masses, can be spread out, so that they are equal. So a small mass of bone blended with a minimum volume of honey would not need to be smaller in volume than the honey.

The second problem is this: if the Stoics say that the wine is spread through three litres and the water is also spread through three litres, why does not that make a total of six litres? The answer is that the wine is able to expand to three litres only because it is *not separate* from the water. In order to obtain *six* litres, we should need the wine and water to be *separate* from each other after their expansion, contrary to the whole idea of interpenetration.

(v) Infinite division

A final complaint by Alexander is premised on an 'if'. If the Stoics' blending is achieved by an infinite division of the ingredients, then they will be in trouble. I have already argued in Chapter 5 that Anaxagoras envisaged his ingredients as infinitely divided. But this was partly for the reason just given, that he did not seem to think of them as blended with each other like fluids. I have also pointed out that Aristotle attacked the view that mixture could be due to infinite division.[94] Alexander wheels out the full range of Aristotelian paradoxes against such a view.[95] Plotinus in an early treatise endorses this line of attack,[96] and in a later one presents it as part of the Aristotelian case.[97] But is there the slightest reason to suppose that the Stoics expected their fluids to blend by any process of division, let alone of infinite division? Chrysippus denies that division can reach an infinite limit; it is merely for ever capable of being continued.[98]

(vi) Aristotle and the Stoics: adjudication

I believe that the Stoic commitment to interpenetration cannot be shown by these arguments to be absurd. It is time, then, to adjudicate: of those who reject the idea of juxtaposition, Aristotle and the Stoics, who has the more reasonable theory? The Stoics' case for their ingredients remaining actual is not unlike that presented in Chapter 5 for the possible interpenetration of billiard balls. Just as the

[94] Aristotle *GC* 1.10, 328a1-5; 15-17.
[95] Alexander *Mixt.* ch. 8, 221,25-222,26: The ingredients would then be divided at every point, and into mere points, which would in turn imply that they could be built up out of mere points. Or if the infinite division were into units with a positive size, the stuff so divided would be infinitely large. Meanwhile if the division were infinite only in the sense of for-ever-to-be continued, one would achieve not blending, but the mere juxtaposition of lumps incompletely divided.
[96] Plotinus 4.7.8²(15-18).
[97] id. 2.7.1(12-13;24;32). The last two are in *SVF* 2.478.
[98] Diogenes Laertius *Lives* 7.150; cf. Plutarch *Com. Not.* 1079B-C (*SVF* 2.483).

distinctive stripes of the two billiard balls might be visible, so the Stoics argue that the distinctive qualities of the ingredients still 'show forth together' (*sunekphainesthai*) in the mixture. Just as the billiard balls are recoverable, so the Stoics argue that the ingredients and their qualities can be 'separated out again' (*apokhôrizesthai, khôrizesthai palin*).[99] They use both arguments to show that the qualities and the ingredients persist in the mixture.

Such evidence is not, of course, conclusive. In the case of the billiard balls, for example, there was a question whether the stripes might now belong to a *tertium quid*. And there was a question whether the billiard balls, or Aristotle's ingredients, needed to be *reconstituted* before recovery. In that case one would recover entities of the same kind, but not the very same entities. I argued that this would be an uneconomical hypothesis in the case of the billiard balls. And it was not a hypothesis readily suggested by ancient atomic theory, because that did not recognise a level of molecules which would need to be reconstituted from constituent atoms.[100] To argue the need for reconstitution in the case of the Stoic example of wine mixed with water, it might be necessary to look at the technique used for recovery, whether it be an oil-drenched sponge (a method to whose efficacy I can testify),[101] or evaporation. Is the technique for recovery better explained by a theory of reconstitution, or of simple reseparation? But that was not discussed, and in the absence of discussion, the Stoic theory remains reasonable.

What might be objected to the Stoics is that the original qualities do not always *show forth*, and they may not always be recoverable either. Aristotle trades on the fact that the qualities of elemental fire do not show forth, but are replaced in the mixture. But even here the Stoic theory is reasonable. For we are inclined to affirm the presence of ingredients on the basis of their *powers*, even when their qualities do *not* show forth: quite indirect effects on health can lead us to suspect the presence of mercury in the salmon we eat. That is what helps to make modern juxtaposition theory attractive. Because our health is affected, we welcome a theory which allows affecting substances, such as mercury, to be present in actuality in our food, even when their qualities do not 'show forth'. Juxtaposition is one such theory; the Stoics' is another.

How does Aristotle's theory compare? He runs into an opposite difficulty that he cannot easily accommodate the Stoic examples, like

---

[99] The first point is made at Arius Didymus fr. 28, *Dox. Gr.* 464,1-6, translated above, the second at Arius Didymus loc. cit. and Alexander *Mixt.* ch. 3, 216,32-217,2, translated above.

[100] As remarked above, the nearest equivalent to a molecular theory was provided by Plato's distinction of elementary triangles and corpuscles, *Timaeus* 53C-57D

[101] Arius Didymus fr. 28, ap. Stobaeum *Eclogae* 1.17.4, *Dox. Gr.* 464,4-6 (*SVF* 2.471); Alexander *Mixt.* ch. 15, 232,2-3. I am grateful to Connie Meinwald and Wolfgang Mann for confirming its efficacy for me in front of witnesses.

wine and water, in which the original qualities *do* show forth, thus testifying to the *actual* existence of the ingredients. I doubt is this is outweighed by the advantage he may have in explaining the increase of volume produced by admixture. And meanwhile there was the question whether he could give context to his view that *all four* elements persist in each compound potentially. On balance it may be thought that his theory is *less* reasonable than the Stoic postulation of interpenetrating ingredients in exactly the same place.

(vii) Individuals do not interpenetrate

It remains to consider a major restriction which the Stoics placed on their theory of interpenetration. They rightly saw that the identity criteria for individuals, such as persons, are different from those for kinds of stuff, and they denied that such individuals can interpenetrate. David Sedley has revealed the meaning of a discussion by Chrysippus which had previously seemed unintelligible.[102] It concerns two persons Theon and Dion, or for memorability we might say Sorab and Sorabji. Sorabji is a familiar person, but Sorab needs some introduction. He is Sorabji with his right foot ignored. Sedley's innovation is to see that they are not separated persons. So far, there is no threat that Sorab is in *exactly* the same place as Sorabji, for they differ by a foot. But suppose Sorabji is run over and the foot amputated: will Sorab and Sorabji then be in exactly the same place? Chrysippus' answer is 'No': only one of them will survive the amputation. The text chooses Sorabji, or in the original Dion, without giving a reason, and Philo complains that, as Sorab (or Theon) has not been touched, he can hardly have been killed. But Sedley ingeniously suggests that the person hopping around shouting 'My foot, my foot' must be Sorabji, since Sorab never had a foot. In concluding that one of the two must have perished, Chrysippus uses it as a premise that two individuals marked out by unique qualities (*idiôs poioi*) cannot occupy the same substance (*ousia*), that is, the same matter, jointly, and it is this premise that constitutes the restriction on interpenetration.

As a matter of fact, Sedley points out, Chrysippus would have had ways of avoiding another premise, that Sorab and Sorabji were ever two persons. For the one property which distinguishes Sorabji from Sorab, his possession of a second foot, does not meet Chrysippus' rather stringent requirement for individuality, that distinguishing characteristics should be of life-long duration. Aristotle too would have avoided the premise that there were ever two persons, but by a

---

[102] Philo *Aet.* 48 (SVF 2.397), discussed by David Sedley, 'The Stoic criterion of identity', *Phronesis* 27, 1982, 255-75.

different means. For he denies that one substance is ever *part* of another.[103]

The puzzle of Theon and Dion and analogous puzzles were known to the Middle Ages, and have provided a feast for modern philosophers. One conclusion has been that Theon and Dion are the same person, but different lumps of flesh,[104] another that Theon is not a person.[105] A third is that we should think of Theon and Dion as four-dimensional sums made up of temporal parts, parts which are identical after the amputation, but merely overlap before it.[106]

Chrysippus' own purpose in raising the puzzle need not concern us, since the present relevance lies only in the premise restricting interpenetration. But it must now be noticed that the restriction is only partial, because Chrysippus denies only that two individuals can occupy the same *substance* (*ousia*), that is, the same subject (*hupokeimenon*), or matter. This would not itself exclude two persons occupying the same *place*, so long as the different *matters* could still be distinguished. I can only say that no examples are presented, even by the Stoics' opponents, of two individuals occupying the same place. For anything like that we must wait for the Neoplatonists.[107]

---

[103] Aristotle *Metaph*. 7.16, 1040b5-10; 1041a4-5.

[104] Peter Geach, *Reference and Generality*, 3rd ed. only, Ithaca N.Y. 1980, sec. 110, draws for the puzzle on the thirteenth-century logician William of Sherwood.

[105] David Wiggins, 'On being in the same place at the same time', *Philosophical Review* 77, 1968, 90-5, draws for the puzzle on Peter Geach. Reply by Harold W. Noonan, 'Wiggins on identity', *Mind*, 1976, 559-75, at 570-3.

[106] Mark Johnston, 'Is there a problem about persistence?', *Proceedings of the Aristotelian Society* supp. vol. 61, 1987, 107-35, refers to Judith Jarvis Thomson, 'Parthood and identity across time', *Journal of Philosophy* 80, 1983, 201-20.

[107] In writing this chapter, I have benefited from talking with Victor Caston, David Sedley, Robert Sharples, and above all from discussions with Eric Lewis surrounding the topic of his Ph.D. dissertation, in preparation, on Stoic and Aristotelian natural philosophy.

CHAPTER SEVEN

Neoplatonists and Christians:
place and bodies in the same place

Plotinus rejects the Stoic belief in interpenetration. In the early treatise *Enneads* 4.7.8[2], he simply endorses the Aristotelian objections. But in *Enneads* 2.7.1-2, he sides against the Stoics only after a thoughtful discussion of Stoic and Aristotelian views. After this verdict, it is surprising to find his successors allowing interpenetration after all.

In 2.7.2, Plotinus grants that the example of a soaked papyrus certainly looks like a case of stuffs in the same place. But the swelling of the papyrus, he says, reveals that the water is not absorbed into exactly the same place as the papyrus. If it were, we should expect the two original quantities to diminish each other, just as two blended *qualities* dim each other. We encountered this argument in Chapter 6.

In 2.7.2, he also asks a new question: what *prevents* interpenetration? It is not the prime matter, nor simply the qualities as such, for matter and qualities are both incorporeal. It is either the *number* of qualities in a body, or a special quality of density (*puknotēs*). Both questions remained problematic for the Neoplatonists: what prevents bodies from interpenetrating, and are all bodies prevented?

Concerning the first problem, what stops bodies interpenetrating, Simplicius is tempted to give the responsibility to qualities rather than to matter. At any rate, he says that there is nothing strange in two matters interpenetrating, nor in two enmattered extensions doing so, provided that one of them is qualityless (*apoion*) and incorporeal (*asômaton*).[1] His enemy Philoponus prefers matter. Extensions can indeed interpenetrate, for example place and body. But then place lacks matter,[2] and it is *matter* that makes things act, or be acted on, and divide each other rather than interpenetrating. The reason why bodies cannot interpenetrate is that they are not first and foremost extensions, but are matter and form.[3] Elsewhere Simplicius seems to

---

[1] Simplicius *in Phys.* 623,11-18.

[2] Corporeal extension has matter, spatial extension and vacuum do not, Philoponus *in Phys.* 577,10-16; 687,29-35.

[3] Philoponus *in Phys.* 559,9-18; 561,3-12.

switch to Philoponus' side, if he is the author of the *de Anima* commentary: it is only *enmattered* bodies that cannot interpenetrate.[4] Aristotle was commonly interpreted as having offered a third answer, neither matter nor qualities. He was seen as holding that it is in virtue of being *extensions* that bodies cannot interpenetrate.[5] That interpretation was criticised in Chapter 5. The position of Syrianus and his pupil Proclus will become clear later. Syrianus supported matter. At any rate, he is at pains to say that those bodies which do interpenetrate are immaterial.[6] The idea of immaterial bodies may sound contradictory, but it is freely used, and I shall consider its meaning below. Proclus agrees with Syrianus,[7] although we shall see that his position is more difficult to determine.

The second problem for the Neoplatonists was that they wanted to allow exceptions: certain bodies do interpenetrate after all. The exceptions include not just stuffs, as with the Stoics, but even individuals. Porphyry may provide an example, when he says that certain corporeal demons 'offer no resistance',[8] while Syrianus holds that the immaterial bodies (*aüla sômata*) which serve as the vehicles of our souls can interpenetrate other bodies.[9]

These vehicles are the subject of elaborate theories. According to Proclus, we have both luminous and pneumatic vehicles. 'Pneumatic vehicle' might equally be rendered 'spiritual vehicle', for *pneuma* (spirit) was still commonly thought of as a body in Greek thought, as it was later in the conception of animal spirits. Proclus classes both luminous and pneumatic vehicles as 'immaterial' bodies,[10] and he agrees that these vehicles can interpenetrate with the celestial body.[11] He does not allow, however, that they can interpenetrate with 'material' bodies. They will not then, presumably, interpenetrate, as Alexander had objected they would,[12] with what he calls, following Plato, the 'oysterlike' body, that is, the human body which carries the soul like a pearl.[13] Simplicius is uncertain: perhaps for our souls the oysterlike body is enough, and they do not need a second body, the pneumatic vehicle, as well. Or if they do, perhaps the two bodies need not strictly interpenetrate.

Simplicius does not, however, deny the interpenetration of bodies in

---

[4] Simplicius (?) *in DA* 134,6. The authorship of this commentary is assigned to Priscian of Lydia by F. Bossier and C. Steel, 'Priscianus Lydus en de *In de anima* van Pseudo (?) Simplicius', *Tijdschrift voor Filosofie* 34, 1972, 761-822.

[5] Themistius *in Phys.* 134,1-135,8; Philoponus *in Phys.* 687,7-12; Simplicius *in Phys.* 622,19-21; 682,5-7.

[6] Syrianus *in Metaph.* 85,27-8; cf 16; 19; 25.

[7] Proclus *in Rempublicam* (Kroll) 2.162,26-8.

[8] Porphyry ap. Proclum *in Timaeum* (Diehl) 2.11.10-13: *ouden ekhousin antitupôs.*

[9] Syrianus *in Metaph.* 85,27-8.

[10] Proclus *in Tim.* 3.297.25.

[11] Proclus *in Remp.* 2.162.20-4; 2.163.1-7.

[12] Alexander ap. Simplicium *in Phys.* 964,19-23.

[13] Plato *Phaedrus* 250C.

general.[14] He holds that the celestial spheres penetrate all the way down to the centre of the earth, coinciding with each other, with the stars and with the sublunary elements on the way.[15] He cites as a parallel the view of the *Chaldaean Oracles*, that the extra empyrean and aetherial firmaments, which they view as immaterial bodies and postulate over and above the 'material' cosmos of Plato's *Timaeus*,[16] interpenetrate that material cosmos.[17]

The light of the sun provides another exception. Proclus, taking it to be a body, holds that it interpenetrates the celestial body. It does not, however, interpenetrate the moon.[18] Nor does it reach us through the air by interpenetrating it, but by being minced up and passing through passages in the air.[19]

These are some of the exceptions which the Neoplatonists allow. But a word of warning is needed about the theory of elemental mixture, which does not constitute a genuine exception. The four elements, earth, air, fire and water, are never found in separation from each other, but are always mixed together,[20] and for this reason they are described as 'going through each other'.[21] Light is the purest form of the element fire,[22] so it too is described as 'going through all things'.[23] It would be uneconomical to take this as conflicting with Proclus' claim that solar light does *not* interpenetrate the moon or the air. Fortunately, we need not take the theory of mixture to imply that ingredients strictly interpenetrate. Aristotle, as noted in Chapter 5, thinks that ingredients in a mixture do not interpenetrate, since they are present only potentially, not actually, and what is present actually is not them, but a *tertium quid*.

The theory of elemental mixture raises a second problem about how the compounds produced by mixture can themselves interpenetrate. The mixture of four elements is found even in the heavens. The idea that the heavens are a mixture is endorsed by Simplicius, whose

---

[14] Simplicius *in Phys.* 966,3-14. Careful interpretation is needed of *hopote* in line 8: 'it is not necessary for one to be in the other in such a way that body would go through body, <not that that would be so bad> since (*hopote*) I do not even think it absurd that ...' Or we can translate: 'while <in any case> I do not even think it absurd ...' An alternative would be to read *amêkhanon* or *adunaton* in place of *anankaion* in line 7: 'it is not *impossible* for one to be in the other ..., since I do not even think it absurd ...'

[15] Simplicius *in Phys.* 531,3-9; 616,23-617,2; 623,32-624,2; 643,18-26; 966,11-12.

[16] Proclus *in Tim.* 2.57.10-17.

[17] Simplicius *in Phys.* 616,25-9.

[18] Proclus *in Remp.* 2.163.1-7.

[19] Proclus ap. Simplicium *in DA* 134,5-8.

[20] Proclus *in Tim.* 2.8.13-2.13.14, esp. 2.8.21; 2.9.1-16; similarly Simplicius *in Cael.* 85,21-31.

[21] Proclus *in Tim.* 2.9.2-4; similarly Simplicius *in Cael.* 85,24-7.

[22] Proclus *in Tim.* 2.8.22-7; similarly Simplicius *in Cael.* 12,28; 16,20-1; 130,31-131,1; but at *in DA* 134,13-20, if he is the author, he doubts whether light is a body.

[23] Proclus *in Tim.* 2.9.3-4; similarly Simplicius *in Cael.* 66,33-67,5; 85,7-15.

account has been well analysed by Philippe Hoffmann.[24] What I want to add is that before Simplicius the idea is already in Proclus.[25] Not only the heavens, but also solar light and[26] our pneumatic immaterial vehicles are composed of the four elements. How, then, can such bodies interpenetrate? Would not the earth mixed into each of them create resistance to interpenetration? Earth is said to be solid (*stereos*),[27] which means that it can offer resistance to touch (*antereidein pros haphên*).[28] It withstands pressure (*hupomenei tên antereisin*),[29] and is resistant (*antitupos*),[30] and it gives its own solidity to the bodies it is mixed with.[31] The answer is probably that the earth in each of the interpenetrating bodies retards their interpenetration, without actually preventing it. Such talk of light as being retarded would involve a rejection of Aristotle's view that light is not the sort of thing that travels.[32] Light for Aristotle is not fire, but the presence of things like fire in a transparent medium.[33] Plato by contrast thinks of what provides light to the eyes as a kind of fire that comes off flame.[34] Proclus agrees that light is a kind of fire, and there are signs[35] that he, and perhaps Iamblichus before him, think of it as travelling.

Such are the complications introduced by elemental mixture. I now come to the most interesting of the examples of interpenetration allowed by the Neoplatonists. It emerges from Proclus' theory of place as supra-celestial light. Proclus speaks of a very special sort of light which is described in the myth of Er in Plato's *Republic*. It is not solar light, but is supra-celestial. He approves of Porphyry describing it as the luminous vehicle of the world soul,[36] and he himself uses Plato's word for implanting a soul in a vehicle: *embibazein*.[37] But in doing so, he goes further. It is not just the cosmic soul, but the whole spherical cosmos that we should imagine implanted in a sphere of this light.[38] Most startlingly of all, he identifies this supra-celestial light with the *place* of the cosmos.[39] It forms a kind of absolute place against which

[24] Philippe Hoffmann, 'Simplicius' polemics', in Richard Sorabji, ed., *Philoponus and the Rejection of Aristotelian Science*, London and Ithaca N.Y. 1987. For the heavens as a mixture of pure elements with fire predominating, see Simplicius *in Cael.* 66,33-67,5; 85,7-15; 360,33-361,2; 379,5-6; 435,32-436,1.

[25] Proclus *in Tim.* 2.43.20-2.44.24. The preceding notes will reveal how closely Simplicius follows Proclus on elemental mixture.

[26] Proclus *in Tim.* 3.297.24-8.

[27] ibid. 2.10.17-26; 2.43.23; 2.44.3-5.

[28] ibid. 2.12.21-2.

[29] ibid. 2.17.14.

[30] ibid. 2.13.8; 2.17.14.

[31] ibid. 2.10.17-26.

[32] Aristotle *DA* 2.7, 418b20-6; *Sens.* 6, 446a26-b2; 446b27-447a11.

[33] Aristotle *DA* 2.7, 418b13-17.

[34] Plato *Tim.* 58C5-6: *apo tês phlogos apion*.

[35] See Simplicius *in DA* 133,31-5; 134,6-20 who disagrees.

[36] Proclus *in Remp.* 2.196.22-30; ap. Simplicium *in Phys.* 615,34.

[37] Plato *Tim.* 41E1.

[38] Proclus ap. Simplicium *in Phys.* 612,29-33.

[39] Proclus *in Remp.* 2.198.25-9; ap. Simplicium *in Phys.* 612,27-9.

the cosmos can rotate and other things move.[40] Finally, as we shall see, it is argued that place, as well as the cosmos, is a body,[41] and so we have a case of two bodies interpenetrating.[42]

I must pause to express two disagreements with Pierre Duhem and others. First, Proclus seems to have been the only person in antiquity to have made place a body. This at any rate is the opinion of Simplicius.[43] I am not persuaded by the view that the passage in Syrianus which we shall shortly consider already treats place as a body.[44] Nor do I agree that Damascius subsequently adopts Proclus' view of place.[45] Damascius, who defines place as a certain kind of measure, is explicitly said to consider it incorporeal.[46] Insofar as it is an ideal unit of measurement rather than an instrument of measuring,[47] place is even unextended.[48] In the passage which has given rise to a different impression, Damascius is talking only of solar light, I believe, not of supra-celestial light, and saying that when the sun is said to reside in its light, the light is an example of what would ordinarily be called a place. (It is offered as an example of a place *superior* to the body it contains.)[49]

My second disagreement is with the idea that the doctrine of place as supra-celestial light is an afterthought of Proclus, motivated by the need to go beyond the heavens in order to find something immobile. It may well be true that, in appealing to supra-celestial light, Proclus laid the foundation for later views of the empyrean as the place of all bodies in the universe.[50] But I do not on the other hand see an afterthought here; only one integral doctrine. Proclus cannot intend, if he is consistent, to go *outside* the heavens. For he insists that his light does not project beyond (*huperanekhein*), the universe,[51] and that the sphere of light is equal in volume (*onkos*) to the cosmos.[52] So we must construe the words carefully, when he says that it is other than and *prior* to celestial things (*heteron, allo, pro*), when he speaks of *supra*-celestial light as above (*huper*),[53] and when Simplicius describes it as being other than (*allo*), above (*huper*), superior (*huperteron*) and prior

[40] Proclus *in Remp.* 2.197.17-198,4; ap. Simplicium *in Phys.* 612,32-5.
[41] Proclus *in Remp.* 2.198.8-15; ap. Simplicium *in Phys.* 611,35.
[42] Proclus ap. Simplicium *in Phys.* 612,29-35.
[43] Simplicius *in Phys.* 611,13.
[44] Pierre Duhem, *Le Système du Monde*, vol. 1, 1913, 348; S. Sambursky, *The Concept of Place in Late Neoplatonism*, Jerusalem 1982, 19.
[45] Sambursky, op. cit., 94.
[46] Damascius ap. Simplicium *in Phys.* 601,16-19; 642,15.
[47] Simplicius *in Phys.* 634,11-31; 636,7-8; 636,27-637,18.
[48] Damascius ap. Simplicium *in Phys.* 601,17-19.
[49] Damascius *Dub. Sol.* (Ruelle)=*in Parmenidem* 2.219.18-220.1.
[50] These are the ideas of Pierre Duhem, *Système*, vol. 1, 341-2.
[51] Proclus *in Remp.* 2.199,12.
[52] Proclus ap. Simplicium *in Phys.* 612,31.
[53] Proclus *in Tim.* 2.80,5-19; *in Remp.* 2.195,25-197,12.

to (*pro*) the aetherial and empyrean firmaments of Oriental Theology.[54] If Proclus is to be consistent this must mean not that his light is *spatially* further out, but only that it is *superior*.[55]

It is time to consider the main passages leading up to Proclus' view on place as supra-celestial light, and I shall start with his master Syrianus. Syrianus succeeded Plutarch (of Athens, not of Chaeronea) as head of the Athenian Neoplatonist school, succeeding also to the house,[56] which archaeologists claim to have identified.[57] He had Proclus as a pupil, living in the same house, but only for a few years, perhaps from 432 to 437.[58] Following a practice that became characteristic of late Neoplatonism,[59] he encouraged his pupil to write up one of his lecture courses, and the result was Proclus' *Timaeus* commentary, written by the time Proclus was 27.[60] Following another characteristic practice, Syrianus sought to keep the headship in the family, intending his relative Aedesia as a wife first for Proclus, and then, in the face of a divine warning, for another pupil Hermeias, who was to become head of the Alexandrian school, and to have his own son Ammonius installed as successor.[61]

In the relevant passage, Syrianus starts by reporting two theories on extensions interpenetrating, without yet committing himself to their truth. According to the first theory,[62] place is an immaterial spherical extension (*diastêma*), distinct from mathematical extension. Place interpenetrates with body, because the only alternative is that one of them should be parted or divided by the other. But division would be a kind of change, and place is exempt from change. According to the second theory,[63] actual bodies, provided that they are simple and immaterial, can pass through each other without dividing each other, or suffering obliteration (the Stoic word *sunkhusis* is used). In this, they would resemble interpenetrating lights, a simile which can be appreciated even by those who deny that lights are bodies. (Aristotle is one who denies that, precisely on the ground that light can penetrate bodies.)[64] Finally, in agreement with the second theory, Syrianus gives it as his own view that the vehicles of our souls are indeed immaterial

---

[54] Simplicius *in Phys.* 613,4-5; 616,2-3.

[55] This is the interpretation of Philippe Hoffmann, 'Simplicius: Corollarium de loco', in *Astronomie dans l'antiquité grecque*, Actes du colloque tenu à l'université de Toulouse-le-Mirail, 21-23 Octobre 1977, Paris 1979, 143-61.

[56] Marinus *Life of Proclus*, ch. 29 Boissonade.

[57] Alison Frantz, 'Pagan philosophers in Christian Athens', *Proceedings of the American Philosophical Society*, 119, 1975, 29-38.

[58] H.D. Saffrey and L.G. Westerink, *Proclus, Théologie platonicienne, Livre 1*, Paris 1968, ix-lx (vie et oeuvres), at pp. xiv-xvi.

[59] M. Richard, '*Apo phônês*', *Byzantion* 20, 1950, 191-222.

[60] Marinus *Life of Proclus*, ch. 13.

[61] Damascius *Life of Isidorus* fr. 124 Zintzen.

[62] Syrianus *in Metaph.* 84,27ff

[63] ibid. 85,15ff.

[64] Aristotle *DA* 2.7, 418b13-18.

bodies which can interpenetrate with body.[65] The notion of immaterial is connected, in the report of the second theory, with the notion of simple,[66] which suggests that the meaning is: not a compound of matter and form, i.e. lacking prime matter. So here at least, if not always below, an immaterial body is one that lacks prime matter. The passage runs as follows:[67]

> Since Aristotle directed us to the puzzles set in Book B as well, it must be replied to what is said there that it is not impossible in all cases for a couple of three-dimensional solids (*sterea*) to be in the same place. In attending to this point, one should look not to the Stoics who allowed even material volumes (*onkoi*) to pass through each other, but should look rather to those who postulate that an extension (*diastêma*) goes through the whole world and receives into itself the whole of corporeal nature. They say that it neither cuts nor is cut in such a way as to be divided up along with the air and other bodies. Rather, it is steady, firm and unmoved, exempt from all change, and extends through the whole world, supplying the contents of the perceptible world with space, a receptacle, a boundary, an outline and everything of that sort. They deny that place and extension of this sort is simply mathematical body, but say that it resembles mathematical body in being immaterial (*aülos*), without motion, intangible, free of resistance (*antitupias eleutheros*), and at the same time pure from all affective quality. It is like mathematical body again, in that this has its being in the *pneuma* and in the imagination in the *pneuma*, when the corresponding form (*logos*) is entertained in discursive thought (for when the form of a sphere, for example, comes within reach, at the same time the imagination sees the mathematical sphere with its immaterial volume.) In just the same way the spherical extension in the universe has its being concurrent (*sundromos*) with the will and thinking of the World Soul.
>
> (85,9) That Soul makes the extension spherical, because it is looking towards the Intellect; but because it is contemplating all the Forms (*eidê*), it makes the extension accommodate all bodies, each one separately, and all as a single whole. This is because its thoughts are both continuous and successive, since the intelligible Forms which are in the universal living creature are both united and separated, even though contained within a single nature. The world carries imitations of these Forms and receives them extended, not deprived of magnitude.
>
> (85,15) But the reason for which this whole discussion was set going is that neither these men, nor those who hold that simple immaterial bodies pass through each other without dividing each other, take it to be impossible for a couple of three-dimensional solids to come together. What they say is that it is absolutely impossible for two material and resistant (*antitupa*) bodies to occupy the same place, but that immaterial bodies resemble the light emitted from different lamps which goes right through the whole of the same building, the light of each lamp passing through that of the other without being obliterated (*asunkhutôs*) or

[65] Syrianus *in Metaph.* 85,27-8.
[66] ibid. 85,16-17; 25.
[67] ibid. 84,27-86,7.

divided. Even if someone wants to call the light incorporeal, it still has the same extension as bodies and stretches with them over the three dimensions, and nothing prevents the light of each lamp occupying the same place as that of every other and of the various bodies. The reason is none other than that the light of each lamp is simple and immaterial and is not divided or parted, but is united with its source and so attached to it that it exists when the source is shining, and departs when the source leaves.

(85,27) In fact, nothing prevents the immaterial bodies attached to our souls from doing the same.

(85,28) This much, so that we shall not be frightened by any startling claims from physicists who transfer what is confined to material, affectible and resistant bodies to the whole of extended reality. For we have already said that to do so makes no contribution to the original intention of our text. Nothing will be said against Aristotle on this question by the Pythagoreans or Platonists, for he is not trying to correct any of their opinions. Nor does Aristotle himself have any quarrel with the present champions of extension, but only with those who locate other bodies within perceptible ones, these other bodies being solid, because extended in three dimensions, but mathematical. We know of no reputable philosopher who supports this hypothesis. The five shapes mentioned in Plato's *Timaeus* and borrowed for giving form to the cosmic elements are explained in mathematical terms, but they allude to active and creative powers of nature. Again, if anyone declares that the luminous vehicle in us is extended in three dimensions and unresistant, he will not be affirming that it is simply a geometrical body. For how could anybody rank as a motionless geometrical object something which is full of life and motion and which of all the things in us is the most susceptible to the kind of motion recognised by Aristotle?

Syrianus' discussion is relevant to Proclus in at least four ways. First, Syrianus introduces as examples of interpenetration both light and vehicles, even though he does not for present purposes insist on the view that light is a body. Secondly, he suggests that interpenetrating bodies are immaterial and simple. Thirdly, he reports the view that place is spherical and that it interpenetrates with body. Fourthly, he reports the view that this is made possible by place being exempt from change and hence from division.

In all these respects, Proclus follows Syrianus, and the changes he makes may seem small, but they make an important difference – a fact which may qualify the impression of unoriginality which has been so much stressed by recent critics.[68] He goes beyond Syrianus in two

---

[68] This impression is strongly emphasised both by E.R. Dodds, Proclus, *The Elements of Theology*, Oxford 1933, xxiv-xxv, and by John Dillon in the introduction to his and Glenn Morrow's translation of Proclus *in Parmenidem*, Princeton 1987. The above borrowings from Syrianus can be added to those listed by E.R. Dodds, loc. cit. Marinus' example of Proclus' originality, the distinction of a kind of soul which can contemplate different Forms simultaneously (*Life of Proclus* ch.23) need not necessarily be viewed as trivial. The difficulty for humans of contemplating simultaneously rather than successively is often mentioned, and if Proclus argues that it is possible, even for beings

passages, one commenting on the relevant part of Plato's *Republic* and one preserved in quotation by Simplicius.[69] One way in which Proclus goes further is by arguing that place is body. The argument in the Simplicius extract is that place is equal in volume to the body it houses, and that only things of the same kind can be equal to each other.[70] In the *Republic* commentary, the argument proceeds by elimination. Place cannot be incorporeal and separable (*khôriston*) from body, for a thing of that kind would exist everywhere in totality (*holon pantakhou*). Nor can place be incorporeal and inseparable from body, because then it would be in a body and in the place of that body. It remains that place is corporeal.[71] I shall have to consider below in what sense he can think it a body.

Although place is a body, Proclus argues that it is immaterial. The arguments in the two passages are similar, so I shall take the slightly fuller version in the Simplicius extract. The argument comes in three steps. First, place is immobile, since, if it moved, this would presuppose that it had a place, and a place of place is impossible, as Aristotle and Theophrastus confirm.[72] Secondly, place must in that case be indivisible, or its parts would move, when they were parted; moreover the intruding body which divided place would need a further extension, to house itself.[73] Thirdly, place must in that case be immaterial, so that it may avoid suffering any change, including division.[74] What does Proclus mean by 'immaterial' here? He cannot mean merely 'tenuous' or 'flimsy', for what is flimsy is *easier* to divide. He makes it quite explicit, when discussing interpenetration by solar light and psychic vehicles, that the sense in which interpenetrating bodies are immaterial is that they lack prime matter:[75]

> We showed some while ago both that body can go through body and that this does not happen to all bodies, but that immaterial (*aüla*) ones go through material (*enula*) ones, and immaterial through immaterial. By matter I mean the ultimate matter which underlies the ultimate bodies.

---

who cannot contemplate all Forms in a simple intuition, Marinus could have selected this for mention not because he could find no better example of philosophical analysis in Proclus, but because he thought it important in relation to our spiritual progress.

[69] Proclus *in Remp*. 2.197.16-199.21; ap. Simplicium *in Phys*. 611,10-613,20. I am grateful to Larry Schrenk for discussion of these passages. Both are translated into French (the second in an appendix) in A.-J. Festugière's translation of Proclus *in Remp*. There is an English translation of the Simplicius passage in S. Sambursky, *The Concept of Place in Late Neoplatonism*, 64-81. An Italian translation of Simplicius *in Phys*. 4 by Cecilia Trifogli is in preparation and will include the Simplicius passage. Further English translations are in preparation (ed. Sorabji) which will cover all citations of Simplicius and Syrianus in this chapter.

[70] Proclus ap. Simplicium *in Phys*. 611,35.

[71] Proclus *in Remp*. 2.198.8-15.

[72] Proclus ap. Simplicium *in Phys*. 612,2.

[73] ibid. 612,10.

[74] ibid. 612,16-25; 613,17; *in Remp*. 2.163.7-8.

[75] Proclus *in Remp*. 2.162.24-8.

But how does the absence of this ultimate matter make bodies exempt from change and hence from division? The idea of matter was originally introduced by Aristotle, precisely in order to analyse what happens in change: the same matter first lacks, and then acquires, a given form.[76] Perhaps Proclus is inferring that what has no matter cannot undergo change and hence not motion or division. We shall see, however, that Proclus also uses the concept of immateriality in a different way, to locate something further away from matter in the Neoplatonist hierarchy, and closer to the One. Insofar as he has this other concept of immateriality in mind, his argument will need a different basis. Perhaps he would view it as a general principle, although it is not a self-evident one, that what is further from matter is less subject to change.

Having argued that place must be body, yet immaterial, Proclus argues that what exactly meets these requirements is the light described in the *Republic*'s myth of Er, which he regards as supra-celestial. For it fits the description of being more immaterial (*aüloteron*) and most simple (*haploustaton*),[77] The Simplicius extract concludes by imagining one body (the cosmos) implanted in the other (place), so that they interpenetrate each other:[78]

> So, to summarise all that has been proved, place is an immobile, immaterial, indivisible body. But if so, it is clearly more immaterial (*aüloteron*) than any other body, more so than moving bodies, including the immaterial ones amongst them. So if among other bodies the simplest (*haploustaton*) is light (for fire is more incorporeal than the other elements, and light is the most incorporeal kind of fire), evidently place will be light, the purest of bodies.
>
> Let us then (he says) conceive two spheres, one made of a single light, the other of many bodies, the two equal to each other in volume. But seat one concentrically with the other, and on implanting (*embibazein*) the other in it, you will see the whole cosmos residing in its place, moving in the immobile light. You will see the cosmos not moving as a whole, so that it may imitate its place, but moving in respect of its parts, so that in this way it may be inferior to place.

Having described Proclus' theory, I will now raise some problems about it. First, if 'immaterial' means lacking prime matter, how can Proclus use the comparative and superlative of it, as he repeatedly does, when he says that place or supra-celestial light is more or most immaterial and simple?[79] For lacking prime matter is an all-or-nothing affair, which should not admit of degrees. The answer must be the one

[76] Aristotle *Phys*. 1.7-9: the reference is not yet to prime matter in this passage.

[77] Proclus *in Remp*. 2.198,25-9; ap. Simplicium *in Phys*. 612,27-9.

[78] Proclus ap. Simplicium 612,24-35.

[79] Most immaterial (*aülotaton*) *in Remp*. 2.198.23; ap. Simplicium *in Phys*. 615,26; more immaterial (*aüloteron*), ap. Simplicium *in Phys*. 612,26; most simple (*haploustaton*), ap. Simplicium *in Phys*. 612,27; more simple (*haplousteron*) *in Remp*. 2.196.29; 199.9.

already indicated, that there is another sense of 'immaterial' and of 'simple', perhaps: higher up in the Neoplatonist hierarchy away from matter and closer to the wholly simple One.

Aristotle had got no nearer to allowing degrees of immateriality than saying that fire is the only element connected with form, because its natural movement is to the perimeter of the sublunary world, and a thing's shape or form lies in its perimeter.[80] But the Stoics had gone further: they treated some things as more material (*magis silvestres*) than others.[81] And Proclus' usage goes beyond that again. For example, he calls the heavens 'simple', even though he thinks them composed of four elements.[82] And he describes as more or as less material: eyes,[83] types of fire or earth,[84] and many other things.[85] I conclude that Proclus uses the terms 'immaterial' and 'simple' both to locate something in the hierarchy and to exclude prime matter, without noticing that these two usages do not fully harmonise with each other.

A second problem also concerns immateriality. For in the most recently quoted passage, Proclus allows that some immaterial bodies, including presumably the celestial body and perhaps psychic vehicles and light, are capable of motion.[86] Even if the rotating heaven does not change its position, so Proclus explains at the end of the passage, its *parts* have to change position. But how can Proclus say that any immaterial bodies move, when he insists that the immateriality of the spatial body makes it immune to motion?

There is a third problem, if Proclus' reason for postulating immunity to motion is, as I conjectured, that motion and all change presupposes matter. Certainly, a body that is created or destroyed must, in Aristotle's view, possess some underlying matter. But for a body merely to move, what is required is not that it should *possess* matter, but that it should serve as matter, or serve, like matter, as a subject (Aristotle puts it both ways)[87] characterised now by one set of characteristics, now by another, in this case by different spatial positions. Motion, then, does not require a body to *possess* matter, and that is how Aristotle is able to envisage, however fleetingly, that the celestial substance might move without having matter.[88]

> Perhaps some things do not have matter, or have matter not of that type, but subject only to change of place.

[80] Aristotle *GC* 2.8, 335a18-21.
[81] Calcidius *in Tim*. ch. 289.
[82] Proclus *in Remp*. 2.50.3.
[83] ibid. 2.196.12.
[84] Proclus *in Tim*. 2.8.22-7; 2.9.12; 2.9.27-2.10.5; 2.11.20.
[85] See e.g. Proclus *in Tim*. 2.45.14; 2.46.17-18.
[86] Proclus ap. Simplicium *in Phys*. 612,26-7.
[87] Matter: *Metaph*. 12.2, 1069b7-9; subject: *Phys*. 1.7.
[88] Aristotle *Metaph*. 8.4.1044b7-8.

The tradition inspired by this remark was described in Chapter 3 (n. 69).

A fourth question for Proclus' theory can be handled more quickly. If place is light, why does not place collapse when the light goes out? The answer is that place is supra-celestial light, visible not to the eyes, but only[89] through the luminous vehicles of our rational soul. It is not affected, therefore, when visible light goes out.

A more subtle question is how Proclus' endorsement of inter-penetration can cope with Aristotle's objection to the idea of interpenetration,[90] which came to be illustrated by the example of the sea in a cup. If you could really fit a ladle full of sea water into a cup full of wine once, without increasing the volume of the liquid, then (absurdly) you should be able to repeat the operation, until the whole of the sea was fitted into the cup. Proclus replies that this objection to the interpenetration of bodies does not affect his supra-celestial light, because the light cannot be divided, and so cannot be separated off into a ladle or cup, but would rather pass right through it.[91] The appeal to indivisibility lay at the root of Proclus' explanation of interpenetration: it is because place, or space, cannot be parted by a barrier that it goes right through it. There could not be a greater contrast with the earlier idea that the interpenetration of bodies is made possible by the infinite division of those bodies.

Of the five questions so far raised, Proclus would have answers to several. But there are two questions which he does not consider, and which would, I think, be hard for him to answer. First, we have seen that supra-celestial light is meant to serve as an absolute place against which the cosmos rotates and its parts move. But is this idea coherent? When it was suggested in the nineteenth century that the ether might serve as an absolute place, it was hoped that the motion of the earth through the ether would be visually detectable. But Proclus abjures any corresponding hope from the outset. For, as we have also seen, he declares that supra-celestial light is not detectable by the senses, but only through the luminous vehicle which houses the rational soul. For something as tied to the sensible world as motion, is it not legitimate, we may ask, to expect a sensory means of detecting it?

Finally, Proclus wishes to insist that, though immaterial, his supra-celestial light is a body. But if it interpenetrates everything without resistance or increase in volume, in what sense can it still be thought of as body? The Stoics allowed wine and water to interpenetrate with each other, but at least these stuffs were *selectively* resistant, since they did not interpenetrate with the spoon which stirred them. Perhaps Anaxagoras allowed all stuffs to

---

[89] Proclus *in Remp.* 2.195.9-14; 196.20; 199.21. The vehicle is luminous rather than pneumatic, or 'pneumatique', as Festugière renders it at 2.195.9-14.

[90] Aristotle *Phys.* 4.6, 213b5-12.

[91] Proclus ap. Simplicium *in Phys.* 613,15.

interpenetrate, but if so, at least each would add to the volume of the whole. Proclus' supra-celestial light seems neither to add to volume, nor to be resistant in any relation or degree. How, then, can we accept his description of it as body? One suggestion would be that he means to describe it only as a *mathematical* body, that is, as a geometrical solid. Credence may be lent to this suggestion by the form of his argument for its corporeality. For when he points out that a place is *equal* to the body which it houses, and that equality holds between quantities of the same kind, he gives the mathematical examples of lines and surfaces as well as bodies.[92] But in fact he cannot mean that place is a *mathematical* body, because mathematical bodies in his view exist only in the mind.[93] An alternative suggestion would be that in calling place a body, he means no more than that it is three-dimensional. But, unless the context were one (which it is not) of arguing against the Aristotelian conception of place as a two-dimensional surrounding surface, its three-dimensionality would not require the argument that he gives it. I cannot see, then, how Proclus can make good sense of his claim that place is body.

I have now considered the Neoplatonists' views on what prevents the interpenetration of bodies, and their admission of exceptions in which interpenetration is allowed, notably the case of Proclus' supra-celestial light. Although the views of Simplicius and Philoponus on these issues have already been canvassed, it remains to say a final word about these two late Neoplatonists. Can Simplicius explain how the heavens are able to penetrate each other and other bodies? What stops interpenetration in other cases, in his view, is either prime matter or qualities. But he allows that the heavens have both of these. Admittedly, the matter is different from any down here, being subject only to motion,[94] and the qualities also are quite unlike those down here.[95] But none the less, in a suitable sense the heavens do have heat and colour.[96] Perhaps it is the difference of matter which makes interpenetration possible, for if the matter is subject only to motion, not to division, it cannot be parted by what it encounters.

Simplicius certainly stresses indivisibility in another connexion, for he follows Proclus' reason for rejecting the 'sea in a cup' objection to interpenetration. The heavens are indivisible, and so would pass right through a cup rather than being separated off into it.[97] On this his opponent, Philoponus, would disagree – he thinks it consistent with the nature of the heavens that they should be divided.[98] Among things beneath the heavens Simplicius thinks that interpenetration is never

---

[92] Simplicius *in Phys.* 611,38-612,1.
[93] Proclus *in Eucl.* 54,8-12.
[94] Simplicius *in Cael.* 133,24-134,9.
[95] ibid. 87,29-31; 88,19-21; 89,15-19; 89,20.
[96] ibid. 83,17; 86,13; 117,30.
[97] Simplicius *in Phys.* 531,8-9.
[98] They are divisible *tôi idiôi logôi*: Philoponus ap. Simplicium *in Phys.* 1333,6; 12; 16.

found, as shown by the fact that, when you mix two such things together, they take up more space than either did singly.[99] The implication is that the heavens do not add to the volume of the things they penetrate. But this draws attention to a difficulty which Simplicius has to face in common with Proclus. If the heavens neither add to the volume of anything, nor offer resistance to anything, by what right can they still be thought of as *body*? Simplicius cannot say, any more than Proclus, that they are merely a *mathematical* body, in other words, a geometrical solid, because he shares Proclus' view that geometrical solids reside in the thought and imagination of the geometer.[100] Nor can he claim that on his view to have three dimensions is to be a body because he views *place* as three-dimensional, but not a body.[101] This, then, is a problem for which, like Proclus, he appears to have no answer.

As regards Philoponus, his belief that it is prime *matter* which prevents the interpenetration of bodies creates a problem for him in his early *de Anima* commentary. For he had taken it as a sign of the incorporeality of light that it could penetrate the air. But he then has to meet the objection that light could penetrate, even if it were a body, just so long as it were a *matterless* body.[102] He replies with the 'sea in a cup' objection to such penetration, using the different example of the heavens in a grain. And he further adds that, if it were a matterless body, light would pass not only through air, but also through earth, and furthermore would thicken the air on its way.

Philoponus does not to my knowledge explicitly recognise any cases of bodily interpenetration. But he does believe that our souls have special bodies as their vehicle,[103] and we saw what difficulty Simplicius had over the question whether the vehicles of our souls would have to interpenetrate with our own 'oysterlike' bodies. This is a problem for Philoponus too, but it is one more problem that appears not to have been tackled.

The picture that has emerged of the Neoplatonists is that Plotinus sided judiciously with the Aristotelians in their rejection of the Stoic belief in interpenetration. But his successors ran into problems in saying what stopped interpenetration, and in admitting cases of interpenetration as exceptions. Moreover, the cases they admitted were problematic in ways that landed them in far greater difficulties than the Stoics.

[99] Simplicius *in Phys.* 531,3-9.

[100] Simplicius *in Phys.* 512,19-26; 621,33; 623,15-16; Simplicius (?) *in DA* 233,7-17; 277,6-278,6.

[101] Simplicius *in Phys.* 623,17-18; 19-20.

[102] Philoponus *in DA* 328,13-33.

[103] Philoponus *in DA* 18,24-8 postulates pneumatic and luminous vehicles; *Opif.* 26,8-9 appears to drop the luminous.

Christian theology

The issue of bodies interpenetrating and the Greek discussions of mixture were important for Christian theology. From the time of the Council of Chalcedon in A.D. 451, it was an orthodoxy that Christ had two natures, one human and one divine. The proceedings of the Council, with its cries of 'chuck him out', its armed guards, and its fears of lynchings back home were recorded at the time and have recently been described again.[104] The doctrine of two natures was one of the doctrines that Philoponus later resisted, and concerning which he declined, in a surviving letter a summons to explain his view before the Emperor Justinian.[105] The relation of the two natures of Christ was described by many Christian writers in terms of the Stoic theory of mixture.[106] The Stoics, unlike Aristotle, viewed the soul as a body, and applied their conception of mixture to body and soul. This was an analogy favoured by orthodox Christians, because, appealing to a different tradition, they could claim that Aristotle regarded body and soul, like other instances of matter and form, as two natures.[107] For the character of the mixture involved, one would expect orthodox believers in two natures to draw on Stoic, rather than Aristotelian theory. For the ingredients in a Stoic mixture persist actually, not potentially, and one can be dominant, as the divine nature was supposed to be, without obliterating the other. None the less, the dominance is described by Gregory of Nazianzus and Gregory of Nyssa in terms more reminiscent of Aristotle's obliteration of a drop of wine.[108] Such obliteration of one ingredient is better suited to the view of the monophysites, who assign to Christ only one nature. And in Theodoret of Cyrrhus' dialogue between a Eutychian monophysite and an orthodox, the monophysite's wording strongly suggests the Aristotelian idea that the subordinate ingredient, Christ's human

---

[104] By Geoffrey de Ste Croix, paper in preparation. The Proceedings are recorded in E. Schwartz, ed., *Acta Conciliorum Oecumenicorum* II i-vi, 1932-8. The resulting creed of Chalcedon is translated into English in Henry Bettenson, *Documents of the Christian Church*[2], Oxford 1962, 51-2.

[105] Philoponus, *Letter to Justinian*, Syriac with Latin translation, no. VI in A. Šanda, *Opuscula Monophysitica Ioannis Philoponi*, Beirut 1930; Latin translation by G. Furlani, 'Una lettera di Giovanni Filopono all' imperatore Giustiniano', *Atti del Reale Istituto Veneto di Scienze, lettere ed arti* 79, 1919-20, 1247-65; extracts in German in W. Böhm, *Johannes Philoponos: Ausgewählte Schriften*, Munich, Paderborn, Vienna 1967. For some of the crucial arguments from Philoponus' *Arbiter* or *Diaitêtês*, see Šanda p. 77, with German translation in Böhm, pp. 423-4.

[106] H.A. Wolfson, *The Philosophy of the Church Fathers*, Cambridge Mass. 1956, 364-403; cf. Henry Chadwick, 'Origen, Celsus and the Stoa', *Journal of Theological Studies* n.s. 48, 1947, 34-49.

[107] Aristotle *Phys.* 2.1; Wolfson pp. 364-72.

[108] Gregory of Nazianzus *Orat.* 29,19 PG 36,100A; Gregory of Nyssa *contra Eunomium*, PG 45,693A; 697B-C; *adversus Apollinarem, ad Theophilum*, PG 45,1276C-D; Wolfson pp. 397-8.

nature, is obliterated, even though it leaves certain properties behind.[109]

If Christ's two natures are not corporeal, it may seem surprising that Christians felt the need to support their conception by reference to theories of how bodies mix, and this observation is actually made by Nemesius and Theodoret of Cyrrhus.[110]

Greek theories of mixture also left their mark on Christian descriptions of union with God in mystical experience. But it was the *examples* of mixture, rather than the theories, which were exploited here.[111]

Interpenetration is also relevant to resurrection bodies and to the virgin birth of Christ. Doubting Thomas, according to St John's Gospel, had to be allowed to *feel* Christ's wounds before he believed in his resurrection.[112] But Eutyches, anxious to deny a distinct human nature in Christ, argued that after that demonstration, Christ's resurrection body would have been impalpable.[113] The impression that it could pass through closed doors may be fostered by the statement made twice in the same passage of St John's Gospel that Christ appeared to the disciples *when the doors were closed.* This does not in fact exclude its materialising within the room. But none the less Thomas Aquinas quotes Augustine with approval, when the latter connects Christ's passing through closed doors with his birth's leaving his mother's virginity intact. Thomas concludes that the divine, as opposed to the human, nature in him made the latter possible.[114]

In 1277, the interpenetration of bodies was turned into a test of God's omnipotence. Bishop Tempier, condemning propositions that limited the power of God, insisted that God could put bodies in the same place. In this he was followed in the next century by William of Ockham, who said that in the bread of the sacrament Christ's body is in the same place as the parts of the matter.[115]

---

[109] Theodoret of Cyrrhus *Eranistes* Dialogue 2, PG 83,153C-D; Wolfson 445-6.

[110] Nemesius *On the Nature of Man* PG 40,601B; Theodoret *Eranistes*, Dialogue 2, PG 83,156A. Cf. Gregory of Nazianzus, letter 101 to Cledonius, translated in Nicene and Post-Nicene Fathers, 2nd series, vol. 7, pp. 440-1 (I am grateful to Robert Gregg for this reference).

[111] See J. Pépin, appendix to his *Ex Platonicorum Persona*, Amsterdam 1977, reprinted from *Miscellanea André Combès=Divinitas* 11, 1967. Thomas Aquinas rejected all mixture analogies: *Summa Theologiae* 3, q.2, a.1, *responsio; Compendium Theologiae* ch.206; *contra Gentiles* 4, ch.35. I thank Christopher Hughes for the references.

[112] John 20:19 29.

[113] Gregory the Great, *Morals on the Book of Job* 14,72-4.

[114] Thomas Aquinas *Summa Theologiae* III, q. 28, a. 2, quoting Augustine *On John*, tract. 121.

[115] Mentioned in Chapter 3 above. See Edward Grant, 'The condemnation of 1277, God's absolute power and physical thought in the later middle ages', *Viator* 10, 211-44. pp. 217, 224, 232 cite Tempier's proposition 141 (renumbered by Mandonnet and in the translation by Ernest L. Fortin and Peter D. O'Neill in Ralph Lerner and Muhsin Mahdi, *Medieval Political Philosophy: a source-book*, New York 1963, 337-54), and Ockham, *Quotlibeta septem: Tractatus de sacramento altaris*, Strasbourg 1491 repr. Louvain 1962, quotlibet 1, q. 4, sig. a4v, col. 1.

Part I: Matter

Interpenetrating beings in literature

Besides the relevance of interpenetration to theology and science, the idea was to influence literature. The belief that the vehicles of our soul can interpenetrate other bodies seems to be reflected in Dante's astonishment at being carried through the very body of the moon.[116] But I will finish with Milton's angels, who surely correspond to Porphyry's demons. One pagan Neoplatonist instructs his Christian pupils that when he speaks of demons they should think of angels.[117] What Milton tells us is that angels, unlike humans, make love by total interpenetration, and

> Obstacle find none
> Of membrane, joint or limb, exclusive bars.
> Easier than air with air, if spirits embrace
> Total they mix.[118]

[116] Dante *Paradiso* 1, 31-9.

[117] Olympiodorus *in Alcib.* 21,15-18. Porphyry himself, however, regards angels as higher than demons (Augustine *City of God* X 9).

[118] See Milton *Paradise Lost* 8, 614-29. Thomas Aquinas, however, denies that angels can be in the same place, *Summa Theologiae* 1, q.52, a.3. I am grateful to Wolfgang Detel, Robert Gregg, Andrew Harrison, David Konstan, John Maybury, Larry Schrenk, Avrum Stroll, Zeno Vendler and Christopher Williams for discussions connected with this chapter.

Part II
Space

Archytas' problem: is there an edge to the universe? Woodcut from Camille
Flammarion, *L'atmosphère*, 1886.

CHAPTER EIGHT

Is there infinite or extracosmic space?
Pythagoreans, Aristotelians and Stoics

Is space infinite? Does it extend beyond the boundaries of the physical cosmos? These questions are not the same. The question of extracosmic space has no application to Epicurus, whose infinite space is studded throughout with cosmoi. It applies to those who thought that physical matter[1] was congregated in a single finite cosmos, with the earth at the centre and the stars in a sphere at the edge. It is then natural to ask if there is space beyond the stars. Some ancient arguments bear only on this issue, some only on the infinity of space, but I want to consider both.

The most compelling argument ever produced for the infinity of space was devised by Plato's friend, the Pythagorean Archytas. Archytas challenges the attempt to imagine an outermost edge, by asking whether you can stretch your stick out from the supposed edge. He thinks it would be absurd (*atopon*) if you could not. The thought experiment is illustrated in a well-known woodcut, which does indeed reveal something beyond the edge.[2]

> Archytas, according to Eudemus, put the question this way: if I came to be at the edge, for example at the heaven of the fixed stars, could I stretch my hand or my stick outside, or not? That I should not stretch it out would be absurd (*atopon*), but if I do stretch it out, what is outside will be either body or place – (it will make no difference, as we shall discover). Thus Archytas will always go on in the same way to the freshly chosen limit (*peras*), and will ask the same question. If it is always something different into which the stick is stretched, it will clearly be something infinite.[3]

Eudemus of Rhodes, Aristotle's pupil, who reports the argument, would not in fact accept its un-Aristotelian conclusion. None the less he

---

[1] I speak of *physical* matter, to make clear that I have no interest in this chapter in *prime* matter.
[2] See facing illustration.
[3] Simplicius *in Phys*. 467,26-32.

elaborates it, and borrows an anti-Aristotelian argument cited by Aristotle himself,[4] to show that what lies outside will not be empty, but filled with body. It is the infinity of *body* against which Aristotle directs most of his attack on infinity, and Eudemus' motive is probably to show that Archytas is committed to this discredited thesis.

> If it is a body, the thesis has been demonstrated, while if it is a place, and place is that in which body exists or could exist, and in the case of eternal things we must treat what could exist as actually existing, then on this option too body will be infinite, and so will place.[5]

The Epicureans and Stoics were also to elaborate Archytas' argument, and it is still found in yet other versions in Locke and Newton.[6] According to the Epicurean and Stoic elaboration, if you can stretch out, there will be space outside, while if you cannot, there will be a body outside obstructing you. So either way, you have not reached an edge. To this version someone might respond by asking why it need be a three-dimensional body rather than a two-dimensional surface that formed the obstruction. In the face of that, the infinitist would still have an answer, but he might find himself falling back on something more like Archytas' simple appeal to intuition: would it not be 'absurd' if there were no space beyond the two-dimensional barrier, even if you were physically prevented from reaching that space? Archytas' intuitive appeal may thus present the argument in its most persuasive form.

Modern physics would answer Archytas by arguing that space could be finite without having an edge, so that no question about edges need be faced by the finitist. I shall return to that answer in Chapter 10. But what was the best answer devised in antiquity? I think it was formulated by the greatest expositor of Aristotelianism, Alexander of Aphrodisias, at the beginning of the third century A.D.:

> He will not stretch out his hand; he will be prevented, but prevented not as they say by some obstacle bordering the universe (*to pan*) on the outside, but rather by there being nothing (*to mêden einai*). For how can anyone stretch something, but stretch it into nothing? How can the thing come to be in what does not even exist (*to mêde holôs on*)? In the first place, nothing would have any desire to stretch any of its limbs in nothing, for such is the nature of what has no existence.[7]

Alexander wants us to take seriously the idea of absolutely nothing, and to see it as different not only from body, but also from empty space

---

[4] Aristotle *Phys.* 3.4, 203b28-30.

[5] Simplicius *in Phys.* 467,32-5.

[6] Lucretius 1.968-83; Stoics ap. Simplicium *in Cael.* 284,28-285,2; Locke *Essay Concerning Human Understanding* 2.13.21; Newton *de Gravitatione*, translated by A. Rupert Hall and Marie Boas Hall, Latin 101, English 133.

[7] Alexander *Quaestiones* 3.12, 106,35-107,4.

with its three-dimensional, receptive structure. To stretch is by definition to stretch into *something*. One could not even want to stretch without wanting to stretch into something. So Archytas is wrong that it would be absurd if a man could not stretch, and the Epicureans and Stoics are wrong that what prevented him would need to be a body. A man at the edge would be prevented – prevented not by a body, but by the logical implications of the fact that there would be nothing, not even empty space, to stretch into.

Simplicius slightly alters Alexander's answer, when he repeats it. He does not think it right to say that nothingness *prevents* a man from stretching or makes him *unable* to do so. Nothingness neither repels nor accommodates a hand:

> Perhaps this argument will create a keen problem also for us who say that there is nothing outside the heaven, since the body of the cosmos, whose limit is the heaven, has matter as its space (*khôra*). Because of that, if someone came to be on the back of heaven and stretched out his hand, where would it stretch? It could not stretch into nothing (*mêden*), for no existent thing is in what does not exist (*to mê on*); but neither will the man be prevented from stretching it out, for one cannot be prevented by nothing.[8]

> But the answer which Alexander himself presented is a better one, that if the cosmos is the whole universe (*to pan*), and nothing is outside the universe, the supposed situation is like someone trying to stretch his hand into the non-existent (*to mê on*). If he were to stretch it out, it will be a place that receives his hand, and not the non-existent, while if he were unable to stretch it out, there will be an obstacle. So the supposition is absurd (*atopos*), since in imagination it assumes in advance what it seeks to prove, that there is something, whether empty or solid, outside the universe.[9]

Is Alexander's reply adequate? His opponents may still want to know how this third option, neither body nor empty space but nothing, is a possible one. If there is an edge to matter, how can there fail to be empty space beyond? If Alexander wants to argue merely defensively for the Aristotelian system and its belief in finitude, he can point out that that system has an answer. For Aristotle argues[10] that there is no such thing as space, conceived as a three-dimensional extension which goes right through things. All there is is place, which he defines as the

---

[8] Simplicius *in Phys.* 467,35-468,3.

[9] Simplicius *in Cael.* 285,21-7.

[10] At *Phys.* 4.4, 211b20-5. Aristotle attacks the idea of place as a portable three-dimensional extension (*diastêma*), with arguments intended to show that that would make many places coincide. At *Phys.* 4.8, 216a26-b16, he attacks the idea of a three-dimensional empty space filled by body, by arguing that it would be a redundant addition to a body's volume (cf. Sextus *M* 10.27), and again that if two such things could be in the same place, so could any number. These arguments are further described in Chapter 5 above.

two-dimensional inner surface of a thing's surroundings. The universe as a whole has no surroundings, and so there is no place available there for a hand to stretch into.

For the Aristotelian system, then, Alexander's answer follows that beyond the physical world there is nothing.[11] So he succeeds in defending that system from Archytas' point. On the other hand, he has not shown how to go on the offensive and convince his opponents that on their own principles they too should admit there would be nothing beyond the edge of matter. Aristotle's attacks on the three-dimensional conception of space are too obscure and unpersuasive for that.

Archytas' stretching argument is one of a family of arguments which are well known to mediaevalists thanks to the work of Edward Grant.[12] They might be called intrusion arguments because they all involve the idea of something intruding beyond the present edge of the physical cosmos. One is an expansion argument, first associated with another Pythagorean, Xuthus. According to Xuthus, if there were no vacuum, when I expanded a kettle full of water into steam, a bulge would have to appear on the outer surface of the cosmos, to accommodate it.[13] If this seems amusing, it is no worse than the theory of compensation (later: *antapodosis*) mentioned in the same breath by Aristotle, that every time I boil a kettle, there must be a compensating contraction elsewhere, for example of cloud into rain. The next appeal to cosmic expansion comes in the Stoic Posidonius (c.135-c.55 B.C.), and the currently neglected, but once influential,[14] Cleomedes, who collected arguments from Posidonius and from earlier Stoics at some time in the first one and a half centuries A.D.[15] Posidonius exploits the

---

[11] The same answer, that there is nothing outside space, would follow also from modern accounts of how space may be finite, although these accounts differ in not allowing that space has an edge.

[12] Edward Grant, *Much Ado About Nothing, Theories of Space and Vacuum from the Middle Ages to the Scientific Revolution*, Cambridge 1981 (hereafter *Ado*); id., 'The condemnation of 1277, God's absolute power, and physical thought in the later middle ages', *Viator* 10, 1979, 211-44.

[13] Aristotle *Phys.* 4.9, 216b22-8; 217a10-12. Beware the Loeb translation's substitution of 'to use Xuthus' expression', for 'as Xuthus said'. Cf. Philoponus *in Phys.* 626,12-17; 671,10-672,17; 674,26-675,6.

[14] For Cleomedes' influence on Patrizi and Gassendi in the sixteenth and seventeenth centuries, see E. Grant, *Ado*, 201, 210. Cleomedes was not available to the Middle Ages.

[15] Cleomedes *de Motu Circulari Corporum Caelestium* or *Kuklikê Theôria*. *SVF* gives almost none of the relevant arguments, which may have contributed to their neglect. See R. Goulet's French translation of Cleomedes with commentary *Cléomède: théorie élémentaire*, Paris 1980, p. 34 for the alternative title, pp. 11-15 for Cleomedes' earlier sources (Posidonius is acknowledged repeatedly and most fully at the end of Cleomedes' work), and see W. Schumacher, *Untersuchungen zur Datierung des Astronomen Kleomedes*, diss. Cologne 1975, for the dating. All are reported by Robert B. Todd, 'Cleomedes and the Stoic concept of void', *Apeiron* 16, 1982, 129-36, notes 3 and 4. Todd is preparing an English translation and revised text, which he has kindly allowed me to see.

fact that the Stoics, following certain Pythagoreans, believed that the history of the cosmos would be endlessly repeated. Some said it would be repeated exactly, in an infinity of cycles. They added that at the end of each cycle, the entire cosmos would be dissolved into fire. What Posidonius points out is that fire takes up more room than the other elements, and he infers that there must be a finite amount of extracosmic space to accommodate the conflagration.

> Posidonius said that what is outside the cosmos is not infinite, but large enough to accommodate its dissolution.[16]

Cleomedes, who elsewhere argues that the extra space must be infinite, develops Posidonius' point as follows.

> And if furthermore all substance is reduced to fire, as the most gifted physicists think, it must occupy a place more than ten thousand times as great, just like solid bodies when they are vaporised into smoke. So the place that is occupied by dissolved substance in the conflagration is now void, seeing that no body fills it.
> But if someone says that the conflagration does not happen, such a claim is no objection to there being void. For even if we were only to conceive (*epinoêsai*) of substance being dissolved and expanding further, since nothing can get in the way of its so expanding, that into which we conceive it moving when it expands would be void, just as what it now occupies is no doubt void that has been filled.[17]

Cleomedes thus converts Posidonius' idea into a thought experiment, an approach that is repeated in Islamic thought by the so-called 'philosophers' whom Ghazâlî (A.D. 1058-1111) reports in his *Destruction of the Philosophers*. God could build extra layers onto the world, so there must be the space in which he could do it,[18] an argument which reappears in the Latin West.[19]

A third argument, the shifting argument, is also introduced by Cleomedes. The possibility of imagining the whole physical cosmos shifting shows that there is extracosmic void.

> Furthermore we can conceive (*epinoêsai*) the cosmos itself moving out of the place which it happens to occupy now. And together with its motion, we shall conceive the abandoned place as being empty and the place to

[16] Posidonius ap. Aëtium *Placita* 2.9.3 (fr. 97 Edelstein-Kidd=*Dox. Gr.* 338.17).

[17] Cleomedes book 1, ch. 1, Ziegler p. 6, lines 11-25 (partly in *SVF* 2.537).

[18] Ghazâlî *Tahâfut al-Falâsifa* I, in Averroes *Tahâfut al-Tahâfut*, translated S. van den Bergh, p. 51.

[19] Bradwardine *de Causa Dei contra Pelagium*, p. 177, discussed by E. Grant, *Ado*, 137. A more complex version in Gassendi *Syntagma Philosophicum*, in *Opera Omnia*, Lyon 1658, vol. 1, p. 183, col. 1, discussed in *Ado*, 208-9, will be mentioned below. Accepted by Isaac Barrow, *The Usefulness of Mathematical Learning*, tr. by John Kirby, London 1734 from the Latin of 1664, 10th lecture, p. 169. I am grateful to Peter Alexander for information on Barrow throughout.

which it has moved as being occupied and held by it. This latter place would be a filled void.[20]

Aristotle would have responded that void would have the opposite effect: it would actually make cosmic motion impossible. For one thing, if there were extracosmic void, why should the cosmos move in this direction rather than that, and why should it stop here rather than there? Would it not disintegrate in every direction rather than moving as an intact whole?[21] The style of argument was later associated with Buridan's ass, which died of indecision, being equally hungry and thirsty, and placed between equidistant quantities of food and drink. But in fact the example is Aristotle's although it involves a man, not an ass.[22]

Some people would think that a stronger point than Aristotle's could be made against Cleomedes' shifting. Does it even make sense to speak of the whole cosmos moving? Leibniz, in his letters to Newton's associate Clarke, appears to combine two different types of objection.[23] In a theological version of Aristotle's argument, he protests that there is no sufficient reason for God to shift the universe from one position to another. But he also argues that the two supposedly different positions of the universe would not be different because they would be indiscernible – that is, there would be nothing to differentiate them.[24]

The last argument would suggest that we cannot make sense of the idea of the physical universe shifting as a whole. But might we not do so by imagining that we suddenly experience the forces normally associated with moving off? We have the familiar sensation in our tummies, the books slide off our desk, but as we look out of our window, the trees are keeping pace with us, their leaves and branches swept back by similar forces. And evidence comes in from all over the universe that, while everything maintains the normal distances from everything else, everything has none the less experienced the forces associated with moving off.

It seems to me that this sort of experience would give a sense to the idea that the physical universe was moving off as a whole. But it would be an impoverished sense, because there would still be no answer to the questions 'whence?' and 'whither?' Some people would think the sense very impoverished indeed. They would say that to talk of moving here would be to say no more than that the forces described had been

[20] Cleomedes 6,6-11.

[21] Aristotle *Phys*. 4.8, 214b28-216a26, esp. 214b17-19; 214b28-215a1; 215a19-24.

[22] Aristotle *Cael*. 2.13, 295b32-4. See Steven Makin, 'Buridan's ass', *Ratio* 28, 1986, 132-48.

[23] The Leibniz-Clarke correspondence, written in 1715-16, ed. by H.G. Alexander, Manchester 1956, with Clarke's English translation of Leibniz's letters, 4.13 and 5.29.

[24] David Hirschmann plans in the *Proceedings of the Aristotelian Society* 88 (1987-8) to relate these arguments more closely than I have done (see his 'The kingdom of wisdom and the kingdom of power in Leibniz', pp. 147-59).

inexplicably experienced.[25] I am more inclined to say that the forces are explained by the occurrence of a genuine motion, which however lacks the customary whence and whither. This would in turn carry with it the idea of a space through which the motion takes place, but a space which contains no positions.

On behalf of Newton, Leibniz's opponent Clarke produced a theological version of Cleomedes' argument. God could shift the whole material world, and that power presupposes a genuine extracosmic space.[26] He was not the first to argue this way. In 1277, Bishop Stephen Tempier of Paris, intent on magnifying the power of God, condemned 219 philosophical and theological propositions, including the proposition that God could not shift the world, or he would leave a vacuum behind.[27] The condemnation provides one instance – belief in the harmony of Plato and Aristotle is another[28] – in which an arbitrary and even preposterous dogma proved very fruitful in the history of philosophy. Edward Grant has shown how it helped natural philosophers of the fourteenth century to rethink Aristotle's views.[29] Some of them then argued on Cleomedes' side (although not under his influence) that, since God could shift the world, there must be extracosmic space.[30] Others were led to work out how God could shift the world without the prior existence of extracosmic space.[31]

There is a fourth intrusion argument, connected this time with multiple worlds. Plato and Aristotle both maintained that there could not be two or more physical cosmoi existing simultaneously.[32] Tempier condemned this proposition too: God could make as many worlds as he chose.[33] There is an obvious connexion with the question of extracosmic space, although it is not the connexion taken up by Aristotle. For it

---

[25] This is in effect, I think, the view of Lawrence Sklar, *Space, Time and Space-Time*, Berkeley and Los Angeles 1974.

[26] Clarke's Letter 3.4 in Alexander.

[27] The 49th condemned proposition, translation by Ralph Lerner and Mushin Mahdi, eds, *Medieval Political Philosophy*, Ithaca N.Y. 1963.

[28] See Chapters 4, 10, 12 and 15.

[29] Edward Grant, 'The condemnation of 1277, God's absolute power and physical thought in the late middle ages', *Viator* 10, 1979, 211-44.

[30] Thomas Bradwardine, John de Ripa and Nicolas Oresme are cited by Edward Grant, 'The condemnation', 230-2 and *Ado*, 131, 136, 413. The argument is reported by Isaac Barrow, *The Usefulness of Mathematical Learning*.

[31] Richard Middleton, Walter Burley and John Buridan are cited by Edward Grant, 'The condemnation', 229-31.

[32] Plato *Tim.* 31B; 32C; Aristotle *Cael.* 1,8-9; *Metaph.* 12.8, 1074a31-8. Further arguments in Philo Judaeus *Aet.* 21; Proclus *in Tim.* 2.65,14-66,14. Origen uses this (*On First Principles* 2.1, transl. Butterworth pp. 79-80 and fr. on Genesis preserved in Eusebius *PE* 7,335a-b), to argue for creation *ex nihilo*. For if the cosmos had been made out of *prior* matter, only chance would have prevented some being left over to form a second mass of matter.

[33] Tempier, proposition 34.

may well seem, and it did seem to some later thinkers, that if there were two physical cosmoi existing simultaneously they would have to be housed in an all-embracing space that stretched outside each of them.[34]

In fact, the argument is not compelling. For one thing, it is logically possible that there should be two worlds bearing no spatial relationship to each other. A person could pass from one world to the other and back only by snapping his fingers and saying 'abracadabra', thus disappearing from one world and appearing in the other. The worlds, if sufficiently alike, might even be temporally related. For the clocks in the other world might enable someone to resume the thread of his life there, with little more difficulty than the owners of a country cottage experience on arriving from their town house, although there would have been no journey in between.[35] A less fanciful possibility can be developed out of the idea of modern physics that the space we live in may be closed. That is to say, it may be finite, without having a boundary. In that case, there could presumably be a second closed space, neither space being sufficiently extensive to link up with the other. Of course, on either of these suppositions there would be a space *other* than the one we are currently in. But it would not be extracosmic, that is *outside* our physical world, because it would bear no spatial relation to our physical world at all.

Aristotle's own argument against extracosmic space runs in an unexpected direction. He spends most of two chapters arguing that it is impossible for there ever to be any physical matter outside our cosmos. From this he draws a conclusion. What is place (*topos*) but something in which it is possible (*dunaton*) for there to be body, and what is void (*kenon*) but something in which it is possible (*dunaton*) that there should *come* to be body? How, then, can there be extracosmic space or void, things which by definition *can* receive body, if it has just been proved that body *cannot* ever exist extracosmically to be received?

> At the same time it is clear that there is neither place nor void nor time outside the heaven. For in any place it is possible (*dunaton*) for there to be body, and they say that void is that in which there is no body, but it is possible (*dunaton*) that there should come to be body, while time is the number of motion. Now there is no motion without natural body, and it has been shown that outside the heaven neither is there, nor is it

[34] For subsequent history, see E. Grant 'The condemnation', esp. 223-4, 242, who names Robert Holket, Walter Burley, Francisco Suarez, Bartholemaeus Amicus and Thomas Hobbes.

[35] A comparable possibility is discussed by Anthony Quinton, 'Spaces and times', *Philosophy* 37, 1962, 130-47.

possible (*endekhetai*) that there should come to be, body. So it is evident that there is neither place nor void nor time outside.[36]

To this argument Cleomedes offers a stunning reply. You might as well say that there cannot be a water vessel in an impenetrable desert. For by a water vessel we mean something that *can* receive water and by an impenetrable desert let us mean a desert that water *cannot* reach. Cleomedes' analogy is not entirely fair, but it is enough to point to the flaws in Aristotle's reasoning:

> Aristotle and the people of his school (*hairesis*) do not allow void outside the cosmos either. For, they say, void must be a container (*angeion*) of body, and outside the cosmos there is no body. So there is no void either. But this is simple minded, and very much like saying, 'since it is not possible (*hoion te*) for there to be water in dry and waterless places, it is equally impossible for there to be a utensil capable (*dunamenon*) of receiving water'. It should be recognised, then, that a container of body can be meant in two ways, either as something containing body and filled by it, or as what is capable (*hoion te*) of receiving body.[37]

I believe that Aristotle has made a mistake. The incapacity of physical matter to be received extracosmically does not imply a corresponding incapacity of anything so to receive it. It implies only that nothing will have the *opportunity* so to receive it. It implies no more than would be implied, if physical matter were prevented from moving further out by an adamantine wall. Then it would not have the opportunity of moving further out, and the implication would be the same: that nothing would have the *opportunity* of receiving it extracosmically. There could still, however, be extracosmic place, for that should be thought of as something with a capacity, but not necessarily with an opportunity, for receiving physical matter. It is not immediately clear whether Aristotle has here overlooked the distinction between capacity and opportunity, or whether he recognises it, but thinks that the capacity for receiving matter has been removed, as well as the opportunity. In favour of the former suggestion is a recent study which suggests that in his earlier work (though not in his later) he unwittingly treats capacity as if it implied opportunity.[38] In favour of the latter suggestion is the fact that Aristotle would not think there could be an everlasting capacity that

---

[36] Aristotle *Cael.* 1.9, 279a11-18. In what immediately follows, Aristotle discusses two things, the physical heavens which (279b2) are in ceaseless motion and the divine intelligences. The latter are superior (*huper*) to the heavens (279a20), unchanged (a20-1), the first and highest divinity (a32-3), which has nothing superior that moves it (a33-4). I should now correct my *Time, Creation and the Continuum*, London and Ithaca N.Y. 1983, 125-7: in view of the argument in the passage translated, the intelligences appear to be timeless in a stronger sense than that which applies to the moving heavens.

[37] Cleomedes 10,6-15.

[38] Harry Ide in a Cornell University Ph.D. dissertation in progress contrasts the early *de Interpretatione* with the later *Metaphysics* 9.

was never actualised, and some scholars have suggested that his whole argument turns on this,[39] but I cannot myself see that he breathes a word about it.

Whatever his mistake, it appears not to be an isolated one, for there is an analogous mistake in his treatment of time. He there argues that there would be no time, if there were no consciousness. For time is defined by him as something countable (*arithmêton*), that is, as something which *can* be counted. The definition is plausible, only if he means that it has a capacity, not necessarily an opportunity, to be counted. He then complains that if there were no consciousness, nothing could be counted. To put the point in its strongest form, we might say that everything in the universe would be *incapable* of counting (although that itself might be a contingent fact). Unfortunately for Aristotle, the incapacity of things to count does not imply a corresponding incapacity of things to be counted, but only once again a lack of opportunity to be counted. Commentators have been amazed at his argument here, but I think they have been wrong to try to reinterpret the plain meaning of his words.[40]

Not only does Aristotle's mistake occur in more than one context, but also it is not a stupid one. Given two correlative modalities, it is difficult to decide when the absence of one implies the absence of the other.[41] If evolution removed all eyes from the universe, so that there were no beings left which could see, would logs still be visible? They would not at least be visible to us, and there are many other cases apparently favourable to Aristotle's point of view. If evolution made us permanently immune to cobra bite, would cobras still be lethal to us? Surely not. Again, if evolution shrank us into dwarfs, would all the mountains we had once climbed still be surmountable? Again, no. In each of these we get Aristotle's result, that the loss of one capacity implies the loss of the correlative capacity, although we may need to understand that correlative as visible *to us*, lethal *to us*, surmountable *by us*. Even the water vessel in the impenetrable desert might be described as incapable of receiving water *there*. Of course, it could be moved, but then Aristotle might protest that there is no corresponding possibility of moving extracosmic space, so why should it be capable of receiving physical matter *at all*?

It might be thought that Aristotle could draw comfort too from modern physics, which has reasons far more complicated than that of an adamantine wall for thinking that space may possibly be finite. I shall try to describe in Chapter 10 how a certain distribution of

[39] David E. Hahm, *The Origins of Stoic Cosmology*, Ohio State University 1977, 106; F.H. Sandbach, *Aristotle and the Stoics*, Cambridge Philological Society, supp. vol. 10, 1985, 42.

[40] Aristotle *Phys.* 4.14, 223a21-9; Richard Sorabji, *Time, Creation and the Continuum*, 89-93; cf. 266.

[41] cf. Thomas Aquinas' discussion of whether, when someone's virginity can no longer be saved, God has lost the power to save it, and Richard Sorabji, loc. cit.

physical matter in the universe would mean that a body travelling in a straight line, so far from getting for ever further away from its starting point, would eventually return to base. In this situation, the size of space would indeed be related to the greatest possible trajectories. We could not protest that there must be space beyond the straight trajectory, for what could lie further out than the uninterrupted path of a straight line?

Unfortunately, Aristotle's reasons for denying that physical matter can be received further out contain nothing as sophisticated as the idea of a straight line that returns upon itself. They are not as crude as the postulation of an adamantine wall, but I think they have the same defect as such a postulation, that they cannot show why there should not be something further out capable, though denied the opportunity, of receiving matter. Admittedly, Cleomedes' example is not completely apt. For one thing, his water vessel is a piece of pot, and he has no argument to show that that piece of pot could not be in the desert, only that it could not serve as a vessel there. It has also been noted that the pot could be moved and could then serve as a vessel. (A parallel would be that conscious life, even if absent, could always be created, and would then be able to count time.) But if Cleomedes is wrong to treat the case of extracosmic space as being as obvious as that of the water vessel, he has none the less drawn attention to the fallacies in Aristotle's argument.

A fifth and final intrusion argument appears only after antiquity, so far as I know.[42] But it is the mirror image of an argument of Aristotle's. Aristotle, having ruled out extracosmic space, draws a conclusion about the shape of the cosmos. It cannot, for example, be a cube, for its outer part, the heaven, rotates, and the corners of a rotating cube would need more room for their rotation than the rest. That room would be unavailable without extracosmic space. Aristotle concludes that the shape cannot be rectilinear, and, more rashly, that it cannot be that of an egg (as Empedocles had suggested) or lentil, but must be spherical (he overlooks the possibility of an egg rotating on its axis).[43] The argument could be reversed by those who, like certain Pythagoreans,[44] think that the universe has a rectilinear shape. For then, if it rotates, there will be a proof of extracosmic space.

Such are the intrusion arguments, and such is the related debate between Pythagoreans and Stoics on one side and Aristotelians on the other. But the intrusion arguments were by no means the only arguments for infinite or extracosmic space. In the *Physics* Aristotle gives a list of five reasons which seduce people into believing in infinity.[45] Of the two that are relevant one looks like a reference to

[42] It is alluded to by Isaac Barrow, *The Usefulness of Mathematical Learning*, 172.
[43] Aristotle *Cael*. 2.4, 287a14-22. For Empedocles, see Aëtius 2.31.14.
[44] Plutarch, *de def. orac.* 422B: the universe, on this view, consists of 183 worlds arranged in a triangle.
[45] Aristotle *Phys*. 3.4, 203b15-30, answered 3.8, 208a5-23.

Archytas' stretching argument. The other is the argument that if
something is limited, it must be limited *by* something further out, and
that will rule out a furthest limit. Aristotle's reply is that the logic of
'limited' is different from the logic of 'touched'. Admittedly, if
something is touched, it must be touched *by* something, but the limited
need not be limited *by* anything.[46]

Unfortunately, he put his reply so briefly that it was ignored.
Epicurus repeated the 'limited by' argument for an infinite universe (*to
pan*), as if Aristotle had never spoken, and we are told that it was the
Epicureans' chief argument.[47] Similarly, Cleomedes presents a Stoic
version, in arguing for the infinity of void.[48] Consequently it fell to
Alexander to make Aristotle's point inescapable, that 'limited' does not
imply 'limited by something'. He does so in his *Problems and Solutions*
by offering three criteria of limitedness, none of which involves being
limited by something further out.

> If things are limited (*peperasmenon*) (i) when they have a limit (*peras*),
> and (ii) when they can be divided into equal [segments], and (iii) when
> they are composed of limited parts (*peperasmenôn merôn*), none of which
> consists essentially in having something outside – then people will be
> taking something which is not included in the essence of limitedness,
> when they take it that anything limited has its limit in something
> further. They will be taking an accidental feature which belongs to those
> limited things which are parts of a totality.[49]

How is this argument to be understood? Alexander follows and
develops Aristotelian thought, and does not always feel it necessary to
point out the Aristotelian background of his arguments. Perhaps this
is even more likely to be true of the *Problems*, if these correspond to
seminars held with Aristotelian students. At any rate, we have
encountered another example from the same treatise.[50] I therefore
suggest that the first criterion for limitedness (they have a limit)
trades on Aristotle's definition of a limit as 'the first [thing] outside
which you cannot take anything' (*hou exô mêden esti labein prôtou*).[51]
The first criterion of limitedness (a sufficient, rather than a necessary,
condition) will then be: having a first [rim] outside which you cannot
take anything. Quite clearly, so far from implying that there is
something further out, this criterion explicitly excludes there being
anything further.

The third criterion of the limited (composed of limited parts) may

---

[46] ibid. 3.8, 208a11-20.

[47] Epicurus, *Letter to Herodotus* 41; Lucretius 1.958-67; Epicureans according to
Alexander ap. Simplicium *in Phys.* 467,1-3.

[48] Cleomedes 14,13-16,12.

[49] Alexander *Quaestiones* 3.12, p. 104, lines 24-9.

[50] Alexander omits to explain, again in *Quaestiones* 3.12, why, on Aristotelian
principles, there is literally nothing for Archytas' man to stretch his hand into.

[51] Aristotle *Metaph*. 5.17, 1022a4.

appear to be circular, for it mentions limitedness, and even presupposes it twice over. It means that a thing is limited, if it is composed of a *limited* number of *limited* parts. But these two occurrences of the concept of limitedness are innocuous. For Alexander is looking for a criterion of limitedness that will apply to the *whole*. He can allow that the *parts* are limited according to a *different* criterion. He can even allow that they are limited according to his *opponents'* criterion of being *bordered* by something. He could no longer agree, if they insisted that the limited must be bordered on *all* sides, for that would not be true of every part of his cosmos, but such a further requirement would not necessarily occur to them.[52] As regards limited *number*, that would seem to Alexander to be a pleonasm. By a number he means a whole number specifiable in principle, although the number of parts would differ according to different ways of dividing the whole into parts. In sum, then, Alexander's third criterion for limitedness of the whole is that it is composed of a number, specifiable in principle, of parts that are limited according to some other criterion, for example, his opponents'. Once again such a criterion carries no implication that a limited whole need be limited *by* anything.

It is the second criterion of limitedness (what can be divided into equal segments) that is most puzzling. It is puzzling, not because it implies something further out – it clearly does not, any more than the other criteria – but because it is hard to see why it provides a criterion of limitedness at all. For it may seem that a non-limited expanse could also be divided into equal segments. One way of doing this, dividing it into an *infinity* of equal segments, would be ruled out by Aristotle, because he denies that an infinite division can be completed.[53] But why should we not in imagination divide an infinite expanse surrounding us simply into *two* equal segments, by thinking that part of it is to the left of us and part to the right? I think that Aristotle, and hence Alexander, would think this excluded for at least two reasons. First, the supposed two segments would themselves be infinite, so that the infinite expanse would contain infinite parts, which Aristotle considered impossible.[54] Secondly, Aristotle believed that ratios cannot hold between infinites, but only between finite quantities.[55] The two infinite segments could not therefore be regarded as equal.

Whatever we think of the last piece of argument, Alexander is to be congratulated for settling the issue by offering *definitions* of the limited. But the commendable treatment of limits in space is not matched by the Aristotelian treatment of limits in time. Aristotle argues that any instant is *both* the beginning of some future *and* the

[52] To acknowledge that something bordered on only *one* side might still be infinitely large is to open the door to the thought, which the Greeks found unacceptable, that some *part* of an infinite quantity might be infinite.
[53] Aristotle *Phys.* 3.5-7.
[54] ibid. 3.5, 204a20-6.
[55] ibid. 8.1, 252a13.

end of some past. In other words, it must be bounded by time on *both* sides, and he infers that time neither begins nor ends.[56] His treatment of limits should have suggested to him that the temporal limit, the instant, like the spatial limit, need not necessarily be bounded on *both* sides. We shall see in Chapter 15, however, that he needs a contrast between the infinitude of time and the finitude of space. It is an essential ingredient in his argument in *Phys.* 8.10 for the incorporeality of the prime mover.

A version of the 'limited *by*' argument is preserved in an Arabic translation, and is there said to have been used by Seleucus of Seleucia, an astronomer of the second century B.C., to establish his belief in the infinity of the world.[57] It comes with a reply which reasserts Aristotle's contrast between being limited and being contacted. Moreover, Seleucus' argument has been put into a neat, logical form and arranged to make clear its vulnerability to the Aristotelian point about contact. The original Greek source of the quotation and reply has not been identified,[58] but I would expect it to be an Aristotelian source. One might think of Seleucus' fellow-countryman Xenarchus in the next century, an Aristotelian influenced by Stoicism, who concentrated on Aristotle's natural philosophy and astronomy, or one might think of Alexander himself.

So much for the arguments in favour of infinite or extracosmic space. Let us see who have been the heroes of the debate so far. From among the proponents of infinite, extracosmic space, I would have to select Archytas and Cleomedes with his Stoic sources. But the more difficult task was to defend the idea of finite space from their onslaught, and for that I would have to nominate Alexander with his insistence on nothingness and his definitions of the limited.

If we turn now to the arguments *against* infinite or extracosmic space, we find that they come from Aristotle and his followers. The largest group I shall reserve for the next chapter. These are Aristotle's arguments that there cannot be any void or vacuum (I use the terms interchangeably), from which it follows that there cannot be extracosmic space. Some of Aristotle's other arguments have been encountered already. For example, his most explicit argument against extracosmic space is that it is what *can* receive physical matter, but that extracosmically physical matter cannot be received. He is committed to opposition by two further types of consideration, one being his conception of place. For once he rejects the obvious view that place is a three-dimensional extension he is left with the idea that a

---

[56] ibid. 251b19-28.

[57] Râzî, *Treatise on Metaphysics*, ed. Kraus, Cairo 1939, p. 133, 13-18, translated into French by S. Pines, 'Un fragment de Séleucus de Séleucie conservé en version arabe', *Revue d'Histoire des Sciences* 16, 1963, 193-209, at 197, reprinted in S. Pines, *Studies in Arabic Versions of Greek Texts and in Mediaeval Science*, The Collected Works of Shlomo Pines, vol. 2, Jerusalem and Leiden 1986.

[58] S. Pines, op. cit., 198-9.

thing's place is the inner surface of its physical surroundings. This at once makes it impossible for place to be extracosmic, or infinite. For there cannot in this sense be a place of, or outside, the cosmos, since the cosmos has no physical surroundings. Nor yet can a surrounding surface have a more than finite diameter.

The remaining objections stem from Aristotle's rejecting greater than finite quantities. His arguments are assembled in the *Physics* and *de Caelo*,[59] and most of them are directed against the view that a *body* could be infinite. This is all he needs in the *Physics* for his argument that the cause of unending stellar rotation cannot be housed in a *body*.[60] In the *de Caelo*, his concern is partly to deny the infinity of one particular body, the element which constitutes the stars, and so some proofs are confined to showing that a *rotating* body could not be infinite. But there are arguments of wider import, which would apply to space, and the most significant is an expression of the view that there cannot be infinite sub-collections.[61] As applied to space, this might be put by saying that if space extended for an infinity of miles, then there would be parts of space that were equally infinite, and a part cannot be equal to the whole. Nowadays, we have come to accept that there is no harm in this. The even numbered miles, as measured from here, would be infinite, just like the totality of miles. Does this mean, as Aristotle would object, that the collection of whole numbers is no greater than the collection of even numbers? Elsewhere I have pointed out that there is a sense in which it is greater and a sense in which it is not. It is greater in that only it contains *extra* numbers which are not in the other collection. But if one imagines the two collections stretching away in columns from before one's eyes, with 1,2,3,4 matched against 2,4,6,8, then neither collection will be greater in the sense of *sticking out beyond the far end of* the other. For neither collection has a far end. Distinctions of this kind, however, were not drawn until the fourteenth century.[62]

Aristotle's arguments against a greater than finite quantity are developed by Alexander in a little treatise against the infinity of existent things,[63] where he constructs an argument for finitude out of Aristotelian materials.[64] He claims that Aristotle defines 'whole' and 'all' (he should have said 'whole' and 'complete')[65] as that which has none of its parts missing. But in that case the whole universe must be limited (*peperasmenon*), at least if 'having none of its parts missing' is

[59] Aristotle *Phys.* 3.4-8, esp. 5; *Cael.* 1.5-7.

[60] Aristotle *Phys.* 8.10: if no bodies are infinite, none can contain infinite power, so the infinite power needed for the heaven's endless rotation must come from something *incorporeal*.

[61] ibid. 3.5, 204a20-6.

[62] Richard Sorabji, *Time, Creation and the Continuum*, 217-18.

[63] Alexander *Quaestiones* 3.12,101,10-107,4.

[64] ibid. 3.12,102,15-18.

[65] Aristotle *Metaph.* 5.16, 1021b12; 5.26, 1023b26; *Phys.* 3.6, 207a8-10. This definition of *whole* goes back to Plato *Theaetetus* 205A4-5.

also the definition of the limited.[66] The last move appears to exploit once again Aristotle's definition of limit (*peras*). For according to Aristotle,

> We call a limit the terminus (*eskhaton*) of each thing, i.e. (*kai*) the first [thing] outside which you cannot take anything and the first inside which everything lies.[67]

The case against infinite space is attacked once again by Cleomedes. He objects to people who say that, if there is infinite void outside the cosmos, there will have (absurdly) to be infinite body. There are at least two arguments deriving from Aristotle which purport to show this.[68] But since Cleomedes' argument does not take direct account of either of them, I hesitate to say that he is responding to Aristotle here. In his reply, however, I do believe that he exploits one of Aristotle's arguments for the finitude of body: a body might be defined as that which is bounded by a surface (*epipedôi hôrismenon*), and an infinite boundary is impossible.[69] Cleomedes' point is that there is no conceptual restriction on the extent of void, but that the infinity of body does not follow, because there *is* a conceptual restriction on the extent of body, namely, Aristotle's requirement that a body be bounded by a surface.

> It is also simple-minded of them to say that, if there is void outside the cosmos, it will have to be infinite, and that if the void outside the cosmos is infinite, there will have to be infinite body. For it does not follow from the infinity of the void that body is infinite too. For one's thought of the void does not give out anywhere, whereas in the conception (reading: *ennoia*) of body, being limited (*to peperasmenon*) is immediately included.[70]

---

[66] This is more elaborate than Aristotle's argument at *Phys*. 3.6, 207a14-15 that being complete (*teleion*) implies having a limit (*peras*).

[67] Aristotle *Metaph*. 5.17, 1022a4-5. For a different interpretation, see Robert B. Todd, 'Alexander of Aphrodisias and the case for the infinite universe (*Quaestiones* III 12)', *Eranos* 82, 1984, 185-93, esp. 191.

[68] One argument is that a space *eternally* capable of receiving body will have to receive it eventually. This is put into the mouth of his opponents by Aristotle, *Phys*. 3.4, 203b28-30, and recorded also by Aristotle's pupil Eudemus, ap. Simplicium *in Phys*. 467,32-5. This is the reference suggested by R. Goulet, *Cléomède: théorie élémentaire*, Paris 1980, p. 186, n. 46, and Robert B. Todd, 'Cleomedes and the Stoic concept of void', *Apeiron* 16, 1982, 129-36, at p. 131 and n. 29. The second argument is that if there was only a *finite* amount of body in an infinite void, it would all have dispersed, there being no obstacles to stop it. This argument is drawn by Epicurus' *Letter to Herodotus* (in Diogenes Laertius *Lives* 10) §42 from Aristotle *Phys*. 4.8, 215a19-22.

[69] Aristotle *Phys*. 3.5, 204b5-7. Aristotle is trading on the fact that lines and surfaces are also bounded, by points and lines respectively.

[70] Cleomedes 12,6-12. It is because void has no boundary that (8,7-14) it has no shape. For a different interpretation of Cleomedes on both issues, see Robert B. Todd 'Cleomedes and the Stoic concept of void', *Apeiron* 16, 1982, 129-36.

Cleomedes' assertion that space is conceptually unrestricted receives an answer in a different, but related, content from Philoponus, who argues for once on Aristotle's side. He is parrying an attack on his own un-Aristotelian view that place is a three-dimensional extension, whose essence does not require it to be filled with body. It is complained that this extension could have no limit or boundary (*peras*) and so would have to be infinite. Unlike Cleomedes, Philoponus cannot easily allow such an infinity, given his Christian arguments against an infinity of past time for the history of the universe. But he replies first that we can conceive of a surface which limits or bounds place as opposed to body, and secondly that, even if that were not so, place can or must derive such a limit or boundary from the body which it houses:

But they again raise a difficulty for our view, and say that if there is such a thing as this extension (*diastêma*), empty so far as its own nature is concerned (*tôi idiôi logôi*), in order to receive bodies, it will have to be infinite, for it will have no boundary (*peras*). For the boundary of three-dimensional things is a surface, and what could serve as the surface of such an extension, seeing that surfaces subsist only in bodies. Thus if the whole body of the heavens up to its outermost surface is spread through it, but it itself cannot have a boundary, it must spill outside the heavens infinitely far, which is both irrational in itself, and also sufficiently refuted by Aristotle both in this work and in *On the Heavens*.

(582,28) But I do not know how such a difficulty could even seem to be persuasive. For first, just as one can conceive such a thing as being three-dimensional, so one can conceive a surface appropriate to it. Secondly, even if one could not conceive a surface appropriate to it, it does not follow of necessity that the void stretches to infinity. For since the place (*topos*) of bodies is something that exists, it exists for as great a distance as the cosmic bodies can spread, and has a boundary together with (*sumperatousthai*) the boundaries of those bodies. Compare the interior of a jar: if it is thought of as empty, and various bodies are put in it, not continuous with each other, but merely in contact, so that each occupies a part of the place inside, then each part of the empty space must (*anankê*) have a boundary together with (*sumperatousthai*) the body which it has received, and the entire empty space up to the concave surface of the jar must have a boundary (*peratousthai*), since all the bodies inside have boundaries out as far as that concave surface. It is the same with the whole universe: each of the spheres in it takes up a bounded (*peperasmenon*) part of the empty space, and the whole of empty space has a boundary together with the whole cosmos. It has a boundary (*peras*), only not a boundary of its own. For it is not impossible to conceive of a surface for the whole of empty space, just as there certainly is a boundary for a part of it, namely, the boundary of the bodies which that part includes, and, as I said, we can conceive the surface of the outermost body as coinciding with the boundary of the whole of empty space.[71]

[71] Philoponus *in Phys.* 582,19-583,12.

CHAPTER NINE

Motion in a vacuum: Stoics, Epicureans and Philoponus against Aristotle

The chief remaining reason for Aristotle's rejection of extracosmic space lay in the fact that he denied the possibility of any void or vacuum. A high proportion of his objections in the *Physics* connect vacuum with *motion*, and a special group of these is introduced as turning the tables on his predecessors.[1] They had thought that vacuum was necessary to make room for motion. He turns things round by arguing that vacuum would actually make motion impossible.

I shall take a selection of Aristotle's arguments, and consider them in increasing detail. I shall deal briefly with those that are not concerned with motion, pass on to the motion arguments, and finish with a group that require special attention, if we are to understand the nature of the Stoic response to Aristotle.

There are two notable arguments among those unconcerned with motion. One is that void is simply place conceived as a three-dimensional extension empty of body. But that conception of place as extension has already been rejected by Aristotle.[2] The other argument has already been encountered in Chapter 5. Void would have to penetrate a body that filled it, but would then be a redundant addition to that body's volume. Moreover, if two such things could be in the same place, why not any number?[3] The redundancy argument is surprising, because it is Aristotle himself who explains why some concept of place is thought indispensable. It would not be thought indispensable if no bodies moved. But when we see bodies moving, we think of their place as something static, left behind when they move, and distinct from them.[4] We are surely also inclined, then, to think of it as distinct from their volume.

Aristotle's arguments from motion are much more complex. He first

---

[1] Aristotle *Phys.* 4.8, 214b28-31; 215b21-6.
[2] ibid. 4.7, 214a16-22; 4.8, 214b28.
[3] ibid. 4.8, 216a26-b12; cf. Sextus *M* 10.27.
[4] Aristotle *Phys.* 4.4, 211a12-13; 211b14-17.

complains that the void is meant to be a cause (*aition*) of natural motion. Aristotle himself agrees that natural places have some explanatory power.[5] But the void, he objects, is not suited to cause any of the natural motions of elements up or down.[6] To this Philoponus replies, in a manner that will be more fully discussed in Chapter 12. His point is that the natural motions of the four elements are caused not by the places towards which the motions are directed, but by the tendency of the elements themselves, viewed as parts of the cosmic organism, to adopt the appropriate spatial relationships to each other. A head craned sideways tends to return to its correct position in relation to the shoulders, and this in no way depends on the causal efficacy of any external place.[7]

A second argument against motion in the void is that a vacuum would contain no differences, so not the difference of up and down on which natural motions depend, and without natural motion, there could be no motion at all.[8] Aristotle's argument will work best against *extracosmic* void, because a mere pocket of void within the cosmos ought not to exclude differences of up and down amongst its surroundings. As if to confirm that extracosmic void is what he is attacking, he adds the point, which is also found in *Cael*. 1.7,[9] that an *infinite* expanse would make motion impossible by providing no up and down. But we may still wonder if he has restricted his claim sufficiently. There could not be an up or down direction for the entire cosmos, if it were surrounded by void. But if a rock could be flung out beyond the edge of the cosmos (which in *Cael*. 1.8-9 he denies), why should not the downward direction for it remain, as before, the direction towards the centre of the cosmos? The central part of the argument runs as follows:

> But as regards natural [motion], how will there be any, when there are no differences in the void or in the infinite? For insofar as it is infinite, nothing will be up or down or central. And insofar as it is void, up will be no different from down, since there are no differences in the void, any more than there are in nothing. After all, the void is thought of as something non-existent and as a privation. Natural motion, however, is differentiated, so that it will need natural differences.[10]

Epicurus is obliged to find a way out of this argument, since he believes that atoms fall downwards in an infinite void. But since he also believes in a flat earth, he feels free to argue that the direction down in

---

[5] Natural motion shows that place has a power (*dunamis*), *Phys*. 4.1, 208b11; if the earth were dislodged from the centre, loose clods would fall not towards it, but towards the central place, *Cael*. 4.3, 310b3. See Chapter 12.

[6] Aristotle *Phys*. 4.8, 214b12-17.

[7] Philoponus *in Phys*. 632,4-634,2; cf. 581,18-31.

[8] Aristotle *Phys*. 4.8, 215a1-14.

[9] ibid. 4.8, 215a8-9; *Cael*. 1.7, 276a8-12.

[10] Aristotle *Phys*. 4.8, 215a6-12.

the infinite void is simply the direction from our head to our feet.[11]

A third Aristotelian objection to motion in the void is concerned with the artificially produced motion of projectiles. The argument turns on Aristotle's view that air or some other fluid is needed to make possible the motion of projectiles, and there is no such fluid in a vacuum. His question is what makes a javelin continue to move after the thrower's hand has come to rest. For the first few feet, the javelin is grasped and propelled by the hand. But when the work is no longer being done by the hand, it must, he thinks, be done by the air. He suggests two ways in which the air might operate,[12] one ascribed to 'some people', and one the way which he himself subsequently accepts. In that subsequent discussion, his solution is that the thrower's hand imparts to the air the marvellous power of being an unmoved mover, or at least a no-longer-moved mover. In fact, there is a series of pockets of air behind the projectile which acquire this power in turn, and they move the projectile on.[13] Anneliese Maier regarded this group of Aristotelian ideas as one of the two most significant and most fatal for the rise of classical physics.[14] It was accepted by such major commentators as Alexander,[15] Themistius and Simplicius. But Philoponus was not only to reject it, but to have enormous fun subjecting it to ridicule. If this were the mechanism, an army would not need to touch its projectiles. It could perch one on a thin parapet and set the air behind in motion with 10,000 pairs of bellows. The projectile should go hurtling towards the enemy, but in fact it would drop idly down, and not move the distance of a cubit.

If [the thrower] transmits no power (*dunamin endidonai*) to the stone, and it is only by pushing the air that he moves the stone, or the bowstring moves the arrow, what advantage was there in the stone's being in contact with the hand, or the bowstring with the notch in the arrow? It would have been possible, without touching these, to stand the arrow, for example, on the tip of a stick which would serve as a thin line, and similarly for the stone, and with ten thousand machines to set a great quantity of air in motion from behind. Clearly, the more air was set in motion, and the greater the force, it ought to push it more and shoot it further. But in fact, even if you were to stand the arrow or stone on a line that really lacked breadth, or on a point, and were to set in motion all the

---

[11] Epicurus *Letter to Herodotus* (in Diogenes Laertius *Lives* 10) §60.

[12] Aristotle *Phys.* 4.8, 215a14-19.

[13] ibid. 8.10, 266b27-267a20.

[14] Anneliese Maier, 'Ergebnisse der spätscholastiken Naturphilosophie', *Scholastik* 35, 1960, 161-87, translated into English in Steven D. Sargent, ed., *On the Threshold of Exact Science*, Philadelphia 1982, see pp. 148-9.

[15] Alexander's *Physics* commentary is lost, but the evidence comes from an Arabic text cited by S. Pines, 'Omne quod movetur necesse est ab alio moveri', *Isis* 42, 1961, 21-54, at 30.

air from behind with all your thrust, the arrow would not move even the distance of a cubit.[16]

What Philoponus substitutes for Aristotle's theory will be the subject of Chapter 14. His innovative idea is indicated in the opening words of the quotation, that the thrower transmits a force directly into the projectile, not a power into the air behind it. This impetus is both internal to the projectile and impressed from without. Its introduction has been called a scientific revolution,[17] and for projectiles it was still taken for granted as the correct theory by Galileo eleven hundred years later. In fact, it is not correct because it overlooks the idea of inertia, the idea that a body will persist in the same state of rectilinear motion or of rest without any force at all. Force is needed only to change that state. It has been a matter of controversy who was the first to formulate that new idea of inertia, whether Galileo, Descartes or Newton, or Newton's successors.[18]

In a fourth objection to motion in the void, Aristotle returns from the motion of projectiles to natural motion, such as that of falling bodies, and argues that if resistance to motion were reduced to nothing, as it would be in a vacuum, speed would have to rise, absurdly, so that it exceeded any ratio:[19] in effect it would be infinite. Further, in a vacuum all bodies would have to have the same speed.[20] Interpretations have ranged from saying that in relating speed to resistance Aristotle is for the first time formulating laws of motion,[21] to saying that he does not even believe in the relationships, but is showing what would follow from the mistaken assumptions of his *opponents*.[22] Some of the grounds for dissociating Aristotle from the

[16] Philoponus *in Phys.* 641,16-26. There is a translation of much of 639,3-642,9 in Morris R. Cohen and I.E. Drabkin, *A Source Book in Greek Science*, Cambridge Mass. 1958, 221-3.

[17] Thomas Kuhn, *The Structure of Scientific Revolutions*, Chicago 1962, 2nd ed. 1970, 120.

[18] Richard S. Westfall, 'Circular motion in seventeenth-century mechanics', *Isis* 63, 1972, 184-9.

[19] Aristotle *Phys.* 4.8, 215b22-216a4.

[20] ibid. 216a11-21.

[21] e.g. I.E. Drabkin, *American Journal of Philology* 59, 1938, 60-84. The view is attacked by H. Carteron, *La notion de force dans le système d'Aristote*, Paris 1923, introduction, 11-32; G.E.L. Owen, 'Aristotle: method, physics and cosmology', in C.C. Gillispie, ed., *Dictionary of Scientific Biography*, New York 1970, 250-8; id., 'Aristotelian mechanics', in A. Gotthelf, ed., *Aristotle on Nature and Living Things: philosophical and historical studies*, Pittsburgh 1986. The first is translated into English in Jonathan Barnes, Malcolm Schofield, Richard Sorabji, eds, *Articles on Aristotle 1*, London 1975 (for 'lover' read 'magnet' on p. 169!); the second and third are reprinted in Owen's *Logic, Science and Dialectic*, ed. M. Nussbaum, London 1986. In reply, see now Edward Hussey, *Aristotle's Physics, Books III and IV*, Oxford 1983, Additional Notes B, 'Aristotelian dynamics'.

[22] Thomas Aquinas, *in Phys.* 4, *lectio* 12,536; Michael Wolff, *Fallgesetz und Massebegriff*, Berlin 1971, 27-8; James A. Weisheipl, 'Motion in a void: Aquinas and Averroes', in A. Maurer, ed., *St. Thomas Aquinas 1274-1974: commemorative studies*, Toronto 1974, 1,467-88, repr. in his *Nature and Motion in the Middle Ages*, Washington

claim of infinite speed will emerge in Chapter 15. His objection to infinite speed, though not expressed, is taken by Philoponus to be that the moving body would have to be at the starting point and finishing point simultaneously.[23] But Aristotle would find it equally objectionable if infinite speed took the form of a discontinuous disappearance from the starting point and reappearance without delay at the finishing point. Something of this latter kind may well have been accepted by Epicurus.

In his response to Aristotle, Epicurus acknowledged, probably for more than one reason, that his atoms would travel through the void at equal speed.[24] But instead of infinite speed, he had them move through it with a speed as 'quick as thought'.[25] He might have filled this idea out in various ways. If, as appears, he came to believe in minimal times as well as minimal distances,[26] an atom would in a sense move infinitely fast if it disappeared from one minimal position and reappeared in the next at the next minimal unit of time. For longer journeys it would be feasible to propose a less than infinite speed for atoms of one minimal unit of distance for each minimal unit of time. It is only individual atoms that would move at this speed. Compound bodies, he explains, move more slowly, because their component atoms zig-zag.[27]

Epicurus' reply takes the form of accepting as possible what Aristotle thought impossible, but a more thoroughgoing answer was provided by Philoponus. For one thing, he argues, there must be something wrong with Aristotle's view, because the rotation of the heavens encounters no resistance, yet is still finite in speed. But he also offers a *diagnosis* of Aristotle's error: all motion takes time, he says. What you would achieve by removing resistance is not the necessity for time, but the necessity for *extra* time spent in overcoming the resistance.[28] In both answers Philoponus is followed by Avempace.[29] Philoponus finally examines the details of Aristotle's

---

D.C. 1985. This view adds an extra dimension to the debate in Carteron and Owen, and presents Aristotle's argument as dialectical in a stronger sense than that accepted by Owen.

[23] Philoponus *in Phys.* 681,18-22.

[24] The reasons are not stated, but Furley convincingly suggests that the other reason is Aristotle's point that over a single atomic space a slower body would travel *less* than an atomic space (a contradiction in terms), while a faster body travelled it all: Aristotle *Phys.* 6.2, 232b20-233a12; 233b19-33: David Furley, *Two Studies in the Greek Atomists*, Princeton 1967, 120-1.

[25] Epicurus *Letter to Herodotus* (in Diogenes Laertius *Lives* 10) §61.

[26] Sextus *M* 10,142-154; Simplicius *in Phys.* 934,26; cf. P. Herc. 698, fr. 23*N*, published in W. Scott, *Fragmenta Herculanensia*, p. 290. See Richard Sorabji, *Time, Creation and the Continuum*, 375-7.

[27] Epicurus *Letter to Herodotus* (in Diogenes Laertius *Lives* 10) §62.

[28] For these two answers see Philoponus *in Phys.* 690,34-691,5 (translated in Chapter 15) and *in Phys.* 681,17-30. (Much of 678,24-684,10 is translated into English in Cohen and Drabkin, op. cit., 217-21.)

[29] For the references to Avempace, see below.

argument. That argument attacks the idea that motion through a void would take time, by claiming that motion through a thin enough medium would (absurdly) take the *same* time. Philoponus complains that it is hard to see Aristotle's error, because densities are not measurable.[30] But if Aristotle were right that for a given weight of body the speed of fall varies in direct proportion to the density of the medium, then it would be plausible (*eulogon*) that for a given medium the speed of fall would vary in direct proportion to weight. And at least the latter can be measured and seen false by observation: if you double the weight and drop it from the same height, it will not fall twice as fast.[31] Evidently someone had anticipated the experiments usually attributed to Galileo. But Philoponus does not go quite as far as Galileo, for he does not think that the speed would actually be the same.[32]

The central defence of motion in a void, that removing resistance removes only the necessity for *extra* time, had a long subsequent history. It is next mentioned six hundred years later in the Islamic world, as already indicated, by Avempace (Ibn Bâjja, died A.D. 1138). Averroes (1126-1198), who tells us this, makes no mention of Philoponus,[33] and himself dissents from Avempace's view, but Thomas Aquinas (c.1224-1274) agrees, and so thereafter do several of his followers.[34] In the following century, the Jewish philosopher Crescas (1340-1410) also agrees with Avempace,[35] and applies the idea of *extra* time in addition to an argument of Aristotle's about infinite power that

---

[30] Philoponus *in Phys.* 683,1-4.

[31] ibid. 683,5-25.

[32] ibid. 678,29-679,23; 681,3-17; 683,23-4.

[33] Averroes, *Long Commentary on Physics* 4 (Aristotle text 71, 215a25-b20), Latin in Juntine edition, 1562, repr. Frankfurt 1962, vol. 4, p. 160, translated in E.A. Moody, 'Galileo and Avempace, the dynamics of the leaning tower experiment', *Journal of the History of Ideas* 12, 1951, 163-93, 375-422, at 226-7; *Middle Commentary on Physics* 4, translated from Hebrew version, with other texts on Avempace, by H.A. Wolfson, *Crescas' Critique of Aristotle*, Cambridge Mass. 1929, 403-7. For discussion, see E.A. Moody, op. cit. (abbreviated in P.P. Wiener and A. Noland, eds, *Roots of Scientific Thought*, New York 1957, 176-206); Edward Grant, 'Aristotle, Philoponus, Avempace and Galileo's Pisan dynamics', *Centaurus* 11, 1965, 79-95; James A. Weisheipl, 'Motion in a void: Aquinas and Averroes', in A. Maurer, ed., *St. Thomas Aquinas 1274-1974: commemorative studies*, Toronto 1974, vol. 1, 467-88, repr. in his *Nature and Motion in the Middle Ages*, Washington D.C. 1985.

[34] Thomas Aquinas *Scriptum super libros Sententiarum*, in lib. 4, dist. 44, q. 2, a. 3, ql. 3 ad 2; *in Phys.* in lib. 4, lectio 12,536; *in Cael*, in lib. 3, lectio 8, n. 9, passages translated and discussed in James A. Weisheipl, 'Motion in a void: Aquinas and Averroes'. The followers of Thomas whom he cites are John Capreolus, Domingo de Soto, John of St. Thomas, Cosmo Alamanno. For natural motion Thomas thinks air is not necessary, and he further believes that Aristotle would agree, his arguments to the contrary being merely *ad homines*, Thomas Aquinas *in Phys.* in lib. 4, lectio 12,536, as above.

[35] Hasdai Crescas, in *Or Adonai (The Light of the Lord)*, translated in part in H.A. Wolfson, *Crescas' Critique of Aristotle*, Cambridge Mass. 1929, proposition 1, part 2, p. 185, with notes.

will engage us in Chapter 15.[36] Philoponus' authorship of this style of argument was recognised once again by Pico della Mirandola at the beginning of the sixteenth century.[37] And at the end of the century, in his early *de Motu* (c.1590), Galileo acknowledges Philoponus, Thomas Aquinas and Scotus as recognising finite velocity in a vacuum.[38] But he thinks that Philoponus reached the truth only by faith (*fides*), and it is true that Philoponus' account of velocity in a vacuum falls far short of Galileo's, for reasons which others have brought out.[39] On the other hand, it must be said that at the time of the *de Motu*, Galileo himself had not yet reached the idea of the equal velocity of freely falling bodies in a vacuum and meanwhile the point about *extra* time remains Philoponus' own.

I come now to a final group of objections to motion in a void. They, and the Stoic response to them, will occupy the remainder of the chapter. The arguments turn on the principle, encountered in Chapter 8, of Buridan's ass, or, as it should rather be called, Aristotle's man. In a void why should things move in one direction rather than another? Once moving, why should they stop here rather than there? Sometimes instead of *denying* a preference, Aristotle *suggests* one: the lack of differentiation would make things rest rather than move,[40] and the void's uniform yieldingness would make them disintegrate rather than stay intact. The arguments would be most persuasive, if restricted to the behaviour of the entire physical cosmos in an extracosmic void. But Aristotle implies no such restriction, and at one point suggests that a body moving in the void might be stopped by an obstacle.[41] He cannot

---

[36] Would infinite power in a finite body produce motion in no time at all, as Aristotle protests, *Phys.* 8.10, 266a24-b6? No, says Crescas, it would only obviate the need for *extra* time beyond the irreducible minimum: op. cit. proposition 12, part 2, p. 171, with Wolfson's notes. See Chapter 15 below and H.A. Davidson, 'The principle that a finite body can contain only finite power', in S. Stein and R. Loewe, eds, *Studies in Jewish Religious and Intellectual History presented to A. Altmann*, Alabama 1979, 75-92.

[37] See Charles Schmitt, *Gianfrancesco Pico della Mirandola (1469-1533) and his Critique of Aristotle*, The Hague 1967, 146-9; 154-5.

[38] Galileo *de Motu* tr. I.E. Drabkin, Madison 1960, pp. 49-50. I am not able to comment on the report that a copy of a Philoponus text currently kept in a Swiss bank has annotations in the hand of Galileo.

[39] First, in talking of weight, he thinks of gross weight. For Galileo's concept of *specific* weight, which takes into account the volume of a body and permits a direct mathematical comparison between the falling body and the medium through which it falls, was closed to him by his Aristotelian inheritance. Secondly, he thinks, wrongly, that there will be a marginal difference in speed of fall, according to the weight of the body. So Edward Grant (1965); Michael Wolff, 'Philoponus and the rise of preclassical dynamics', in Richard Sorabji, ed., *Philoponus and the Rejection of Aristotelian Science*, London and Ithaca N.Y. 1986. But on one of Wolff's charges I would defend Philoponus, for in Chapter 12 I shall present his comparison of elemental motion with the motion of an organism's parts not as an ad hoc expedient, but as a considered development in a long tradition.

[40] A similar explanation, says Aristotle, had been given of the earth's resting. Elsewhere, *Cael.* 2.13, 295b12 (cf. Hippolytus *Refutation of all Heresies* 1.6.3) he cites Anaximander. Slightly different is the theory of Plato *Phaedo* 109A (see also *Tim.* 62D).

[41] Aristotle *Phys.* 4.8, 215a21-2.

be thinking of the entire cosmos meeting an obstacle, but must have in mind pieces of matter, such as stray rocks, This invites the question, which I shall argue Chrysippus raised, why the normal causes of motion should not operate on an individual rock to give it its normal direction and destination, if part of its journey took it through a void region. The rock would still have an inner nature and a natural destination at the centre of the cosmos to account for its centripetal passage through a limited portion of void.[42]

It is important to be clear which arguments of this type Aristotle does bring, and which he does not. He is not interested in the question, 'Why should a body *stay still* in the void rather than move?' At most he asks at one point, 'How [not why] will a body placed in the void move *or stay still (menei)*?'[43] But the brief reference to staying still is not what occupies him, because his opponents are people who introduce the void as the cause (*aition*) of *motion*,[44] or as necessary (*anankaion*) for *motion*.[45] And he replies that void would *rather* produce the *opposite* result (*tounantion mallon*),[46] i.e. *staying still*. Because of this, he argues in what follows not merely negatively against motion in the void, but positively for the proposition that a body in the void must stay still (*anankê êremein*).[47]

Aristotle's insistence that a body in the void will *rather* stay still is an insistence that it will stay still rather than move as an intact unit. This does not at all prevent him from raising a *different* question: why does it remain intact rather than disintegrating in every direction.[48] There is a puzzle about this last question, because Aristotle's use of the word *oisthêsetai* might suggest that he is charging his opponents with a view he has earlier declared absurd,[49] that a thing will move (*oisthêsetai*) as an intact unit into all of the void. But it is philosophically unlikely that he means that, or would have been taken to mean it, because such a charge does not look as if it would stick. The puzzle of Buridan's ass might be stated in at least three ways, ways which are all mentioned by Aristotle in another passage.[50] First the ass equally hungry and thirsty and equidistant between equal quantities of food and drink could not move. In that form, the puzzle has some seductive power. Less plausible would be a second claim that the ass would explode in both directions, although such an idea is

---

[42] For Chrysippus ap. Plutarchum *Sto. Rep.* 1055B-C (*SVF* 2.550), see below. The question is raised also by David Furley, 'Aristotle and the atomists on motion in a void', in Peter K. Machamer and Robert G. Turnbull, eds, *Motion and Time, Space and Matter*, Ohio State University 1976.

[43] Aristotle *Phys.* 4.8, 214b21-2; *pôs gar oisthêsetai ... ê menei.*

[44] ibid. 214b15-17.

[45] ibid. 214b29.

[46] ibid. 214b28-31; repeated 216a21-6.

[47] ibid. 214b32.

[48] ibid. 215a22-4.

[49] ibid. 214b19.

[50] Aristotle *Cael.* 2.13, 295b15-17; 296a7-8.

recognised in Aristotle's other passage, and would gain plausibility when applied to the different example of a body in the yielding void. There is, however, as the other passage makes clear, no persuasive force at all in a third suggestion, that the ass would move as an intact unit in both directions simultaneously. I have been reminded of Stephen Leacock's *bon mot*, 'Lord Ronald ... flung himself upon his horse, and rode madly off in all directions.'[51] The relevant arguments run as follows:

(214b17-22) Again, if whenever there is void, there is something like a place deprived of body, where will the body placed in it move to (*pou*)? For it will not move into all of it (*eis hapan*). And the same argument holds also against those who think that place is something separate into which a thing moves. For how will the thing placed in it move or stay still (*menei*)?

(214b28-215a1) Those who say that void exists as something necessary (*anankaion*) for motion get the opposite result rather (*tounantion mallon*), if you look at it, that not a single thing will be able to move, if there is void. For just as some people say that the earth stays still because of a homogeneity, so things would have to stay still (*anankê êremein*) in the void, since there is no place to which (*hou*) they can move more or less than to another, in that, insofar as it is void, it contains no differences.

(215a19-22) Again, no one could say why a thing, once moved, would halt anywhere. For why here rather than there? So either it will stay still (*êremêsei*), or it must move *ad infinitum*, unless some superior obstacle meets it.

(215a22-24) Again, as things are, people believe that things move into the void because it is yielding. But in the void it is like that in every direction (*pantêi*) equally, so things will move in every direction.

Of the arguments in this set, 'Why halt anywhere?' will turn out in Chapter 15 to create difficulties for Aristotle. But it was to be endorsed and exploited for his own purposes by Epicurus,[52] while 'Why not disintegrate?' and 'Why any direction?' were taken up, I believe, by the Stoics. In saying this, I am diverging from the thesis of F.H. Sandbach that the Stoics were ignorant of Aristotle's views. Cleomedes, in a passage to be quoted, cites two Stoic replies to the question of

---

[51] Stephen Leacock, *Nonsense Novels, Gertrude the Governess*. Thanks to David Sedley for the quotation, and to Eric Lewis for the rival interpretation of 215a22-4 and discussion of it. The disintegration argument is not noticed in Aristotle in David Hahm's pioneering book (*The Origins of Stoic Cosmology*, Ohio State University 1977, 119, 121), although his discussion of the Stoics on this subject is by far the most thoroughgoing I know (109-24, 140-3, 165-8, 249-59).

[52] Epicurus *Letter to Herodotus* (in Diogenes Laertius *Lives* 10) §42. For Avicenna's endorsement, see Chapter 15.

disintegration, although in the first, he combines the Aristotelian question of what could hold the cosmos together (*sunekhein*) with the un-Aristotelian question of what could stop it falling down in extracosmic void. Both questions are answered by saying that the only down there is is the centre of the cosmos.[53] The parts of the cosmos do indeed incline (*neuein*) towards this, but this means both that there is no other downward direction in which they can fall in extracosmic void and that the centripetal inclination holds the cosmos together. This Stoic solution is permeated with Aristotelian and anti-Platonic thought. First, the equation of down with centre is Aristotle's distinctive contribution. It is his reply to Plato's complaint that in a spherical cosmos it is geometrically impossible to find a down, and to his further complaint that a falling rock is attracted not so much by a particular place (down) as by a tendency to join the main mass of earth by the principle of like-to-like.[54] Aristotle's ingenious reply to Plato, that the *centre* can serve as down, is *rejected* by Epicurus, so that in accepting it, the Stoics are going back to *Aristotle*. Of course, the Stoics diverge from Aristotle in making all the elements, even fire and air, incline toward the centre. But first this divergence is required in order to deal with the problem posed by Aristotle himself as to why the cosmos does not disintegrate, if (contrary to his view) it is surrounded by vacuum. Secondly, the divergence had already been licensed in the earliest days of Stoicism by the Aristotelian school in the person of Strato.[55] And thirdly, the Stoics give to fire and air not one, but two tendencies, not only a downward one, but also, in conformity with Aristotle, an upward one as well.[56] Cleomedes (10,15-23) puts the Stoic view as follows:

> But, they say, if there were void outside the cosmos, the cosmos would move through it, since it has nothing capable of holding it together (*sunekhein*) and propping it up (*hupereidein*). But we shall say that it is

[53] This answer is a great improvement on the one apparently attempted by Zeno of Citium, according to Arius Didymus fr. 23, ap. Stobaeum *Eclogae* 1.19.4 in *Dox. Gr.* 459-60 (*SVF* 1.99), according to which the centripetal inclination directly explains why the cosmos stays still. See further on this below.

[54] Aristotle *Cael.* 4.1, 308a17-29, answering Plato *Tim.* 62E. Plato had complained that a man walking round the circumference of a sphere would successively call the same point up, when it was where he was standing, and down, when he had reached the opposite side of the sphere.

[55] Strato, frs 50-2 Wehrli (=Simplicius *in Cael.* 267,29; 269,4; Stobaeus *Eclogae* 1.14.1 in *Dox. Gr.* 311). Strato (died c.269 B.C.) was contemporary with the founder of Stoicism, Zeno (died 262).

[56] See Chapter 6, n. 38, and the different analysis by Michael Wolff cited there. The ascription of an upward motion is associated in the sources with the description of fire and air as weightless: Plutarch *Sto. Rep.* 1053E (*SVF* 2.434 and 435); Cicero *ND* 2,115-17; Arius Didymus fr. 23, ap. Stobaeum *Eclogae* 1.19.4, *Dox. Gr.* 459-60 (*SVF* 1.99). The Plutarch passage tells us that in another work Chrysippus further describes air and fire as neither heavy nor light, but this third description fits perfectly well with the others, on the interpretation given in Chapter 6: the two opposite tendencies balance each other, so that fire and air neither crash downwards, nor fly off into outer space. For the claim that fire

impossible for it to move through the void, for it inclines (*neuein*) to its own centre and this, the point to which it inclines, is its down. If centre and down were not identical for the cosmos, it would move down through the void. The above will be demonstrated in the discussion of motion towards the centre.

Cleomedes' second reply to the threat of disintegration invokes a holding power (*hexis*), which holds the cosmos together:

They also say that if there were void outside the cosmos, substance (*ousia*) would be diffused infinitely far through it, scattered and dispersed. But we shall say that this cannot befall it either, for it has a holding power (*hexis*) that holds it together (*sunekhei*) and keeps it fast (*suntêrei*). And whereas the void that surrounds it does nothing, substance employs an overwhelming power (*dunamis*) to keep itself fast, both when it contracts and again when it is diffused in the void in accordance with its natural changes, as it is sometimes diffused into fire and sometimes impelled into forming a [new] cosmos.[57]

These two answers to the disintegration question, the centripetal inclination and the holding power, go back, I believe, to a very much earlier Stoic, Chrysippus. In a direct quotation supplied by Plutarch,[58] Chrysippus gives one of Cleomedes' two explanations of why the cosmos does not disintegrate: its parts incline (*neuein*) towards the centre. Chrysippus does not explicitly speak in this passage of the other idea, the holding power (*hexis*). But in another text discussed in Chapter 6, Chrysippus is credited by name with that same doctrine. The whole of substance (*ousia*) is held together (*sunekhein*) by an interpenetrating gas, *pneuma*.[59] This is the same doctrine, because this gas, according to the Stoics, in one form is God, in another Reason, in another soul, in another nature and in another what we are looking for, namely, a holding power (*hexis*).[60] This is why the holding together function is ascribed in some reports to *hexis* and in others to *pneuma* or to God.[61] The two solutions, the *pneuma* and the centripetal inclination, are closely related to each other. For we are told that there is a centripetal motion in *pneuma* and that it is by this centripetal

---

and air, along with all bodies, also have a *downwards* tendency, see Arius Didymus loc. cit.; Plutarch op. cit. 1054E (*SVF* 2.550); Achilles Tatius *Isagoge* 9 (*SVF* 2.554); Cicero op. cit. 2.45-6; 115-17 (partly in *SVF* 2.549); Philo *Prov.* 2.56 (*SVF* 2.1143); Strabo 17.809.

[57] Cleomedes 10,24-12,5 (*SVF* 2.540).

[58] Plutarch's discussion runs from *Sto. Rep.* 1054B to 1055C. The drift is rather lost in the truncated and scattered excerpts at *SVF* 2.539, 550, 551.

[59] Alexander *Mixt.* 216,14-17 (=*SVF* 2.473).

[60] ps-Galen *Introductio medicus* 9, Kühn vol. 14, p. 697 (*SVF* 2.716); cf. Plutarch *Sto. Rep.* 1053F (*SVF* 2.449); Diogenes Laertius *Lives* 7.138 (*SVF* 2.634); Themistius *in DA* 35,32-4 (*SVF* 1.158); Galen *Comm. 5 in Hippocratis Epidemiarum 6*, Kühn 17B,250 (*SVF* 2.715); Galen *Quod Animi Mores Corporis Temperamenta Sequantur*, Kühn 4.783 (*SVF* 2.787); Galen *Definitiones Medicae* 95, Kühn 19,371 (*SVF* 2.1133).

[61] Alexander ap. Simplicium *in Cael.* 286,10-21.

motion that it holds things together.[62]

My claim that there is Aristotelian influence on the Stoics obliges me to consider Sandbach's important contrary thesis, and I thank him for discussion of his view.[63] His position is that Stoics down to Chrysippus (head of Stoa c.232-c.206 B.C.) were ignorant of the contents of Aristotle's works (except for his early publications as a member of Plato's Academy). They were not influenced by them, nor were they reacting to them. Even at a later date (135-51 B.C.), Posidonius had no more than second-hand knowledge of the contents of these works, apart from the *Meteorology*. I appreciate the provocativeness with which Sandbach expresses his view on the void. Concerning this topic, he considers the supposition of Aristotelian influence on the Stoics so implausible that he hesitates even to discuss it.[64]

I would take Sandbach's suggestion very seriously for some subjects other than physics (for example, for logic, or metaphysics). But in connexion with the void, I think the situation looks very different, when three pieces of evidence are taken into account. One is Aristotle's fullest attack on the void in *Phys.* 4.8, for it is this which I believe to have influenced Chrysippus in the passages quoted by Plutarch. A second is the Stoic defence of extracosmic void in Cleomedes and his sources, which has not received due attention, because von Arnim's collection of Stoic fragments excludes crucial passages. A third is the developments and recantations within the Aristotelian school after Aristotle, which account, I think, for some of the Stoic silences to which Sandbach draws attention. Sandbach has, however, raised the standards for identifying Aristotelian influence, which can never be done with more than probability, and I must explain the indications which I see in Plutarch's text that Chrysippus is indeed responding to the contents of Aristotle's *Phys.* 4.8. There are five reasons altogether, but I do not take it to matter, nor to be ascertainable, whether Chrysippus actually read Aristotle's text first hand, or whether he was responding to a tradition that preserved its contents faithfully.

All five points bear on the text in Plutarch which has already been cited.[65] The first point, however, cannot carry much weight. It is that it was Aristotle who originated the objection which worries Chrysippus, that a body would disintegrate in a surrounding void. But of course, Chrysippus could have encountered the objection elsewhere: the Epicureans thought that their multiple worlds would disintegrate eventually. What may be more significant is the second point, that it was again Aristotle who originated the objection that an infinite expanse contains no up or down, a point which Chrysippus uses against

---

[62] Nemesius *Nat. Hom.* ch. 2, p. 42 Matth. (=*SVF* 2.451).

[63] At a conference held in Cambridge to consider his new suggestions 23-25 May 1986.

[64] F.H. Sandbach, *Aristotle and the Stoics*, Cambridge Philological Society, supp. vol. 10, Cambridge 1985. The void is discussed on pp. 42-6, and the remark cited is on p. 42.

[65] Plutarch *Sto. Rep.* 1054B-1055C.

Epicurus' falling atoms.[66] This time it is less likely that Chrysippus learnt of the objection from Epicurus. For at least in what is extant, Chrysippus shows no knowledge of Epicurus' answer to it.[67] Thirdly, it was Aristotle again who originated the objection that if there were directions in a void, there would be no reason for things to be drawn in one direction rather than another, a point which Chrysippus repeats.[68] The argument had been used neither by Plato nor by Epicurus, who indeed thought that there *would* be a reason for his atoms to take the downward direction in a void. This last point, the lack of preferred direction in a void, makes Chrysippus like Aristotle in yet a fourth way. For like Aristotle, he has an answer to the question why a body surrounded by void does not move, and, at least in Plutarch's excerpts, worries only about why it does not disintegrate. And this is what he is next quoted as tackling by reference to centripetal inclination.[69] In treating immobility in the void as natural, Chrysippus is certainly not following Epicurus who expected *motion* in the void. Nor yet is he following his own predecessor Zeno of Citium, at least on the usual interpretation. For on that interpretation, Zeno had cited the centripetal inclination as one of two explanations of why the cosmos stays still,[70] a hopeless proposal, which is easily ridiculed by Alexander and Themistius.[71] Admittedly, Chrysippus did at one time endorse Zeno's other suggestion about why the cosmos stays still.[72] But here in the lack of preferred direction he has a better answer, which corresponds exactly to Aristotle's idea. Meanwhile, he wisely reserves the centripetal tendency for the different purpose of tackling the problem Aristotle thinks unanswerable, why there would not be disintegration. It looks as if Chrysippus' discussion is sensitive to Aristotle's, in a way that Zeno's was not.

Even now the probable connexions with Aristotle are not exhausted.

[66] ibid. 1054B-C (*SVF* 2.539); Cleomedes 16,13-16 (*SVF* 2.557).

[67] Epicurus *Letter to Herodotus* (Diogenes Laertius *Lives* 10) §60

[68] Plutarch *Sto. Rep.* 1054E (*SVF* 2.550).

[69] ibid. 1054E-1055A (*SVF* 2.550)

[70] Arius Didymus fr. 23 ap. Stobaeum *Eclogae* 1.19.4, in *Dox. Gr.* 459-60 (*SVF* 1.99). Michael Wolff has raised the possibility that the word *monê* in Arius Didymus refers not to the *immobility* of the cosmos and of the earth, but to their *permanence*, or to their *retention of shape* ('Hipparchus and the Stoic theory of motion', in J. Barnes, M. Mignucci, eds, *The Bounds of Being*, Naples, forthcoming), the latter being a meaning attested for Chrysippus by Arius Didymus, fr. 22, ap. Stobaeum *Eclogae* 1.19.3, in *Dox. Gr.* 459,14-15. But although this interpretation has the advantage of freeing Zeno from a bad argument, it does not yet deal with the testimony of Alexander and Themistius, it has to contend with the fact that the passage ends with a cognate word, when it says that the earth remains (*menei, Dox. Gr.* 460,5) in the same place (*epi tou topou toutou*), and the sense of *monê* might need to switch at 459,23 from permanence, which is the meaning more relevant to the cosmos, to retention of shape, which is the meaning more relevant to the earth.

[71] Alexander ap. Simplicium *in Cael.* 286,10-21, and ap. Simplicium *in Phys.* 671,8-12 (*SVF* 2.552); Themistius *in Phys.* 130,15-17.

[72] Achilles *Isagôgê* 4 (*SVF* 2.555): the equilibrium of heavy and non-heavy elements prevents its falling.

A fifth emerges from a further quotation by Plutarch. I earlier raised against Aristotle the objection that an individual rock in transit through a portion of void need not lose all direction. It should have its direction and destination determined by the normal causes. It looks as if it is precisely this point which Chrysippus makes in Plutarch's final quotation.[73] The quotation has been otherwise interpreted. For it has been seen as discounting the need for external causes of motion, such as the squeezing (*ekthlipsis*) upwards of lighter bodies postulated by the Epicureans.[74] But as addressed to Epicureans, Chrysippus' point would have no force, for he would not be able to say to them that, even on *their* principles, lighter bodies should rise, if surrounded by void. By contrast, if Chrysippus is addressing Aristotle, he can fairly claim that even on his *opponent's* views about the causes of motion, a moving body suddenly surrounded by void should maintain the direction it had previously. The quotation runs as follows:

'It is reasonable that the way in which any part moves when it is united in the natural way with the rest should be the way in which it would still move on its own, even if we were for the sake of argument to suppose in thought that it was in some void within the cosmos. Just as it would have been moving to the centre when held together from every side, so it will continue in this motion, even if for the sake of argument void is suddenly created around it.'[75]

The other four connexions with Aristotle are manifested in earlier parts of the same passage, as the following extracts show:

It is often said by him [Chrysippus] that the void outside the cosmos is infinite, and that the infinite contains neither beginning, middle, nor end. And this they [the Stoics] use especially to annihilate the independent downward motion of the atom postulated by Epicurus, by saying that there are no differences in the infinite by reference to which one thing can be thought of as coming to be up and another down.[76]

In the next extract, Plutarch is talking of the rejected idea that bodies move not to the centre of substance (*ousia*), but to the centre of the space (*khôra*) that surrounds substance:

About this [Chrysippus] has spoken often and said that it is impossible and contrary to nature. For there is no differentiation in the void by which bodies might be drawn here rather than there. Rather the coordination (*suntaxis*) of the cosmos is the cause (*aitia*) of motion, with all its parts inclining (*neuein*) and moving from all sides to its centre and middle. To show this, it is enough to quote a text from the second book on

[73] Plutarch *Sto. Rep.* 1055B-C (*SVF* 2.550).
[74] David Hahm, *Origins*, 125. I am most grateful to Hahm for discussion, which has helped me to sharpen my presentation of the evidence.
[75] Plutarch *Sto. Rep.* 1055B-C.
[76] ibid. 1054B-C.

Motion. He remarks that the cosmos is a perfect body, whereas its parts are not perfect because they exist not independently, but in relation to the whole. He expatiates on the motion of the cosmos. By means of all its parts, it moves naturally towards its own continuation (*summonê*) and its holding together (*sunokhê*), not its dissolution and fragmentation. He then adds: 'Thus the whole has a tension (*teinesthai*) and motion towards the same point, and the parts have this motion from the nature of body. So it is plausible that the first natural motion is motion towards the centre of the cosmos for all bodies, for the cosmos which thus moves towards itself and for its parts insofar as they are parts'.[77]

My claim has been that, regardless of whether Chrysippus had Aristotle's text before him, he had, unlike his predecessor Zeno, accurate enough knowledge of its contents, and was responding to them. Sandbach's evidence on the other side consists of two arguments in Aristotle which he believes to be concerned with extracosmic void and to be ignored by the Stoics. One of the two is the argument, recorded again by Aristotle's pupil Eudemus, that if void and place are of infinite extent then body will have to be so too. For on Aristotle's own principles, void and place that *eternally* have the capacity for receiving body will have to receive it, at least eventually.[78] But this argument is not described by Aristotle as an argument against extracosmic void, but as an argument by an opponent in favour of the infinite extent of body. So if it really is ignored, it will still not provide us with an example of the Stoics ignoring Aristotle's arguments against *extracosmic void*. But it must further be added that, on the prevailing interpretation of Cleomedes, the argument is in any case not ignored, but is taken up by him at 12,6-9, although I myself have expressed doubts above about this interpretation.

The other argument which is said to have been ignored is definitely not ignored by Cleomedes and his sources. For it is the familiar argument that void and place can receive body, whereas extra-cosmically body cannot be received.[79] Cleomedes even cites Aristotle as his opponent by name, and replies with his water vessel in the desert, in one of the many passages omitted from von Arnim's Stoic fragments.[80] Sandbach maintains that, if the Stoics did know of this argument, they would ignore it anyhow, because it rests on an assumption they reject, that what is possible must be actual eventually. But in analysing the argument in the preceding chapter, I found no trace of this assumption.

The neglected evidence of Cleomedes is important, even though he himself is too late a figure to be relevant to Sandbach's thesis, belonging as he probably does to some period in the first one and a half

[77] ibid. 1054E-1055A.
[78] Aristotle *Phys*. 3.4, 203b28-30; Eudemus ap. Simplicium *in Phys*. 467,32-5.
[79] Aristotle *Cael*. 1.9, 279a11-18.
[80] Cleomedes 10,6-15.

centuries A.D.[81] His evidence is important, because he is drawing on earlier Stoics. The one repeatedly mentioned is Posidonius, and he declares at the end of his work, In a remark now accepted as his, not a librarian's addition, that his course notes (*skholai*) are not his own, but are drawn from older or more recent works, mostly from those of Posidonius.[82] This acknowledgment appears to cover his whole work, both books of which are called *skholika*.[83] And it is confirmed by independent testimony that one of the arguments cited above from Cleomedes (Chapter 8) is to be found in Posidonius, the argument that extracosmic space is needed to accommodate the periodic conflagration.[84] I have tried to argue that some of the ideas in Cleomedes, those on the disintegration of the cosmos and on the absence of locations in the void, go back further still to Chrysippus. And the anti-Platonic equation of down with centre goes still further back to Zeno. But it would be interesting enough if the attention to Aristotle could be referred with some degree of likelihood as far back as Posidonius.

Not only does Cleomedes explicitly name Aristotle as his opponent in one argument (the one about the water vessel), but at least seven of the arguments which he draws from his predecessors, and perhaps more, appear to refer to Aristotle's ideas. I have so interpreted (i) his discussion of the water vessel (10,6-15), (ii) the objection he meets concerning cosmic disintegration (10,15-12,5, partly in *SVF* 2.540), (iii) his equation of down with the centre of the cosmos (10,15-23), and (iv) his claim that void differs from body in not including the idea of limitation (*peperasmenon*) (12,9-12). I would further so interpret, (v) his point that there are no directions in the void (16,13-16=*SVF* 2.557). His Posidonian argument (vi) for void to accommodate the conflagration (6,11-25, partly in *SVF* 2.537) is directed against those who see the cosmos as surrounded by *nothing* and these are surely Aristotelians. His further argument (vii) that void cannot be limited, because it would need to be limited *by* something (14,13-16,12) is also first recorded by Aristotle. Finally, some see an Aristotelian reference too in (viii) the argument he rejects, that infinite void would imply infinite body (12,6-9).

In spite of all this, Sandbach's salutary doubts should teach us caution. For the Stoics are more likely to have learned the 'limited *by*' argument from the Epicureans whose favourite argument it is said to be, than directly from Aristotle. For like Epicurus, they show no awareness of Aristotle's admittedly brief answer to it.[85] But if this particular

[81] W. Schumacher, *Untersuchungen zur Datierung des Astronomen Kleomedes*, diss. Cologne 1975, accepted by Robert B. Todd, op. cit., n. 3.

[82] Cleomedes 2.7,226,24-228,4. See E. Martini, 'Quaestiones Posidonianae', *Leipziger Studien* 17, 1896, 394-8; O. Rehm, 'Kleomedes (3)', *RE* 9.1, cols 683-5; R. Goulet, op. cit., introduction, 15 and n. 127; Robert B. Todd, op. cit., n. 4.

[83] Cleomedes 2.2,168,20.

[84] Aëtius *Placita* 2.9.3 (Posidonius fr. 97 Edelstein-Kidd=*Dox. Gr.* 338,17).

[85] Aristotle *Phys.* 3.4, 203b20-2, answered 3.8, 208a11-20. Epicurus, *Letter to Herodotus* (in Diogenes Laertius *Lives* 10) §41; Lucretius 1.958-67; Epicureans' favourite according to Alexander ap. Simplicium *in Phys.* 467,1-3.

argument from *Phys.* 3.4 and 8 was known only indirectly, I am less inclined to think the same of the arguments in *Phys.* 4.8. For of these I have suggested that Chrysippus' knowledge is remarkably detailed and accurate.

Although I have disagreed with the claim that the Stoics ignored Aristotle's attack on the void, Sandbach has performed a service in pointing out that they ignored Aristotle's belief in natural places, his postulation of a prime mover to explain motion,[86] and, I would add, his definition of place. But while he puts this down again to ignorance, I would suggest a quite different explanation. Aristotle's successors Theophrastus and Strato began to recant some of Aristotle's views, and this occurred during the lifetime of Zeno the founder of Stoicism. Indeed, the second of these Aristotelians, Strato, died around 269 B.C., only seven years before Zeno. We shall see in Chapters 11 and 12 that Theophrastus assembled powerful doubts about Aristotle's definition of place and his belief in natural place, and he questioned the need for a prime mover to explain motion.[87] Strato was more forthright: he not only rejected the prime mover,[88] but positively denied Aristotle's conception of place[89] and his belief in natural places.[90] On the other hand, he agreed with Aristotle in rejecting void, except perhaps in microscopic interstices.[91] Now philosophy often proceeds dialectically, arguing against those positions which are currently being held elsewhere and ignoring those which are not being pressed. This, I suggest, not ignorance, is the explanation of the Stoics ignoring Aristotle on place, natural place and the prime mover. Indeed, if I am right, there is a selective attention to Aristotle (on void) and ignoring of him (on the other subjects mentioned), which precisely corresponds to the stage that Aristotelianism had reached. This selective attention and neglect is not well explained by a blanket ignorance.

So much for the Stoics. I believe that, in response to the Buridan's ass objections, they provided not only answers, but the main answers. But they were not the only ones to address these arguments. We have seen Epicurus exploiting the 'Why halt anywhere?' argument (215a19-20) for his own purposes,[92] while Philoponus *answers* it by saying that a projectile in a vacuum would halt at the point where the force impressed in it was exhausted – Aristotle himself admits the

[86] F.H. Sandbach, op. cit., 39 and 44.

[87] Theophrastus *Metaphysics* 7b15-23; ap. Proclum *in Tim.* 2,122, 10-17.

[88] Strato ap. Ciceronem *Academica Priora* 2.38.121 (=fr. 32 Wehrli).

[89] Simplicius *in Phys.* 601,24; 618,24.

[90] Strato frs 50-52 Wehrli=Simplicius *in Cael.* 267,29; 269,4; Stobaeus *Eclogae* 1.14.1 (*Dox. Gr.* 311).

[91] Simplicius says he believes (*oimai*) that for Strato space, though void in its own nature, is always filled by body, *in Phys.* 618,20-5. But he may make an exception by ascribing to Strato belief in microscopic void interstices *in Phys.* 693,11-18: see David Furley, 'Strato's theory of the void', in J. Wiesner, ed., *Aristoteles Werk und Wirkung, Paul Moraux gewidmet*, vol. 1, Berlin 1985, 595-609.

[92] Epicurus, *Letter to Herodotus* (in Diogenes Laertius *Lives* 10) §42.

exhaustibility of the *different* force which he postulates in the pockets of air behind a projectile.[93] One might adapt another argument of Philoponus[94] and remind Aristotle that he thinks the heavens would come to a halt without an infinite power to propel them, even though they have no differentiation in their surroundings (indeed, have no surroundings at all).[95] So lack of differentiation should not prevent a body travelling in the void from doing the same. Philoponus has an answer too to the 'Why any direction?' question (214b17-24), for he thinks that his comparison of elemental motion with that of an organism's parts answers both this objection and the objection (214b12-17) against void as an efficient cause of motion.[96]

There is a final curiosity to be noted. Aristotle momentarily deserts his own principle that the infinite could have no centre, when he allows that if the earth were infinite, the reason for its resting would still be that it hugs the centre.[97] It has been suggested that this passage may have influenced some of Zeno's unsatisfactory attempts to explain the immobility of the earth and the cosmos.[98]

In the preceding chapter, I looked for heroes. In this chapter, I have concentrated on Aristotle's case against a vacuum, and the replies made by the Epicureans, the Stoics and Philoponus. The Stoics turned out to offer a much more thoroughgoing response than has been supposed. But the answers of most significance for the history of science were produced by Philoponus. A more general discussion of vacuum would have to take in much more. It would have to consider, for example, the experiments proposed to demonstrate its existence, experiments involving burning candles, siphons, clepsydras and suddenly separated plates. Those experiments would introduce many new figures into the picture,[99] but they would take me too far from my Aristotelian theme.

---

[93] Philoponus *in Phys*. 644,16-22.

[94] The corresponding argument by Philoponus, *in Phys*. 690,34-691,5, involves the lack of *resistance* to celestial motion, rather than the lack of differentiation.

[95] Aristotle *Phys*. 8.10, discussed in Chapter 15.

[96] Philoponus *in Phys*. 630,8-634,2.

[97] Aristotle *Phys*. 3.5, 205b10-18.

[98] David Hahm, *Origins*, 117-18.

[99] e.g. Philo of Byzantium, Strato, Hero of Alexandria, Lucretius, Galen, Themistius, Philoponus. Some of the main experiments, with a few of the Greek references, are recorded by Edward Grant, *Ado*, 77-100.

CHAPTER TEN

Closed space and closed time:
the Pythagoreans

The most severe objection ever put to the finitude of space was examined in Chapter 8. It was formulated already in the fourth century B.C. by the Pythagorean Archytas, who asked, if space were finite, what would happen at the edge. The Pythagoreans, we shall see, were particularly imaginative about space and time. The modern answer is to say that space could be finite without having an edge. In a curious way, Aristotle would agree with that conclusion. It does not help him in answering Archytas, because he has to concede that the physical world has an edge, but *place* does not. Place *is* an edge, the inner edge of a thing's surroundings. Moreover, there cannot be at the edge of the physical world a place which is the outermost of all inner edges. But the way in which modern physics conceives space as possibly being finite without having an edge is by thinking of it as closed. The difficult thing is to show how that is possible. I shall merely try to offer a clear statement of the type of explanation that has been given by others.[1]

Closed space

If space were infinite in every region and direction, a rocket ship that went on travelling in a straight line should get for ever further away from its starting point. Is it conceptually inevitable that that would happen? Let us imagine ourselves stepping into a space rocket erected outside our window in London, or wherever we are. And let us make sure, once under way, that the rocket ship travels in a straight line. We shall have to consider what is meant by 'straight'. According to a definition as old as Plato, the straight line is the line of sight,[2] which we now know to be the line taken by light, although we must guard against refraction or reflection. It is also the shortest route between two points. Let us make sure, then, that our path is straight by all

---

[1] Stephen F. Barker, 'Geometry', in Paul Edwards, ed., *The Encyclopaedia of Philosophy*, New York 1967.
[2] Plato *Parmenides* 137E boils down to this.

three criteria. The rocket ship will emit a coloured wake, and, looking back, we shall note with satisfaction that, in the absence of any wind, the wake corresponds to the line of sight. We shall send out ahead a laser beam, and radio signals will warn us if the space craft deviates from the line of the light ray. Finally, we shall make sure that we are taking the shortest route in the direction of travel between any two points. We shall do so by taking measurements with a light ray gun, equipped to time the return of bounced light signals. We shall want to measure the distances between points successively reached A, B and C. If light signals take one minute to bounce back and forth between B and A, and one minute to bounce between C and B, then they should take two minutes to bounce between C and A, or the route through B will not be the shortest.

Suppose we have assured ourselves that our path is straight by all three criteria. What is there to prevent the cabin boy, who is looking through the front window, from shouting, 'Ahoy there! I see something that looks remarkably like London'? After all, there is nothing but habit to connect the shortest line, the line of sight, or the line of light with the view from the front window. There is no conceptual link which dictates what the view will be for one who has followed such a line. And it is this lack of conceptual link which makes it conceptually possible that a traveller following a straight line should not get for ever further away from his starting point, but should return to it. Whether he would in fact return depends on empirical facts about the universe and in particular on the distribution of matter. Relativity theory does not determine whether he would return, but allows for the possibility. And it was already anticipated by Riemann that his Geometry, in which this possibility is implied, may be true of the space we live in.

But we must be careful how we formulate our description of what we have imagined. The entire case will be thrown away, if we allow the cabin boy to say, 'We have come round in a circle.' Of course a *circle* comes back to where it started, but that only invites us to think that there is space *outside* the circle. The point on which we must insist is that it is a *straight line* that has come back to its starting point. And how would that suggest finitude? Well, if even a straight line does not get for ever further away from its starting point, there is no other line that will get further. The distances in the universe then seem to be finite. And yet they are finite without the universe having an edge (that is what makes space closed). For on our journey we did not come to a boundary where we had to change direction. The journey was finite without that.

The claim to have taken the shortest route needs to be qualified. For between any two points on our track there will be *two* routes, and only *one* of them will be shortest. It may therefore be thought better to say that our track represents the nearest available analogue of a straight line.

There are other descriptions which share the defect of the cabin boy's

exclamation, 'We have come round in a circle.' It is often said that
space may be curved, or that the geometry of such a universe is like
that of a sphere. These statements have their point, but they do not
enable us to see how space could be finite. For they invite the question
what is beyond the curve, or outside the sphere. And that is why it is
better to describe the situation in terms of a straight line, or the closest
analogue to it.

But how much would our imaginary experiment establish about
finitude? Some caveats are needed. We should not rule out the
possibility that space, though finite, might expand indefinitely. Nor, I
believe, should we exclude the possibility that there are spaces which
bear no spatial relation to our space or to each other, and which have
different dimensions from ours. Further, before deciding that our space
is finite, we should need to make sure that it is not finite merely in
certain directions or in certain pockets. But it might be urged that the
experiment suffers from a more radical shortcoming. For why would it
be rational to suppose that we had returned to London itself, rather
than reached an exact replica of London, or, if London had perished in
our absence, of the earth?

I think this could be made the subject of physical experimentation.
Riemannian Geometry is a consistent system which permits the
straight line to return on itself. If we find that the space we travel
through has yet other features of Riemannian Geometry, it will be
more credible that our space is Riemannian than that London and the
earth is duplicated. By sending our astronauts out on little detours
from the main space craft, we could get them to take paths that formed
a triangle with the main path, and we could get them to measure the
interior angles of those triangles. If the interior angles differed from
two right angles in the way that Riemannian Geometry predicts, we
should have confirmation that we were following a straight line in
Riemannian space, and that it was London we had reached.

One feature of the spatial experiment will later become relevant to
the question of closed time, and that is that we can think of London as
both in front of and behind itself. As we departed in our space craft, we
left London behind, but the straight line led us to find it in front of the
city we left behind.

I have deliberately over-simplified by mentioning space-time as little
as possible, and by treating space as a wholly distinct entity. We might
have lived in a universe in which space was indeed a distinct entity,
and in which the notion of space-time had no special utility. My
account of how space could be finite should apply without modification
to such possible universes. But in our actual universe, according to
Relativity Theory, four-dimensional space-time is the fundamental
structure, and it is only on some models of the distribution of matter
that one can separate out such a thing as space, in order to ask if it is
finite. In Gödel's model, for example, which I shall discuss below, one
can separate out time, but not space. It might therefore seem easier to

consider the finitude not of space, but of space-time. But that too would have its complications. For it is only on some models that there would be a unitary answer to the question whether space-time was finite. On other models, there would be different answers corresponding to different frames of reference.[3] I believe that my account of closed space could be adapted to fit those models of space-time in which space is a distinguishable structure. But I shall leave this task to those better qualified and move on to consider time.

Endless repetition

I shall first touch briefly not on closed time, but on a rival theory, also pioneered by Pythagoreans, according to which history is endlessly repeated with the same persons and things recurring in linear time. I have discussed this theory elsewhere.[4] It appealed not only to certain Pythagoreans, but also to the Stoics and to the Neoplatonists. But there were variations according to whether the repetitions of history were thought of as exact. Various Stoics, for example, felt a need to avoid exact repetition, and Aristotle's pupil Eudemus of Rhodes had already objected to the Pythagoreans that the theory of an exact repetition of history would collapse into the acknowledgment of a single occurrence. I presume he was relying on Aristotle's definition of time as the number, or element which can be counted, in terms of before and after in motion. If repetitions were exact, the motions would be the same, hence so would be the countable element in the motions, in other words, time. The conclusion which follows from that is not drawn explicitly in the report of Eudemus, but if the time is the same, there will have been no repetition after all, and history will have occurred but once:

> One might be puzzled whether or not, as some people say, the same time recurs. For sameness is spoken of in different ways, and it does seem that a time the same in kind recurs, e.g. summer, winter and the other seasons and periods. Similarly, motions (*kinêseis*) the same in kind recur, for the sun will accomplish its solstices, equinoxes and other journeys. But if one were to believe the Pythagoreans' view that numerically the same things come again, and I will talk staff in hand, to you sitting like this, and everything else will be alike, then it is plausible, that the time too will be the same. For when the motion (*kinêsis*) is one and the same, and similarly there are many things which are the same, their before and after is one and the same, and hence so is their number. So everything is the same, which means that the time is as well.[5]

The same conclusion, that history occurs once, would follow for those present day philosophers who accept the principle of the identity of

---

[3] I am grateful to Arthur Fine and David Malament for introducing me to these issues.

[4] Richard Sorabji, *Time, Creation and the Continuum*, 182-90.

[5] Eudemus ap. Simplicium *in Phys.* 732,26-733,1.

indiscernibles, as applied to times. If we try to imagine an exact repetition of history, events will be qualitatively the same as before, and the time of their occurrence will also have nothing to distinguish it from the earlier time. The principle of identity of indiscernibles, as applied to times, would then dictate that the time was after all the same, and there was no repetition. For the principle says that in order for two things, for example two times, to be distinct, there must be some feature that differentiates them.[6]

A differentiating feature would be provided by those Stoics who allowed that next time around, I might differ in some minor way, e.g. by having freckles.[7] But other Stoics, it has been argued, would be committed to the time being the same by principles of their own.[8] So long as events are numerically the same (and for some Stoics they are), time too will have to be the same, since time for the Stoics has no existence independent of events. All these arguments for sameness of time turn on purely *logical* considerations. I shall consider later whether any *empirical* test for reduplication could be found, parallel to the measuring of interior angles, which was used above in the discussion of space, to decide whether a traveller had reached a duplicate of London.

If the hypothesis of a finite history endlessly repeated in linear time can be collapsed, we might expect the result of collapse to be a finite history, occurring only once, and having its events arranged in a linear sequence with a beginning and an end. If time depends on the occurrence of events, it too will be finite, and indeed that result would follow from the identity of indiscernible times, or from the other logical arguments used above. There would be no distinct times, according to the identity of indiscernibles, outside the span of history, because there would be nothing to distinguish such times. But could history and time be viewed as circular rather than linear? That is, they would be finite without having a beginning or an end, in the manner of a circle whose length is finite, but which lacks beginning or end. That would be a form of what is called closed time, and A.A. Long has suggested that if the Stoic idea of endless repetition can be collapsed we should think that what the Stoics and perhaps the Pythagoreans were really expressing was the idea of closed time.[9]

---

[6] This application of the principle to *times* is parallel to Leibniz's application of it to *places* in the passage translated by G.H.R. Parkinson under the title 'The Principle of Indiscernibles', in *Leibniz: logical papers*, London 1966, 135.

[7] Alexander *in An. Pr.* 181,25 (*SVF* 2.624); cf. Origen *contra Celsum* 5.20 (*SVF* 2.626). Conversely, some Stoics wondered whether it would still be Socrates, even if there were *no* differences, Origen op. cit. 4.68 (*SVF* 2.626).

[8] Jonathan Barnes, 'La doctrine du retour éternel', in Jacques Brunschwig, ed., *Les Stoïciens et leur logique*, Paris 1978, 3-20, esp. 12.

[9] A.A. Long, 'The Stoics on world-conflagration and everlasting recurrence', *Southern Journal of Philosophy* vol. 23 supp., ed. R. Epp, 1985.

Closed time

I am grateful to Long for re-opening the issue, because I had previously argued that the idea of closed time was conceptually impossible, and a number of others have thought the same.[10] But on the other side, a very simple situation has been described in which we might be motivated to think of time as closed.[11] I shall start by treating the situation as purely imaginary; we can consider later whether it might be our actual situation.

We have already imagined that there might be only a finite sequence of qualitatively distinguishable events, and we have seen that, on the postulate of identity of indiscernibles, the sequence could occur only once. Now suppose that in this single sequence no event looked like a first or last, but on the contrary at every moment there were events which appeared to be caused in a perfectly normal manner by their predecessors. The sequence of events would then appear to form a seamless, closed circle. No event could be assigned to a later or earlier end of the sequence, because the seamless circle of events would not permit the distinction of a beginning or end. And if there was no such thing as a later or earlier end of the sequence, that would mean that time itself, as well as events, would be seen as forming a circle. The talk of a circle, which was unsatisfactory in connexion with space, would not raise the same problems here, because no comparable question arises about what lies outside the circle.

I do not see why this situation should preclude talk of earlier and later, past and future. But each event would be seen as lying *equally* in the past and future, just as someone stationed at a point (say, 3 o'clock) around a clockface, can see 12 o'clock as lying *equally* in the clockwise and the anticlockwise direction (I am deliberately imagining the observer as stationary),[12] and yet for all that, 12 o'clock does not occur more than once on the clockface. Similarly, from the present, your birth would be seen as lying both in the past and in the future direction, but it would occur only once.

I have no proof that time would have to involve a past, present and future, but I know of no compelling reason why it should not. One of my previous objections to the idea of closed time, that each event would be

[10] Richard Sorabji, *Necessity, Cause and Blame*, London and Ithaca N.Y. 1980, 115-19; W.C. Kneale, 'Time and eternity in theology', *Proceedings of the Aristotelian Society* 61, 1960-1, 91-2; J.R. Lucas, *A Treatise on Time and Space*, London 1973, 57-60; Jonathan Barnes, 'La doctrine du retour éternel', p. 19, n. 64; R.G. Swinburne, *Space and Time*[2], London 1981, 141; D.H. Mellor, *Real Time*, Cambridge 1981, 187.

[11] Bas van Fraassen, *An Introduction to the Philosophy of Time and Space*, New York 1970, elucidating Adolf Grunbaum, *Philosophical Problems of Space and Time*[2], Dordrecht 1973; cf. W.H. Newton-Smith, *The Structure of Time*, London 1980, 65-8; Lawrence Sklar, *Space, Time and Space-Time*, Berkeley, Los Angeles 1974, 303-17; David Lewis, 'The paradoxes of time travel', *American Philosophical Quarterly* 13, 1976, 145-52.

[12] It is irrelevant that the hands may reach 12 o'clock more than once: the analogy concerns only the *stationary* observer.

before and after itself, has its analogue in the spatial situation, where London proved to be in front of and behind London. That turned out to be the natural consequence of a perfectly intelligible situation, and its temporal counterpart should not deter us. In talking of past and future, before and after, I am not withdrawing the claim that there would be no earlier or later *end* of the sequence. The point is that there would be no ends in the sequence, and therefore no earlier or later ends, even though any event could be viewed as both earlier and later than any other.

Despite the fact that each event could be seen as lying equally in the future and in the past, I see no reason why there should not be a direction for the flow of time. The static question of where events *lie* in the circle is not the same as the dynamic question of when we *reach* them. Given any three events, say, planting a tree, its sprouting and its shading our house, there would still be a fixed order in which, starting from one (the planting), we could reach the other two. The sprouting would be *reached next*, and only then the shading. The same would apply to causal influence: there would be a fixed temporal order in which we could influence the other two. And besides the temporal order there might be an explanatory order, in that the first event was a necessary condition for the third only *because* it was a necessary condition for the second. The event we were reaching would admittedly lie in the past as well as in the future, and the event we were leaving would lie in the future as well as in the past.

The possibility of temporal direction would be challenged by someone who thought that temporal direction could be reduced to some unidirectional physical process like entropy (a process in which physical differences are increasingly levelled out, like sandcastles on a sea shore). The circle of events would not allow a unidirectional process, because the circle would require that all changes should eventually be reversed. However, I do not myself see that the direction of time can be reduced to any unidirectional physical process[13] any more than the direction of clockwise motion can be reduced to a physical process distinct from itself.[14] Nor would the absence of such a process rule out the talk of past and future, or of events being earlier and later than each other.

The fact just mentioned, that in circular time all changes would need to be reversed eventually, shows that circular time would require a different physics. A tree could grow, so long as it shrank or

[13] The issue is, of course, controversial. John Lucas has argued against the possibility of direction in circular time, and hence against the possibility of circular time itself, on the basis of the absence of suitable physical correlates for direction (*A Treatise on Time and Space*, London 1973, 59-60). And I am grateful to my colleagues in King's for pressing this type of view on me. For discussions of the direction of time, see e.g. Lawrence Sklar, op. cit., 351-414; John Earman, 'An attempt to add a little direction to "the problem of the direction of time" ', *Philosophy of Science* 41, 1974, 15-47.

[14] The clockwise direction of motion does depend on the direction of time, but that does not mean that it depends on a physical process like entropy.

disappeared in time to grow. It could lose a branch, so long as it later had the same branch with the same particles. The particles would need to be assembled from wherever they had got to.

A different physics would deal with the objection that circular time is impossible, because it would lead to self-defeating situations. In a forty-year circle, somebody somewhere, by design or accident, will surely kill his mother. Would not that prevent him from being around to be her killer?[15] The simplest answer is that, since a different physics will anyhow be required, we should expect one which permits his mother to return from the dead in time to give him birth.

A different danger would be that in circular time an event would be caused by itself. For I have imagined that at any moment there are events which are caused by their predecessors, but among an event's predecessors in circular time we should have to include itself. The idea of an event causing itself may seem even more startling than that of an event preceding itself. But it is not an idea which I believe we need accommodate. For elsewhere I have connected cause with explanation,[16] and explanation is not a transitive relation. That is to say, from the fact that an event A is explained by some predecessor B, and B is explained by its predecessor C, it in no way follows that A is explained by C. Nor, then, should it be argued that an event is likely to be *caused* by its remote predecessors. Admittedly, there are other relations which are normally viewed as transitive, such as the relation of being a causally necessary prerequisite, or that of being a causally sufficient guarantee. I do not know how many events have causally sufficient guarantees, but I think that the simplest solution would be to introduce an exception into our normal way of viewing things. The normal understanding that these last relations are transitive should be viewed as overridden in any case in which it would lead us to talk of an event being so related to itself.

Implications for immortality and psychology

Circular time has implications for immortality. Suppose the universe lasted a total of forty years. Then a living being which lasted forty years without having a beginning or an end during that time would be immortal. However, human beings all have a beginning when they are conceived. So how would they fare in a circular system of forty years? They might die within the forty years, or they might dwindle into the

---

[15] See Samuel Gorovitz, 'Leaving the past alone', *Philosophical Review* 73, 1964, 360-71; John Earman, 'Implications of causal propagation outside the null cone', *Australasian Journal of Philosophy* 50, 1972, 222-37; David Lewis, op. cit.; Paul Horwich, *Asymmetries in Time*, Boston Mass. 1987, ch. 6, 'Time travel'; Murray Macbeath, 'Who was Dr. Who's father?' *Synthese* 51, 1982, 397-430. The context of Earman's discussion is the possibility of signals travelling faster than light, for which he supplies a useful bibliography.

[16] Richard Sorabji, *Necessity, Cause and Blame*, ch. 2.

sperm and ovum from which they were generated. In either case, despite the death or dwindling, they could, in a sense, enjoy survival after death. For whenever they died or dwindled, there would still be life ahead of them. Any particles scattered at the time of death or dwindling would need to be in their former positions later on. Of course, death or dwindling would involve an interruption in human life. But for many people, irrational though it may be, the most horrifying aspect of death is the prospect of permanent extinction, of having no more future. And circular time would deal with this, without relying on the need for an infinite future.

To some people an infinite future would be preferable. But others would find a positive advantage in a guaranteed future that was finite. For the length of experience would not then be a source of tedium such as some philosophers have feared in immortality.[17] For the people described would live at most forty years, some of them less. It should not be thought that they would live the same forty (or fewer) years again and again; they would live it only once. For the principle of identity of indiscernibles, which was invoked to establish that there would be only one circle, would equally imply that there was only one *experience* of the circle. Boredom might come from another source, if the fact that all their experience is past (as well as future) leaves them with no surprises. But if their memory or attention is selective, or if for those who die or dwindle it is eradicated, there will be plenty of novelty to surprise them. In general a different psychology would be needed, just as much as a different physics, if life was to be tolerable.

In some respects, circular time would have the same effects as endless repetition. People would have to accept the prospect which Augustine found so repulsive in endless repetition, that any progress would be undone. I have discussed his reaction and alternatives to it elsewhere.[18] The crucifixion would have been in vain, and would need to be repeated. Augustine's youthful activities, including the sex shows which scholars have lovingly sought to reconstruct, would have been renounced in vain, and his progress towards God would be reversed. To be happy in such circumstances one would need aspirations very different from Augustine's. One alternative would be to seek pleasure in activity while avoiding concern for its outcome. Those who lived right through a cycle might try to so pattern their lives that decline in one field was balanced by progress in another.

It would be desirable that memory or attention should be imperfect, not only to prevent boredom, but also in order that a person might avoid terror or guilt at the thought of what he was going to do. Insofar as he did recall unpleasant events, he might need to cultivate an

[17] Bernard Williams, 'The Makropulos case: reflections on the tedium of immortality', in his *Problems of the Self*, Cambridge 1973; discussion in Richard Sorabji, *Time, Creation and the Continuum*, London 1983, 181-2.

[18] *Time, Creation and the Continuum*, 188-90.

attitude of resignation, although pleasant things could be recalled and anticipated gladly. Imperfect memory or attention would also make it possible for a man to deliberate, and it would free him from the inhibition of thinking of future action or inaction as certain, or fruitless, or redundant.

Problems of memory raise the question what form memory would take in a forty-year circle. While watching a tree being planted, a being who lived the full forty years without birth or death might find the planting familiar because he had had this very same experience forty years earlier. The familiarity would be a case of memory because it was due to the experience's being past, despite its also being present and future. He could also remember the forty-year-old remembering, although it would be wrong for him to add forty years to forty years and conclude that the planting occurred eighty years earlier.

Our actual situation

So far I have considered a purely imaginary situation, in order to argue for the conceptual possibility of circular time. But we know that we are not living in a forty-year circle, so can the idea of circular time apply to us? I think it might, but we should have to consider a circle of billions upon billions of years. The 'big bang' from which the current state of the universe started, and the implosion with which it may finish, could not be seen as genuinely first and last events. They would need to be local events in what is sometimes called a ping-pong universe, in which big bangs and implosions alternate.[18a] And entropy also would have to be seen as a local affair: over the longer term, there would be no unidirectional processes, but everything that happened would be undone. Given that known processes all break down in the vicinity of a big bang, and that physicists are very modest about what, if anything, preceded it, these possibilities can hardly be excluded.

An extremely long circle would in many ways be easier to accommodate than a circle of forty years. It would solve problems of memory, because we should die after occupying only a tiny portion of the circle, and could expect our memories to be eradicated when our birth eventually came round. The length of the circle would not reintroduce the threat of tedium, because our lives would occupy only a fragment of the time. There would be ample time for the murdered mother to be reborn. The backwards causation which I shall discuss below would take a less startling form, because in influencing the future, one would be influencing only one's remote, not one's immediate, past.

The great mathematician Gödel sought to introduce a real possibility of circular time by a very different suggestion. He argued that it was

[18a] I have not yet seen the different approach of Stephen Hawking in *A Brief History of Time*, 1988.

allowed for by the equations of Relativity Theory.[19] But he considered the question only in relation to the personal time of individuals, because, as he pointed out, the theory of Relativity does not allow that there is a single time for everybody, regardless of their state of motion. Observers travelling at high velocities in relation to each other do not agree about which events are going on at a given moment, or in what order events occur. Gödel's view was that, on the supposition of a certain distribution of matter in the universe, an individual making use of a rocket ship could find his personal time closed, if only he could attain a high enough velocity. On the other hand, the time of others not so accelerated would not be closed. Moreover, the velocities attainable in practice, given the available fuel, would suffice only to reach the distant past, so that the individual who reached the past would die before meeting his younger self. Even so, this would presumably ensure a kind of survival after death, for the traveller's birth would always lie in his personal future.

Some of the situations rejected by Gödel as impractical would make closed time more like a loop within linear time than like a circle. For if a younger self were able to meet his older self, he might progress in linear fashion until the meeting, and the older self, in his separate body, might progress in linear time thereafter.

Gödel claimed that closed personal time was compatible with the mathematics of Relativity Theory, but is he right that it is also compatible with other things? Much has been done to exhibit the idea of time travel by rocket ship as consistent,[20] but problems remain to be discussed. To mention but one, would there be enough molecules to constitute the rocket ship both on and off the ground? The peculiarities of time travel might require its molecules to be in both places at once, if it made a journey backwards through time, rather than simply disappearing and reappearing. For when it had retrogressed to the time five minutes before its launch, it might by then be far out into space and yet also on its launching pad. Its molecules might also need to shuttle from its landing site at the earlier time to the rock face from which they were mined before the ship was assembled for take off. If such difficulties can be met and Gödel is right, those who seek physical precautions against permanent extinction might consider investing in a rocket. But I shall confine myself to what I regard as the simpler scenario described above. Although molecules have to shuttle in that scenario, there is no question of their being in two places at once. Nor need we make the distinction (although it is an intelligible one and necessary for some purposes) between a personal time and a time common to everyone. Nor need we face the problem which Gödel was so keen to avoid of a person meeting his younger self.

[19] Kurt Gödel, 'A remark about the relationship between Relativity Theory and Idealistic Philosophy', in P.A. Schilpp, ed., *Albert Einstein, Philosopher-Scientist*, New York 1951, 555-62.

[20] See Murray Macbeath, op. cit. For discussion of problems I thank Christoper Hughes.

Simultaneous selves

But although the idea of a person meeting his younger self is a gratuitous complexity, it could be added to the story I told, in order to assess its consequences.[21] For suppose we return to the short cycle of forty years. We may consider whether some individuals might not have a *personal* time of fifty years, although they would have to live out their fifth decade simultaneously with their first in separated bodies. Such a thing would be impossible for immobile things like buildings. For the rusted fifty-year-old building would not get out of the way, but would find itself in exactly the same place as its shiny new self, not to mention an infinity of new selves, if it never perished. This goes well beyond the possibilities of interpenetration that I tried to make sense of in Chapter 5. Nor would mobile things fare any better, if they did not, like living things, change their particles. Unless the planet earth, for example, replaced its particles, there would not be enough particles to constitute both its earlier and its later self. The planet earth and its dwellings, then, are best imagined as never beginning or ending, or alternatively as doing both within the forty-year cycle. But human beings might fare differently, since they are both mobile and living. The postulation of years beyond forty for humans does not violate the original description, according to which there are only forty years in the history of the universe. That requirement is still respected, because the individual would live the first ten years of his life *simultaneously* with his forties in two separated bodies. It would clarify the situation to distinguish his personal time from the time of the universe. His first and fifth decades would coincide with the same decade of the universal time circle. Given a suitably adapted physics, he could have surrendered his earlier particles, in time for the ten-year-old to claim them as his own. The fifty-year-old could meanwhile have recruited (though only temporarily) a succession of particles, ingested from food. That he should perish at some stage would be desirable, or we should need to accommodate an infinity of juxtaposed selves.

We need not doubt the fifty-year-old's claim to be the *same* (person) as the ten-year-old, because he would be linked to him by all the links which in linear time attach a fifty-year-old person to his younger self. There would be not only his memory impressions and his character, but the continuous history of his body. For a detective hired to dog the footsteps of the fifteen-year-old would find himself thirty years later behind the forty-five-year-old. What is more problematic is whether the appropriate concept would still be that of a *person*. For we should have to say that a 'person' could be present in two bodies, standing with one and sitting with the other, and with two minds, each possessed of different information. But our concepts should be capable of revision, and for short we can express the situation by saying that a

---

[21] I am grateful to Lindsay Judson for the point.

fifty-year-old would be the same quasi-person as the ten-year-old.

Worries have been raised in various contexts about whether an older self could kill a younger self.[22] But this will again be possible, if we may imagine a different physics in which a slain younger self is resuscitated, in time to become the killer. As for the suggestion[23] that an older self might have a piece of information only because his younger self was given it by the older self, we do not have to accept that such unexplained circles will occur.

The juxtaposition of selves would produce a very distinctive kind of immortality. Even if the individual died at ninety, his ten-year-old and his fifty-year-old selves would live on until they died at ninety, leaving a ten-year-old and a fifty-year-old self behind. At some stage, the surviving older self would acquire the particles that constituted the departed. There would be a rich variety of vantage points, for the same individual would see different selves simultaneously both from the outside and from the inside, and might also remember his earlier vantage points. There would also be a variety of vantage points on events in the external world. It might be thought that the external world would be lacking in novelty. For a ninety-year-old, though enjoying a different vantage point, would be looking at a world no different from the world he viewed (and was currently viewing) at fifty. But there would be the compensation that the world would be very full of things to explore. For it would contain juxtaposed stages of the same living individuals, for example, the older and younger selves of everyone over forty one could wish to meet. The forty-year cycle would thus present a person with all the development of which the world was capable, in juxtaposed stages. If despite this variety, a person became bored, he could die, while retaining the full benefit of perpetual existence. For there would be no point in the cycle at which he failed to exist.

But the supposition of juxtaposed selves is entirely gratuitous for the purpose of making sense of circular time. I shall therefore return to the simpler model without juxtaposition. For several major questions about it remain.

Backwards causation and the sabotage objection

One question is this: would circular time require causation to work backwards from future to past in an objectionable sense? No; it is true that, in a small time cycle, I could plant a tree tomorrow in order to give shade to my house yesterday. And it has seemed to many philosophers that such a thing would be impossible. But the objections have all been discussed in the context of ordinary linear time. It has therefore been overlooked that, whether or not they are good objections

[22] See n. 15.
[23] David Lewis, op. cit.

in that context, they lose all force in the context of circular time.

An initial point is that it is only in a sense that circular time would involve backwards causation. Since the date of planting can be viewed equally as future and as past, and so can the date of shading, it can be said that a later event is causing an earlier. But it can equally be said that an earlier event is in the normal way causing a later. Moreover I have already argued that there will be only one direction for causation. For given three events, the planting of a seed, its sprouting, and the shading of my house, there will be a definite temporal order for the three events, in that, starting from the planting, the sprouting is *reached next*, and the shading only after that. In this last sense, causation is working, as it normally does, in a single direction. We must, of course, observe the normal constraint that in circular time what is done must be undone. So the tree needs to be felled in time to make room for the planting.

The difficulties of backwards causation were brought into prominence in two articles by Michael Dummett.[24] Even though he was defending the possibility, he drew attention to the most discussed difficulty: the possibility of *sabotaging* backwards causal connexions. I shall devote the following discussion to assessing the supposed force of this difficulty.

If it is suggested that in linear time a certain planting is the cause of a certain earlier shading the reply is offered that, since the shading comes first in time, it would have been possible, and in other similar cases it still is possible, to wait until the shading has occurred, and then to *prevent* anyone planting near that spot. The possibility of such *sabotage* is meant to show that there is after all no causal link between the later planting and the earlier shading in the case we started with. But if it be protested that all attempts to prevent planting might *fail*, the opponent of backwards causation moves in to deliver the *coup de grâce*. The failure of sabotage might indeed show that the planting and shading were causally linked. But systematic failure would call out for explanation, and the best explanation would be that by some

---

[24] The two papers in favour of backwards causation by Dummett have led to the replies listed below. Michael Dummett, 'Can an effect precede its cause?', *Proceedings of the Aristotelian Society*, supp. vol. 28, 1954, 27-44; with reply in the same volume by A. Flew, 45-62; Dummett, 'Bringing about the past', *Philosophical Review* 73, 1964, 338-59; Max Black, 'Why cannot an effect precede its cause?', *Analysis* 16, 1956, 49-58; A. Flew, 'Effects before their causes? Addenda and corrigenda', *Analysis* 16, 1956, 104-10; id., 'Causal disorder again', *Analysis* 17, 1957, 81-6; Michael Scriven, 'Randomness and the causal order', *Analysis* 17, 1957, 5-9; D.F. Pears, 'The priority of causes', *Analysis* 17, 1957, 54-63; William Dray, 'Taylor and Chisholm on making things to have happened', *Analysis* 20, 1959-60, 79-82; Samuel Gorovitz, 'Leaving the past alone', *Philosophical Review* 73, 1964, 360-71; R.G. Swinburne, 'Affecting the past', *Philosophical Quarterly* 16, 1966, 341-7; id., *Space and Time*, ch. 8, London 1968, 2nd ed. 1981; Richard M. Gale, 'Why a cause cannot be later than its effect', *Review of Metaphysics* 19, 1966, 209-34; id., *The Language of Time*, London 1968, ch. 7; D.H. Mellor, *Real Time*, Cambridge 1981, ch. 10. There is an appraisal in J.L. Mackie, *The Cement of the Universe*, Oxford 1974, ch. 7, with reference to further literature.

unfamiliar mechanism the shading was causing the planting. Though surprising, that would in linear time be a case of *forwards* causation, and so the possibility of *backwards* causation would still not have been shown.

I think it can be seen that this, the most prominent objection to backwards causation, loses its plausibility just as soon as it is applied to the case of *circular* time. Its last remark, for example, that the systematic failure of sabotage would show shading to be the cause of planting, plays straight into the hands of the believer in circular time. For in circular time, the shading's causing the planting would be just as much (and just as little) a case of backwards causation as the planting's causing the shading. Indeed, the argument could stop there, but there are so many other defects, I believe, in the objection to backwards causation, when it is applied to circular time, that it is worth bringing them out.

In circular time, it could be conceded that the shading caused the planting, and still argued that the planting might *in addition* cause the shading. Such a suggestion would not have had the same plausibility in *linear* time. For if in linear time an earlier shading caused a later planting, it would merely introduce an unfamiliar mechanism and direction of influence to suggest that in addition the planting might have caused the shading. But in *circular* time that additional suggestion would not introduce an unfamiliar mechanism at all, and even the direction, I have argued, would only be backwards in a sense. Moreover, the additional suggestion could even be confirmed. For if in circular time shading could not be suppressed after planting any more easily than planting after shading, the failure of sabotage ought to show, if it shows anything at all, that each causes the other.

There is more: it is not at all clear that in circular time successful sabotage would prove anything. It might be very easy to wait for shading and then suppress planting, because the suppression of planting might merely open the way to other methods, perhaps more natural and more efficient, of producing shade. It might open the way to windblown seedlings or underground suckers which would have been obstructed, had there been planting. Once again, such a suggestion would have lacked plausibility in linear time. If planting were in *linear* time suppressed after shading, that could not easily open the way to alternative causes of the earlier shade. For in linear time, those alternative causes would have again to work by a wholly unfamiliar mechanism and in a backwards direction. They would have to work backwards to produce shade, or causes of shade. In circular time, these problems are avoided. For the suppression of planting in circular time would open the way to quite familiar mechanisms, seedlings and suckers, which would work in their customary direction, reaching sprouting before shading. Moreover, the direction would be no more from future to past than from past to future.

Admittedly, it would be harder to suppress planting after shading in cases where no alternative causes of shading (seedlings, suckers, etc.) were available. For then the shading would be causeless in these cases, which would in itself be odd (I do not say impossible), and still odder given that shading would be caused in other apparently similar cases. But what follows from this? At worst, we might find that sabotage was frustrated in these cases, in other words, that in these cases we could not suppress planting after shading. But need that frustration lead, as alleged, to the revised conclusion that really shading was in these cases causing planting? We would equally experience frustration, if we tried to suppress simultaneously all possible causes of shading, e.g. planting, seedlings and suckers. The revised objection here would have to be the even less tempting one that shading caused *one or other* of these many things to happen. But is it likely that shading could have such diverse effects? A more plausible explanation of our frustration would be, I think, that nature abhorred causeless shading, and that such things as shading naturally occurred only when paired with a cause. That conclusion would leave intact the claim that later planting caused earlier shading.

It has not yet been explained why the possibility of sabotage is meant to show that in the original example of later planting causing earlier shading there is after all no causal link, and there has been more than one suggestion about this. According to one proposal, the statement that a certain later planting caused a certain earlier shading implies that there is some kind of correlation, not necessarily a perfect one, in *other* cases too between later planting and earlier shading. Or at least it is implied that there would be a correlation, if there were to be enough other cases. Moreover, it is thought that the correlation implied is a *non-accidental* correlation, that is, a correlation that would *resist* being broken down. The idea is that planting is implied to be better correlated than non-planting with shading. So it is implied that we would encounter resistance, if in general we took shaded houses and tried to prevent people subsequently planting trees by them, and took unshaded houses and tried to induce subsequent planting there. What is supposed to refute the original suggestion that a certain later planting caused a certain earlier shading is not our actually carrying out such sabotage in other cases, but the claim that there would be no resistance to such sabotage in other cases, if it were attempted.[25]

But I have argued against such a 'correlation' view of causation elsewhere.[26] Planting, as I have suggested, might be an inefficient method, which occasionally led to shading, but which got in the way of alternative, natural methods, more efficient at producing shade, such

[25] Cf. D.H. Mellor, op. cit. I am grateful to him for explaining his view to me, and hope that I am not distorting it too much in supposing that the present formulation gives much the same sort of view.

[26] Richard Sorabji, *Necessity, Cause and Blame*, ch. 2, esp. pp. 37-8.

as windblown seedlings or underground suckers. In that case, planting would not necessarily be better correlated than non-planting with shading. And so the claim that on some occasion planting did lead to shading implies no such correlation, and is not impugned if any such correlation can be sabotaged without resistance.

There is another suggestion about why the possibility of sabotage would impugn the original claim to have a case of backwards causation. The alternative suggestion appeals not to a general correlation, but to a different principle. A cause, it is said, cannot become fixed or unstoppable later than its effect does so. But the earlier shading becomes fixed or irrevocable at latest by the time it occurs. So the subsequent planting cannot be its cause, at least if it remains open to suppression and therefore unfixed.[27] To this version of the sabotage argument there is an obvious initial reply: the alleged principle still allows the subsequent planting to cause the earlier shading, just so long as it is itself inevitable sufficiently far in advance.[28] But in any case, the alleged principle is itself far from obvious. In circular time, why should it not be left open, after a tree has cast its shade, whether its seed will be *planted* or whether it will be established by *natural* means?

There is an even more fundamental question: in circular time, when would things become irrevocable? On one view, since everything in circular time is past, everything is always irrevocable. On another view, which I shall support, since everything is future, nothing is irrevocable (although some things will be unstoppable for other reasons). On either of these two views, there can be no possibility of ruling out backwards causation by threatening that an effect would become irrevocable before its cause did. For either cause and effect would both be at all times irrevocable,[29] or, as I shall argue, neither ever becomes irrevocable, although they may be unstoppable for other reasons. I shall return to this issue shortly.

For the present I conclude that, insofar as circular time can be said to involve backwards causation, it cannot be impugned by the 'sabotage' argument. I need not therefore discuss any of the more far-fetched answers that have been proposed to that argument.[30] I will only recall that insofar as backwards causation is involved in circular time, it can be made to seem less bizarre if we consider a long cycle of time. In a long cycle, although I could affect things in the past, any direct efficacy would be exerted on the distant past before my birth, and not on anything as recent as yesterday's shading of my house.

[27] J.L. Mackie, *The Cement of the Universe*, 178-83, reporting, not endorsing; Richard Swinburne, *Space and Time*, 2nd ed. only, London 1981, 139.

[28] So Mackie, loc. cit.

[29] This answer has been proposed in discussion by Susan Weir.

[30] Dummett raises the question, much discussed subsequently, whether the evidence might change, after planting had been suppressed, as to whether prior shading had really occurred.

Inevitability

I have touched on the question whether in circular time everything would be fixed and inevitable. That view has been taken about the alternative theory of endless repetition. For many adherents would not think they had reason to accept the theory of endless repetition, unless there were causes making the pattern of events *inevitable*. Why else should that pattern be the same on so many occasions? Admittedly, the theory of endless repetition can have a different rationale, for it has been advocated on probabilistic grounds. Probability led Nietzsche to expect repetition, given a finite number of possible states of affairs and an infinite amount of linear time.[31] But probability ought not to lead us to expect anything as tidy as exact repetition. For on a probabilistic basis states of affairs could be repeated in any order permitted by physics, and protracted to any length of time. One state of affairs, or sequence of states (however long), could be repeated indefinitely often before another was repeated once. There need come no time when these permutations were exhausted and had to be repeated in the original order.[32] So any *good* rationale for accepting exact repetition is likely to presuppose causes which make the pattern of events inevitable.

Would circular time also mean that all events were inevitable? I once thought so,[33] but now propose to retract. The most persuasive reason for expecting events in circular time to be inevitable is probably the consideration that everything is already in the past and the past is irrevocable. But this consideration needs to be balanced by reflecting that everything is equally in the future, so that it should not be *too late* to forestall anything. So far the arguments seem equally balanced and they give rise to a stand off. But I wish to introduce what I may call the 'subjunctive' argument, in order to show that events would *not* be irrevocable. It is a feature of irrevocability in ordinary linear time that we can use the *subjunctive* mood, and say of past events, 'Even if I *were to* act differently in the future from the way I am going to act, it would still make no difference now to the past.' This feature is missing in circular time. Consider a man in a forty-year cycle whose house was shaded yesterday, because he planted trees thirty-nine years ago. He cannot say, 'Even if I *were to* refrain from planting next year, it would still make no difference now to my house having been shaded.' Admittedly, nothing he will *in fact* do will make a difference, but if he *were to* behave differently there *would* be a difference in the past. So the shade he has received is a fact, but it lacks the marks of irrevocability.

[31] See Ivan Soll, 'Reflections on recurrence', in Robert Solomon, ed., *Nietzsche: a collection of critical essays*, Garden City N.Y. 1973.

[32] The point has been made in discussion by Susan Weir. Cf. Richard Sorabji, *Time, Creation and the Continuum*, 190 and n. 81.

[33] Richard Sorabji, *Necessity, Cause and Blame*, ch. 6, 115-19.

This should not be too surprising on further reflection. For in linear time, there would be things a man could have done in the earlier past, if he had wanted to prevent his house from being shaded yesterday. What happens in circular time is that the past is also the future. So anything that a man could have done in the past to prevent his house being shaded is also a thing that he could do in the future, even if he won't. This is not to say that in circular time all things are possible. On the contrary, if the man lives in the depths of the jungle, he may be physically incapable of keeping back the shade. But at least this will not be because he has lost a chance which he once had by leaving things too late.

If there is no irrevocability in circular time, it will be wrong to think of alternative possibilities in terms of branches stemming from a single trunk. The single trunk would represent a past in which there were no longer alternative possibilities, and that is the wrong model for circular time. A better model to represent alternative possibilities in circular time would be parallel lines. That in turn means that it would be wrong to argue in the opposite direction, to assume that the branching model was an appropriate representation of alternative possibilities, and to conclude from its inapplicability in circular time that there were no alternative possibilities there.

A different reason for thinking the pattern of events in circular time inevitable concerns a person's knowledge of what he has done. I have argued elsewhere[34] that if someone else knows what you are going to do, this at most implies that you will not, not that you cannot, do anything else, unless the foreknowledge (unlike human fore-knowledge) is infallible. But the situation is different if you yourself know what you are going to do, and in circular time a person could know this, if he clearly remembered what he had done. When he reflects that his past action or inaction lies also in the future, he may think that, however much he tries, he will not avoid it. Or his memory may tell him that he is not going to try. He will also be able to foresee whether or not he will achieve the desired outcome, and whether that will be because of, or in spite of, his action or inaction. This need not be disagreeable in the case where he welcomes the prospect of the action or inaction and its outcome. But in other cases his knowledge could be extremely inhibiting. He might feel obliged to go along with what he knows will happen, or he might feel that effort is fruitless or redundant. These feelings would impose a genuine constraint on some of his actions, even though they would not affect all. But the constraint would arise not from the fact that such and such was going to happen, but from his *knowledge* that it was going to happen. Such knowledge would be no less inhibiting in linear time, but nothing follows about the man who *lacks* foreknowledge. So long, then, as a person in circular time attends only selectively to what has happened in the past, he will

[34] ibid. 112-13.

be free of this particular source of constraint.

A further argument in favour of his being constrained seems also to be mistaken. It turns on the exact placing of such modal words as 'cannot'. Let the hypothesis be that the person has planted a tree. In circular time, it cannot be that that hypothesis is true and yet that he will not plant a tree in the future. But what is impossible here is only a certain combination: his having planted a tree and his not doing so in the future. It does not follow that the non-planting taken on its own is impossible. Of course, he will not refrain from planting: that is part of the hypothesis. But nothing yet entitles us to move from 'will not' to 'cannot' and to say that he *cannot* refrain from planting.

If the past planting were *irrevocable*, then it would indeed follow that he cannot in the future refrain. But I have argued that irrevocability is missing in circular time. The temptation to misplace the modal words may be greater, although the logic of the situation is no different, in the more complex case where an older self is juxtaposed with a younger self. If the hypothesis is that the younger self plants a tree, will not the older self be unable to stop him? Once again, the modal word has been wrongly placed. All that is impossible is the *combination* of the younger self planting a tree and the older self stopping him. But there is nothing in general impossible about older selves stopping younger selves, and they may frequently do so. Even when they do not, they may have had every possibility of doing so, and merely made no use of it.

Would events in circular time all be made inevitable by my earlier supposition that at every moment there are events which are *caused* by their predecessors? Not at all. First, I did not apply this supposition to every event. Secondly, I have argued elsewhere that what is caused need not be necessitated. I will not repeat my argument for that conclusion. Suffice it to say that it turns on linking cause with explanation, rather than with necessitation, and on divorcing the idea of a complete explanation from that of a necessitating explanation.[35] Some things that are caused are necessitated, no doubt, but they need not all be. Thirdly, in linear time a future effect need not yet be inevitable, even if it is necessitated by its cause. Suppose that cause and effect both lie in the future and that one necessitates the other: you cannot now light a match and avoid causing an explosion (the combination is impossible). But it does not follow that you cannot now avoid the explosion. The explosion would only become inevitable if the match-lighting became inevitable, and that need not happen in linear time, unless and until the match-lighting becomes irrevocable. In circular time, it need not happen at all, if I am right that irrevocability does not apply there.

Other suggested sources of constraint have already been dismissed. A man need not be prevented from killing his mother, given a different

[35] ibid. ch. 2.

physics which allows her to be resuscitated in time to give him birth. Nor does backwards causation imply that a planter who has enjoyed the shade of the trees he is going to plant is *unable* to refrain from planting them.

Circular time versus endless repetition

There is a final question. The case for describing the situation in terms of circular time rather than in terms of endless repetition rests at the moment on the flimsiest of logical considerations. It rests chiefly on a certain application of the principle of identity of indiscernibles. That application will appeal to some philosophers, but it is by no means irresistible.[35a] Can we not imagine an empirical test that would tell in favour of the circular description and against the idea of endless repetition, just as in the spatial case the measurement of interior angles told against the hypothesis of *spatial* repetition. An empirical consideration has recently been suggested by Susan Weir.[36] We should first need some evidence to narrow the correct choice of descriptions down to these two, the circular and repetitive descriptions. We should need evidence, that is, that everything had happened in exactly the same way before. Perhaps inhabitants of circular time could remember things as having happened in exactly the same way before, and even remember remembering. But we must be careful that this first piece of evidence is not of a totally deterministic character, suggesting that everything happens inevitably. For the second piece of evidence we want is something that suggests that a few things at least are *not* required to happen as they do. Taking those two pieces of evidence together, we could reason that it was not likely that world history would repeat itself exactly again and again, unless it was actually required to do so. Since the evidence of indeterminism would show that it was not required to do so, we should have reason to side against the hypothesis of repetition and to prefer the hypothesis of a single circle of time.

Such a conclusion would, I think, be fair, but evidence for indeterminism might be hard to come by. We should have evidence for it, if we found that the same state of the universe occurred twice within a cycle, but was followed in the short term by different sequels. However, even a single occurrence of the same world state might be too improbable in an indeterministic universe, for evidence to become available by this route.

If we cannot imagine any empirical evidence to decide between the circular and the repetitive descriptions, this need not be fatal to the circular one. An approach has been suggested according to which there

---

[35a] For doubts see R.W. Adams in *Journal of Philosophy* 76, 1979, 5-26.

[36] Susan Weir, *An Inquiry into the Possibility and Implications of a Closed Temporal Topology*, Ph.D. diss., Bristol 1985. A rather similar suggestion was put to me in conversation by Malcolm Murchison.

is no evidence at all that would decide. Consequently, we do not have here two rival theories, but only two alternative descriptions of a single state of affairs.[37] On that view, the principle of the identity of indiscernibles should not be taken as having some independent plausibility that gives us reason to prefer the circular over the repetitive description. It would simply be a stipulation that would generate that description rather than the other. We could then once again imagine empirical evidence, perhaps drawn from cosmological physics, but it would not be evidence favouring one description against the other. It would be evidence suggesting the existence of a state of affairs that could be *indifferently* described either way. In that case, what I have said earlier about circular time should still be viewed as legitimate. It would provide a legitimate, but not mandatory, description of the situation.

The Greeks on circular time

Having defended the possibility of circular time, I can return to A.A. Long's suggestion that circular time may be what the Stoics and perhaps the Pythagoreans intended. If they did, they ought not to retain the language, tempting though it is, of endless repetition: of events occurring 'again', or 'recurring', or being 'repeated', nor of a succession of 'worlds' (in the plural). For events, we have seen, would occur only once, and be experienced once. The Stoics would have also to work out a drastically altered physics and set of attitudes to life. They would have to reconsider their views on causation, freedom and responsibility. I see no trace of this in them myself. The Stoic theory that each cycle of history ends with a disastrous conflagration might even preclude their arguing for circular time in the way suggested above. For that particular argument depended on there being a seamless circle of events, none looking like a first or last, and each caused in a normal way by its predecessors.

Nor does Eudemus argue for circular time. He merely collapses into a finite linear time the supposedly endless repetitions of the Pythagoreans, and that in order to refute them, not to express his own view. It is true that the Greeks and their modern commentators do sometimes describe time as circular. But as I have pointed out elsewhere,[38] that normally means something very much less than what I have been discussing. Sometimes the point is no more than that *events* repeat themselves in linear time. But sometimes the reference is to time viewed as a system of measurement. The heavens provide a kind of clock, and the longest period marked by this clock is the period in which all the heavenly bodies return to their original alignments. If we view time as a measure, we ought then to think of it as coming to an

[37] W.V. Quine, 'Comments on Newton-Smith', *Analysis* 39, 1979, 66-7.
[38] Richard Sorabji, *Time, Creation and the Continuum*, 184-5.

end at the end of the cycle and starting up again. But the very word 'again' reveals that there is a further concept of time as not coming to an end, but continuing indefinitely in linear fashion. And this is explicitly acknowledged by Proclus in the lines I have italicised, in some of the very passages where he talks of time as circular.[39]

> The advance of time is not, as it were, a line that is single, straight and infinite in both directions, but finite and circumscribed ...
> The movement of time progresses in accordance with the measures of the temporal monad, fitting end to beginning, *and that an infinite number of times* ...
> The whole of time is contained in a single revolution of the whole universe ...
> The revolution of the whole has as its measures the entire extent and development of time, and no extension is greater – *unless it be by repetition, for in that way time is infinite.*

Despite all of this, however, I now think that there is at least one passage which tries to express the more rigorous idea of circular time, despite its inappropriate use of concepts like beginning, end, revert and again. The pseudo-Aristotelian *Problemata* describes and rejects an argument according to which we are earlier (as well as later) than the people of Troy, because things are circular. Unfortunately, the argument supports the idea of our being earlier by postulating a beginning, and suggesting that we might be nearer to it than are the people of Troy. The pseudo-Aristotelian author then has no difficulty, in the closing sentence, in dismissing the argument by pointing out that a circle contains no beginning. And he looks for no other way in which the circular hypothesis might be true. What is very interesting, however, is that he compares a remark of Alcmaeon, the philosopher of the fifth century B.C., who was at least a friend of Pythagoreans, if not a Pythagorean himself. Alcmaeon says that people die because they are not able to join beginning to end. This remark has been considered hard to understand.[40] And it has been taken to mean either that the soul will eventually be unable to complete its circling motions,[41] or that the body will eventually be unable to complete its physiological cycles.[42] But interesting as are the parallel passages about soul circles and physiological circles, when they are viewed as mere parallels, it needs to be considered that Alcmaeon may be making a fresh point

[39] Proclus *in Tim*. (Diehl) 3.29,3-5; 3.30,30-2; 2.289,14 and 20-3.

[40] Charles Kahn, 'Anaximander and the arguments concerning the *Apeiron* at *Physics* 203b4-15', in H. Diller, H. Erbse, eds, *Festschrift Ernst Kapp*, Hamburg 1958, 19-29 (p. 26); G.S. Kirk, J.E. Raven, M. Schofield, eds, *The Presocratic Philosophers*[2], 1983, 347; Jonathan Barnes, *The Presocratic Philosophers*[1], London 1979, vol. 1, 115.

[41] J. Burnet, *Early Greek Philosophy*[3], 1930, 195; A.E. Taylor, *A Commentary on Plato's Timaeus*, Oxford 1928, 262-3; cf. W.K.C. Guthrie, *A History of Greek Philosophy*, vol. 1, Cambridge 1962, 356.

[42] Charles Kahn, loc. cit.; Charles Mugler, 'Alcméon et les cycles physiologiques de Platon', *Revue des Études Grecques* 71, 1958, 42-50.

here, and one of very considerable philosophical interest. Men would be immortal, if they could only join their beginning to their end, that is, if they could only dwindle into the very seeds from which they came. Of course, the seeds would have to be the very same ones, in order that the ensuing life might be the same life – their own. Alcmaeon need not have worked out what would make that possible, but merely assumed that it was impossible. But the pseudo-Aristotelian author connects Alcmaeon's idea with that of circular time. And the discussion of circular time in the present paper has provided a set of hypotheses under which the immortality which Alcmaeon regards as impossible could come about. For people in the forty-year circle envisaged above could dwindle into the very seeds from which they came, and in doing so, they would secure a kind of immortality. The passage runs as follows:[43]

> How should we take 'earlier' and 'later'? In the sense that the people in the time of Troy are earlier than us, and the people before them are earlier than them, and as you go higher, the people are always earlier? Alternatively, if there is a beginning, middle and end of the universe, and a man too, when he grows old and comes to his limit, reverts again to his beginning, and things nearer the beginning are earlier, what prevents us being closer to the beginning part? If we are, we will also be earlier. Just as in motion there is a circle for the heavens and for every star, what prevents the formation and destruction of perishable things being such that the same things [reading: *t'auta*] are always being formed and destroyed? They also say that human affairs form a circle like that.
>
> Now it is simple-minded to think that those being born are always numerically the same people, but we could more readily accept that they are the same in kind. Thus [sc. on the mistaken view] we ourselves would be earlier, and someone could posit that the order of the series was such as to turn back again to the beginning and to make things continuous and to remain for ever in an identical state. (In fact Alcmaeon says that people die simply because they are not able to join beginning to end – a clever remark, if you take him to be speaking impressionistically and do not want to make the saying precise.)
>
> Now if there is a circle, and a circle has neither beginning nor end, people could not be earlier by being closer to a beginning, neither we earlier than they, nor they than us.

Space and time compared

If this passage does express the idea of closed time, there will be an asymmetry in Greek thought, for no one, I believe, expressed the idea of closed space. But this is not surprising. I myself needed the aid of a far from obvious treatment of straight lines in order to make sense of space being closed.

There is a parallel asymmetry in Greek thought concerning

[43] ps-Aristotle *Problemata* 17.3, 916a18-39.

reduplicated worlds. Reduplication over time became a commonplace. Reduplication in space was rarely considered. Of the three Presocratic authors, Anaximander, Anaxagoras and Democritus, who appear to talk of multiple worlds,[44] the reports on Anaximander are highly obscure, while Democritus' point is merely that, given the play of chance, one might expect the formation of some worlds indistinguishable from ours, as well as of many quite unlike it. The only one of the three philosophers who might be interpreted with any plausibility as offering a spatial analogue of the temporal reduplication of later theories is Anaxagoras. In fr. 4 (Diels-Kranz), which is preserved by Simplicius, he is describing the formation of a cosmos:

> Since this is so, we must think that there are many things of all sorts in all the things that are being concentrated, including seeds of all things with all kinds of shapes, colours and pleasurable sensations. And humans are put together and all the animals that have life. And the humans also possess inhabited cities and cultivated fields, as with us (*hôsper par' hêmin*). And they have a sun and moon and other things, as with us (*hôsper par' hêmin*). And the earth grows many things of all kinds for them, of which they collect the most useful into their homes and use them (*khrôntai*). So this is what I have said about the process of separating things out: things would be separated out not only with us (*par' hêmin*), but also elsewhere (*allêi*).

Could this passage be describing a simultaneous reduplication of similar worlds separated not in time, but in space?

There have been many alternatives to this interpretation. Some have thought that the other men resided in other parts of the earth, moon, or planets. But there would not be room there for the sun and moon (not described as our sun and moon) which they are said to have.[45] It has been suggested to me that the farmers and their sun and moon might be microscopic ones.[46] But when Anaxagoras says elsewhere that there is no limit to how small things come, I take him to be referring to stuffs, not to individuals. An alternative interpretation has represented Anaxagoras as saying only that similar worlds *would* be formed elsewhere, if similar circumstances *were* to arise, his point being to emphasise the inevitability of the one cosmogenesis that has in fact taken place. But to this the reply is that Anaxagoras says the farmers do use their harvests (*khrôntai*), not that they *would* use

---

[44] Anaximander (for a collection of passages, see e.g. W.K.C. Guthrie, *A History of Greek Philosophy*, vol. 1, Cambridge 1962, 106-15; Anaxagoras fr. 4 Diels-Kranz, taken from Simplicius *in Phys.* 34,28ff. with Simplicius' comments 157,9ff; Democritus ap. Ciceronem *Academica Priora* 2.17.55.

[45] For this reply, see Simplicius *in Phys.* 157,9ff and H. Fränkel, *Wege und Formen frühgriechischen Denkens*, Munich 1955, 286-92.

[46] I thank Andrew Harrison for this suggestion.

them.[47] A fourth suggestion would be that Anaxagoras is referring only to a chronological *succession* of worlds. But he describes the reduplication as occurring elsewhere (*allei*), not at other times (*allote*). Finally, Simplicius expresses tentatively, but with doubts, the thought that Anaxagoras might be referring to a Platonist ideal world rather than to a second physical world.[48] But this anachronistic proposal must be seen for what it is, part of the Neoplatonist attempt to view the tradition of pagan Greek Philosophy as unified.[49] The elimination of rival interpretations might seem to open the way for the view that Anaxagoras has indeed got in mind a spatial, not a temporal, reduplication of our world. But there are difficulties about that too. For two ancient reports list Anaxagoras as a person who believes that there is only *one* physical world.[50] One at a time is presumably what is meant, since Anaxagoras is bracketed with Empedocles, who allowed *successive* worlds. The report appears to be confirmed by the fact that a cosmos is formed by matter being separated out in a whirl and only one whirl is envisaged in the fragments of Anaxagoras.[51] Furthermore, even if Anaxagoras did envisage a spatial reduplication of our world, there is no suggestion at all that the reduplication is *exact*. Neither the things, nor *a fortiori* the events, need be exactly similar in the other worlds.

I conclude that the Greeks offered no spatial analogue at any rate for the *exact* reduplication of worlds which some of them envisaged for time. This asymmetry is not altogether surprising, because there is no spatial analogue to the Great Year to encourage speculation on spatial reduplication. The Great Year is the long period within which the heavenly bodies are supposed to return to the alignments they had before. The expectation of such repetition in the heavens encouraged belief in a temporal reduplication for the whole cosmos, but not, of course, in a spatial one.[52]

---

[47] The proposal is Fränkel's, the reply G.E.L. Owen's reported by Colin Strang, 'The physical theory of Anaxagoras', *Archiv für Geschichte der Philosophie* 45, 1963, 101-18, repr. in R.E. Allen and D.J. Furley, eds, *Studies in Presocratic Philosophy*, vol. 2, London 1975.

[48] Simplicius *in Phys.* 157,9ff.

[49] See my introduction to the first volume of translation in the series *The Ancient Commentators on Aristotle* (Philoponus *Contra Aristotelem*, translated by Christian Wildberg, London and Ithaca N.Y. 1987).

[50] Aëtius 2.1.2 (=*Dox. Gr.* 327=Diels-Kranz A 63); Simplicius *in Phys.* 178,25.

[51] Anaxagoras, frs 12 and 14 Diels-Kranz. The point is made by Gregory Vlastos, 'One world or many in Anaxagoras?', review of Fränkel in *Gnomon* 31, 1959, 199-203, repr. in Allen and Furley, op. cit.

[52] I am especially grateful to Susan Weir for her paper replying to my Readwell-Tuck lectures and to Christopher Hughes for discussion. I have also benefited from discussion with Alexis Belash, Victor Caston, Nicolas Denyer, Peter Gibbins, David Hirschmann, Charles Kahn, Clive Kilmister, Alan Lacey and Michael Stokes. John Lucas and Bill Newton-Smith have been kind enough to tell me of their latest views, although I have left it to them to make these known.

CHAPTER ELEVEN

The immobility of space:
Theophrastus on Aristotle

Even though only two fragments on place survive from his work, I believe that Aristotle's friend and successor Theophrastus had a major impact on Ancient Greek treatments of the subject. First, he collected doubts which proved to be seminal about Aristotle's definition of place. Then he offered his own reinterpretation of the idea of natural place. In this chapter, I shall consider the first of these two contributions. To do so, I must first outline Aristotle's views.

Aristotle offers at least four views on place. In the early *Categories*, he classifies place as a quantity and seems to think of it as coextensive with the body that occupies it.[1]

In *On the Heavens*, Aristotle stresses *natural* place. In his geocentric cosmos, the natural place of fire is up above at the periphery of the sublunary world with only the heavens above it, and that of earth is down below at the centre. The other two sublunary elements, air and water, are assigned intermediate positions. If *per impossibile* the earth were dislodged from the centre, loose clods would fall not towards the main mass, but to the place where they fall now.[2] Mere vacuum would contain no differences, and hence not the up and down by which natural motions are defined,[3] nor could it act as a cause (*aition*) of motion.[4] Aristotle evidently wants natural places to play an explanatory role in the natural movement of the elements towards them, for he says that such motion shows that place has power (*dunamis*).[5] The most likely explanatory role, although he never says

---

[1] Aristotle *Cat*. 6, 5a8-14: the parts of a place are occupied by (*katekhein*) the parts of a body, and share a common boundary with them. Averroes notes that this conception differs from that in the *Physics* and the difference is now the subject of an article by Henry Mendell, 'Topoi on topos: the development of Aristotle's concept of place', *Phronesis* 32, 1987, 206-31.

[2] Aristotle *Cael*. 4.3, 310b3.

[3] Aristotle *Phys*. 4.8, 215a9-11, cf. 214b17-19; 214b28-215a1. Upper and lower parts would not be *actually* present, in his view, if there was nothing to mark them.

[4] Aristotle *Phys*. 4.8, 214b12-17.

[5] ibid. 4.1, 208b11.

this, is as a final cause or goal (not consciously sought) of motion.[6]

In biology, Aristotle offers a definition of up, down, front, back, left and right in terms of biological function. Up is defined by the intake of food, so that the upper parts of a plant are its roots, on his theory. Right is defined by the initiation of motion, and front by the direction of the gaze. These ideas are applied not only to plants and animals, but also to the cosmos as a whole, where the initiation of motion can be located by reference to the poles.[7]

Finally, in the *Physics* Aristotle defines place roughly speaking, as a thing's surroundings. A thing's immediate place is the inner surface or boundary (*peras*) of the body that surrounds and contacts it,[8] as the air surrounds and bathes a person. In representing a thing's place as a two-dimensional surrounding surface, rather than as a three-dimensional extension or interval (*diastêma*) that goes right through it, Aristotle quite deliberately[9] rejects the three-dimensional view of place, which seems to have been his in the *Categories*. And his rejection of the idea of an interval or three-dimensional extension helps to show why 'place' (*topos*), rather than space (*khôra*), is the word he chooses for the concept he is getting at. On the other hand, his requirement that a thing's place be *equal* in size to it[10] shows that he is not thinking merely of its *position*, something which we might define in terms of distance from other points, but also of the *space it fits into*.

He sees no conflict between his definition of place as a surrounding surface or boundary and the idea of natural place.[11] We shall see in Chapter 12 that several Neoplatonists later objected that a mere surrounding surface could not account for natural motion in the way that natural place is supposed to do.

The most perplexing dilemma in Aristotle's account of place, it has been said,[12] and one that greatly exercised the Middle Ages, arises from his adding to the requirement of immediate contact the further requirement that the contacting surface which constitutes a thing's

[6] The denial at Aristotle *Phys.* 4.1, 209a20 that place can serve as any of the four causes, or four modes of explanation, is merely part of a puzzle or *aporia*.

[7] See Aristotle *Cael.* 2.3, 285a15f; b3ff; *Inc.* 4, 705a28-b18; 706a13ff; *PA* 2.10, 656a13; 4.7, 683b19-25; 4.8, 683b35; *Juv.* 1, 467b31-4; 468a2-5; *Phys.* 4.1, 208b15, with G.E.R. Lloyd, 'Right and left in Greek Philosophy', *Journal of Hellenic Studies* 82, 1962, 56-66, and James G. Lennox, 'Theophrastus on the limits of teleology', in W. Fortenbaugh, ed., *Theophrastus of Eresos, on his Life and Work*, Rutgers University Studies in Classical Humanities 2, New Brunswick 1985.

[8] Aristotle *Phys.* 4.4, 211a24-b4; 212a2-7 (on some readings); 212a29-30; cf. 4.5, 212b19.

[9] ibid. 4.4, 211b9-29.

[10] ibid. 211a2; a27-9.

[11] The two ideas are presented in the same contexts. See Aristotle *Cael.* 4.3, 310b7-11; *Phys.* 4.1, 208b11; 4.4, 211a3-6; 4.5, 212b29-213a10 (a passage missing, however, from the text read by Themistius); 4.8, 214b12; 215a8-11.

[12] Edward Grant, 'The medieval doctrine of place: some fundamental problems and solutions' in A. Maierù and A. Paravicini Bagliani, eds, *Studi sul XIV secolo in memoria di Anneliese Maier*, Rome 1981, 57-79.

place should also be *immobile*. Aristotle himself recognises the problem that the water surrounding a boat and in immediate contact with it may *not* be immobile. And so on the standard interpretation, he withdraws his original insistence on immediate contact, and redefines a thing's place as the first (i.e. the nearest) *immobile* surface or boundary of the surrounding body. The place of the boat would not after all be the inner surface of the water that immediately contacts it, but the inner surface of the whole river, that is (on the standard interpretation) of the river banks and bed:

> And as a vessel is a portable (*metaphorêtos*) place, so place is a motionless (*ametakinêtos*) vessel. And that is why when something moves and changes within a thing that is [itself] moving, like a boat in a river, it uses its surroundings [sc. the water] as a vessel rather than as a place. Place, however, ought to be immobile (*akinêtos*), which is why it is rather the whole river that is place, because the whole is immobile. So (*hôste*) *the first immobile boundary* (*peras*) *of the surroundings* is what place is.[13]

A difficulty for the standard interpretation is that Aristotle does not after all abandon the contact requirement, upon setting up the immobility requirement, for he goes on to repeat it at least once in the immediate sequel.[14] And it would be hard for him to give it up, because one of the opinions about place which he lists as correct (*alêthôs*) and as constraining any satisfactory definition is that a thing's immediate place is equal to it in size.[15] In fact, whichever of the two requirements he prefers, he will be faced with philosophical difficulties. There is an advantage, which he does not himself point out, of switching to the immobility requirement: if the place of a boat were, as he first seemed to suggest, the inner surface of the water that contacted it, then a boat moored stock still in a moving river would change its place as the water moved on. For new water would bring a new water surface and hence a new place. Conversely, a boat drifting with the water would not change its place. These unwelcome results will be avoided, if he makes a boat's place the surface of the river banks and bed. But in avoiding one set of problems, he will expose himself to another. First, a ferry plying between the two banks will not change its place (it is always between the same banks). Secondly, two ferries between the same banks will be in the same place as each other. Thirdly, if the tide stops, the nearest immobile surface (and hence the place) will suddenly switch from being that of the banks to being that of the water in contact with the boat.

There would be a way of retaining both the contact and the immobility requirements. For the inner surface of the water

---

[13] Aristotle *Phys*. 4.4, 212a14-21.
[14] ibid. 4.4, 212a29-30; cf. 4.5, 212b19.
[15] ibid. 4.4, 211a2; a27-9.

immediately surrounding a moored or moving boat might be described as immobile, on the ground that it does not exist long enough to move. Admittedly, if we consider only a *portion* of that water surface, it comes into being when the water is parted by the boat, moves in relation to the side, and goes out of existence when the waters close again. But if we think of the *entire perimeter* of the water surface round the boat, we can say that it exists as a whole only instantaneously, being continuously replaced by new perimeters. This would indeed allow Aristotle to treat the inner water surface as the boat's immediate and immobile place, but only at a price. For it would give the result already rejected above, that the moored boat continuously changes its place as new water surfaces come into being.

A much more promising solution has been offered by Myles Burnyeat, who tucks his most original interpretation into a footnote.[16] First, he provides an alternative sense in which the whole river can qualify as immobile: not only do the banks and bed differ from the water in not flowing, but also the river as a geographical entity does not change course. Secondly, he invites us to think of a boat-sized hole within this immobile geographical entity, and to think of the boat's place as the rim (*peras*) of the hole. Thirdly, around a moored boat the water may be for ever changing, but the rim can be said to endure, though constituted by ever different water surfaces. As an interpretation of Aristotle, this suggestion has great advantages. It enables him to retain the requirements of immediate contact and equal size, and, if successfully worked out, it would free him from the philosophical difficulties mentioned above. Unfortunately, however, it is not without its own difficulties, and these begin to emerge when we reflect that Aristotle insists on the immobility not merely of the *river*, but also of the *peras* or boundary,[17] here construed as the *rim* of the hole, and indeed it is not the river, but the rim, that we need to be immobile. These requirements do not come to the same thing. Let us first consider a *moving* boat: the rim surrounding a moving boat might naturally be thought of as moving with the boat.[18] It makes no difference that the river is immobile, for the rim still seems to be moving. Aristotle must, however, deny that the rim moves with the boat, since that would prevent the moving boat from changing its place. But what immobile rim can he find around a boat as it moves? He might try appealing to a series of *instantaneous* rims, none of which lasts long enough to move, because each new water surface round the moving boat constitutes a new rim. But that appeal would fail, because it was explicitly denied by the present interpretation that new water surfaces constitute new rims, at least in the case of the *moored* boat.

---

[16] Myles Burnyeat, 'The sceptic in his place and time', in Richard Rorty, J.B. Schneewind, Quentin Skinner, eds, *Philosophy in History*, Cambridge 1984, n. 15.

[17] Aristotle *Phys.* 4.4, 212a20.

[18] On moving holes, cf. Chapter 1 on the immobility of space, and Chapter 5 on whether the vacuum in a thermos flask can be penetrated by body.

Moreover, the idea that Aristotle has in mind *instantaneous* rims is not supported by any hint in the text that he is thinking of instantaneous entities. What immobile place is there, then, which the moving boat can be said to leave behind? If there is no immobile water rim available, we may be forced back to treating a portion of bank as its place.

If we now turn to the *moored* boat, the problem is not one of *finding* an immobile rim to surround the boat. It is rather one of specifying the *sense* in which the rim is immobile, although this may seem perfectly obvious. Thomas Aquinas supplies a sense in which the successive water surfaces surrounding a moored boat are immobile. But then Thomas recognises that he has to go beyond Aristotle in order to supply it. He defines the place of the boat as the spatial relation (*ordo*) of the successive water surfaces that contact it to the whole immobile river, which is in turn fixedly related to the heaven with its immobile centre and poles.[19] Edward Grant has discussed this concept of place as a relation or distance, namely the fixed distance of successive water surfaces (sometimes construed as a single surface or rim) from fixed points. There are anticipations of it in Grosseteste and Roger Bacon, and by the time of Buridan it is called *formal* place, as opposed to material place, which is mobile.[20] But the idea of fixed distances from fixed points goes far beyond anything in Aristotle. He would in any case have refused to put place in the category of relation. And once the idea of fixed distances from fixed points is used, it soon becomes apparent that the immobility of the river itself, on which Aristotle concentrates, is irrelevant. All one needs is a fixed distance between the rim of the hole and some fixed points somewhere in the universe. Nor is Aristotle any closer to Descartes' variant on Thomas, the idea of a boat's extrinsic place. This is a surface immediately surrounding the boat, which, however, belongs no more to the water than to the boat, and which remains the same, so long as the boat retains the same relation to other bodies, for example to the banks, which are *regarded* as unmoving.[21]

Aristotle's own way of giving sense to the idea of motion or rest in connexion with surfaces is different. The motion or rest of surfaces is *derivative* from that of the bodies to which the surfaces belong. This view does not apply to every kind of immobility (not to the immobility of *instantaneously* existing surfaces), but it does apply to rest, where rest implies the *possibility* of motion. Aristotle has to say that the rest of places is derivative for more than one reason: in its own right a place is incapable of retaining or changing its place, because he thinks we cannot say it has a place, on pain of regress.[22] Moreover, it cannot have

[19] Thomas Aquinas *in Phys.* in lib. 4, lectio 6, translated by Richard J. Blackwell, Richard J. Spath and W. Edward Thirlkel, London 1963, paragraphs 468-9.

[20] Edward Grant, op. cit.

[21] Descartes *Principles of Philosophy* 2.15.

[22] Aristotle *Phys.* 4.3, 210b22-7; 4.5, 212b27-9.

a place, if a place is defined as the two-dimensional surface of a thing's immediate *surroundings*. For the two-dimensional surface of the banks and bed immediately surrounding a river cannot itself be *immediately* surrounded by a further two-dimensional surface. Aristotle does think of the bank's surface as being at rest. But this is because he allows that points, and so he would allow that surfaces, can be in motion or at rest *derivatively*, or as he says *accidentally*, by belonging to a body which is in motion or at rest, like a sailor in a ship.[23] So the bank's surface is supposed to derive its immobility from the *bank*.

Aristotle's belief that the immobility of boundaries is *derivative* is manifested in the very passage under discussion, when he says that the whole river is immobile and then draws a conclusion (*hôste* – so) about the immobility simply of the boundary (*peras*) within it:

> It is rather the whole river that is place, because the whole is immobile. So (*hôste*) the first immobile boundary (*peras*) of the surroundings is what place is.[24]

This move from the immobility of the river to that of the boundary will have been plausible, so long as Aristotle was thinking of the bank and its surface: a bank surface does indeed share in the immobility of the bank. But why would it appeal to him, if he were thinking of a river with a fixed course and of the rim to a hole in it? Such a rim, we have seen, may move, regardless of the immobility of the river. Quite generally, it is unsafe to treat the motion and rest of surfaces as derivative, once we leave the reassuring example of banks, and turn to other examples. The instantaneous water surface surrounding a boat moored in a flowing river does not have time to share in the water's movement. The ebb and flow of the rim to a pool of water does not imply that the pool itself is on the move. These problems are disappointing. They mean that the 'hole in the river' interpretation cannot after all explain how Aristotle makes place immobile, as he must, if the moored boat is to be still and the propelled boat is to move. On the other hand, if Aristotle's appeal is to the *banks*, he runs into opposite problems, by abandoning the contact requirement.

We have seen that there are at least two components in his concept of place, the idea of an exactly equal (*isos*) space into which a thing fits, and the idea of *position*. For analysing the *latter*, the notion of fixed distances from other points, found in Aquinas and Descartes, was to prove more fruitful. And given the notion of fixed distances, the notion of a surrounding surface becomes otiose, at least for the purpose of defining position. Insofar as we do pay attention to surrounding surfaces in assessing position or change of position, we often attend only to some *part* of the surroundings, and discount other parts as

[23] ibid. 6.10, 240b17; cf. *DA* 1.3, 406a6-12.
[24] Aristotle *Phys.* 4.4, 212a19-21.

irrelevant. In determining whether a pedestrian is still or moving, we may attend to the ground under his feet, but we ignore the air which surrounds the major part of him. Similarly Aristotle's discussion of the place of a boat ignores the air above it, and in denying that a sailor seated in a ship is moving, except accidentally, he discounts both air and shore.[25] On the other hand, he would be happy to allow that a hand or arm was moving, even though securely anchored at wrist or shoulder to part of its surroundings. These cases are quite different from each other, but the common point is that the notion of surrounding surfaces does not hold the key to them.

I have concentrated on problems concerning the *immobility* of place, which were to prove the most intractable. But it is time to say that Aristotle's immediate successor Theophrastus assembled a whole battery of enduring problems, only two of them concerned with immobility, two concerned with the place of the heavens and one with the two-dimensionality of Aristotelian place. Simplicius summarises the five problems as follows:

> Remember that in his *Physics* even Theophrastus raises puzzles (*aporei*) such as the following against the definition of place given by Aristotle:
> (i) A body will be in a surface.
> (ii) Place will be in motion.
> (iii) Not every body will be in a place – not the sphere of the fixed stars.
> (iv) If the spheres are taken together, the heavens as a whole will also not be in a place.
> (v) Things in a place will no longer be in a place if, without their changing themselves at all (*mêden auta metakinêthenta*), their surroundings (*periekhonta*) are removed.[26]

Let us consider these puzzles in turn. There is a clue to the meaning of the first, because shortly after his report, Simplicius raises in his own person an objection, which is found also in Philoponus, and which may be related.[27] We have seen that one of the data which Aristotle's definition claims to respect is that a thing's immediate place is equal to it in size. But Aristotle has assigned a two-dimensional surface as the place of a three-dimensional body, and that cannot be equal to the body in size, but at best only to the two-dimensional surface of the body. Newton was later to devise a variant objection: bodies equal to each other in volume ought to have their places equal to each other, but their *Aristotelian* places will not be equal. For if one body is for example a sphere, while the other has a dented configuration, the *Aristotelian* place of the dented body will cover a larger area.[28] Aristotle might well reply to either version of the first puzzle that he

[25] Aristotle *DA* 1.3, 406a6-12; *Phys.* 6.10, 240b17.
[26] Simplicius *in Phys.* 604,5-11.
[27] ibid. 604,33-605,2; cf. 604,21-8; Philoponus *in Phys.* 564,3-14.
[28] Newton, Scholium to definitions at the end of the first section of *Principia* III. My thanks to Peter Alexander for the reference.

has sufficiently accommodated the common-sense notion that a thing is equal to its place and fits it exactly.

Theophrastus' first puzzle, then, concerns the two-dimensionality of Aristotelian place. His second concerns its *immobility* once again. We must recall that in Aristotle's geocentric cosmos, the elements are arranged as a nest of spheres, with earth at the centre, partly surrounded by water, then by air, then fire, and with the transparent heavens in a series of up to fifty-five celestial spheres round the outside. I shall speak of the celestial spheres as the heaven in the singular, when I wish to speak of them collectively. The celestial spheres are normally thought of as hollow spheres, and for convenience I shall here ignore the interpretation, described in Chapter 7, according to which they extend continuously to the centre of the cosmos, penetrating each other and the earth on the way. Theophrastus' second puzzle arises when we reflect that the cosmic components appear to be far from immobile. Simplicius and Philoponus develop the point.[29] None of the natural places in the cosmos is immobile, for the fifty-five celestial spheres rotate, and so do their inner surfaces. So also does the sphere of elemental fire beneath them, and air and water are also fluid and mobile. Aristotle himself seems to expect no difficulty about the heavens. At any rate, immediately after introducing the requirement that place be immobile (*akinêton*), he says that this is why (*dia touto*) the centre of the cosmos and the inner surface of the entire heaven are thought of as being down and up respectively, because the former stays still (*menei*), while the latter stays the same way (*hôsautôs ekhon menei*).[30] This shows that he considers the inner surface of the entire heaven sufficiently immobile to serve as a place, and according to some manuscripts, he goes on a little later to say that it is at rest (*êremoun*).[31] How can it be, we may wonder, if the heavenly spheres are rotating?

One answer is supplied by something that Aristotle says in a different connexion. He argues that a rotating body moves only in a sense, for it does not as a whole change its place simultaneously; only its parts move in that way.[32] This is the defence explicitly offered by Eudemus,[33] and it is a view taken, in a different context, by Alexander of Aphrodisias and Themistius.[34] I shall suggest below a positive reason why, in making place immobile, one might indeed not want to exclude its rotating. However, this solution does not deal with the case of water and air, which shift without rotating.

[29] Simplicius *in Phys.* 603,28-604,5; 605,26-35; 606,32-607,9; Philoponus *in Phys.* 564,14-565,1.

[30] Aristotle *Phys.* 4.4, 212a21-4.

[31] ibid. 4.5, 212b20.

[32] ibid. 212a31-b1; *GC* 1.5, 320a21-4.

[33] Eudemus ap. Simplicium *in Phys.* 595,8-9.

[34] Alexander ap. Simplicium *in Phys.* 589,5-8; 595,20-1; 602,31-5; Themistius *in Phys.* 120,17-19.

For the case of water and air, the solution of Burnyeat might be canvassed again. The air rim round a flying or hovering bird is immobile, because the air is immobile, and the air is immobile because the entire mass of it always keeps within the same place, that is, within the inner surface of the heavens which are immobile. This may seem to threaten a regress of immobilities, for what makes the heavens immobile? But Burnyeat satisfactorily puts a stop to this regress. The heaven as a whole is immobile, not through having and retaining a further immobile place, but through having no further surroundings and hence no place at all. By an irony its lack of a place, which Theophrastus is to present in his third and fourth puzzles as a possible source of weakness, is here a source of strength. Burnyeat is further able to show that Aristotle repeatedly speaks as if the heaven put an end to a related regress: your immediate place is on the earth, which is in the air, while the air is in the heavens, but the heaven as a whole has no further place.[35]

The obstacle to accepting Burnyeat's attractive solution lies not in any regress, but in the kind of point that has already been made: it is doubtful that the air rim around a *moving* bird could be called immobile,[36] while to specify the sense in which the air rim around a *hovering* bird was immobile, Aristotle would need to borrow the idea of distances from many centuries later.

Theophrastus' third and fourth puzzles also concern rotation. They were repeated by Sextus,[37] and provided the Middle Ages with the second of the two major perplexities that Edward Grant discusses. But this time Aristotle had anticipated them. In his cosmos of concentric spheres, the heaven as a whole, and still more clearly the outermost of the celestial spheres in it, have no surroundings, and so have no place in the sense defined by Aristotle. Yet it seems that they need a place, since in rotating they change place. Aristotle's main reply is given in *Physics* 4.5. It applies to the heaven as a whole and to its outermost sphere, and would apply, he says, to anything analogous, if there could be such, for example, to a mass of water with no outer surroundings. His answer comprises several distinct points. First, in rotation a body does not change place as a whole, and so the heaven as a whole needs no place in order to rotate.[38] This answer is accepted by Eudemus, Alexander and Themistius, as noted above, but rejected by Simplicius and Philoponus,[39] who insist that Aristotle repeatedly classifies rotation as a

[35] Aristotle *Phys.* 4.4, 209a32-b1; 211b28-9; 212a21-4; 4.5, 212b17-22; *Cael.* 4.3, 310b10-11.

[36] It has been put to me that it may be this last point, rather than the one I suggested, that constitutes Theophrastus' second puzzle: a flying bird takes its Aristotelian place with it. But then the point would be unduly restricted to *moving* bodies, and more narrowly to bodies that move by cutting a path through a medium that closes in behind (it would not apply to a worm burrowing through a rigid pipe). Moreover, it is in the other direction that Simplicius and Philoponus develop the anti-Aristotelian puzzles.

[37] Sextus *M* 10,30-6.

[38] Aristotle *Phys.* 4.5, 212a34-b1; cf. *GC* 1.5, 320a21-4.

[39] Simplicius *in Phys.* 595,16-26; 602,35-603,17; Philoponus *in Phys.* 565,1-8; 14-16.

change of place. We might protest against Aristotle that, even if the sphere which carries the stars does not, in rotating, change its place, any *part* of that sphere will change its place, for example, the segment in which the star Sirius is embedded, not to mention Sirius itself. Indeed, Aristotle himself allows that the parts of a rotating body move and change place,[40] and even points out that the outer parts of a spinning top have a different speed from the inner.[41] Yet he has a further problem how the fixed parts, outer or inner, of any moving body can be said to change place, since they all preserve the same relationships to *each other*. His answer is that the fixed parts of a moving body change their place only in an 'accidental' sense, like a sailor seated in a ship, in that they are part of some larger body which is changing its place.[42] But this solution seems inapplicable to the case of a rotating body, once he says that a rotating body does *not* change its place. This may be why we find some effort in Aristotle to provide a sense in which the whole rotating body does change its place after all.[43] But that only brings us back to Theophrastus' problem that the outermost celestial sphere has no place.

This strand, then, in Aristotle's answer seems unsuccessful. Another strand is the claim that at least the fifty-four *inner* spheres in the heaven have a place and change their place. Each has a place because (*gar*) one surrounds another, and each can move because (*dio, gar*) it has a place.[44] But now Aristotle is up against a double difficulty. First, to serve as a place, a surrounding sphere ought to be *immobile*. Secondly, the outermost of the celestial spheres still has no place, because it has no external surroundings.

A third strand in his answer is the claim that in certain special senses the heaven as a whole does have a place, but none of these senses turns out to throw any light on its rotation. For one thing, the heaven can be said to be in place 'accidentally', because virtually all its parts (the inner spheres) are in place, and it is in place in virtue of them.[45] For another thing, it can be said, by adapting an idea in Anaxagoras and Plato, to be 'in itself'.[46] But this latter location is not presented as one which changes, and so it too throws no light on how the heaven rotates. Moreover, the sense is admitted not to be primary (*prôtos*): a jar of wine can be said to be in itself merely because of an

[40] Aristotle *Phys.* 4.5, 212a34-b1; 6.9, 240a33-b1; 6.10, 240b13-17; *GC* 1.5, 320a21-4.
[41] Aristotle *Phys.* 6.10, 240b13-17.
[42] Aristotle *DA* 1.3, 406a6-12; *Phys.* 6.10, 240b17.
[43] Aristotle *Phys.* 6.9, 240b1-7.
[44] ibid. 4.5, 212a32-4; b12-14.
[45] ibid. 212b11-13. Eudemus and later commentators fill out the point by saying that the heaven is in its parts in the sense that it cannot be separated from them, and its parts are (most of them) in place: Eudemus ap. Themistium *in Phys.* 120,2-3 and ap. Simplicium *in Phys.* 595,14-16; Themistius *in Phys.* 120,31-2; Simplicius *in Phys.* 592,35-593,3; Philoponus *in Phys.* 603,8-10.
[46] Aristotle *Phys.* 8.6, 259b26; 4.3, 210a25-b22. Cf. Anaxagoras, ap. Aristotelem *Phys.* 3.5, 205b1-24; Plato *Parmenides* 138A-B; 145B-E.

ambiguity according to which the expression 'jar of wine' may refer
now to the wine and now to the jar. Finally, the outermost sphere can
be assigned a left, a right, an up and a down, on the basis of Aristotle's
functional definition of 'right', according to which right is the direction
from which motion starts.[47] But this does not help either.

A new solution, however, is added to the discussion by Themistius,
who argues that the outermost sphere is in a place, though not strictly
or without qualification, because it is 'as it were' surrounded 'in a way'
on its *inner* side by the surface of the next sphere in.[48] This defence is
rejected by Simplicius and Philoponus.[49] The latter complains that
Aristotle describes a thing's place as being outside it, as surrounding it
and as being equal to it. Themistius' proposal might still be presented
as a possible *revision* of Aristotle's view. It raises the problem of
something having a place when it is not surrounded on all sides. But
that is no worse than the problem already encountered that Aristotle
often ignores part of the surroundings, in assessing change of position.

Theophrastus' fifth and last puzzle is associated with our difficulties
about river banks and comes close to a puzzle which is said to have
been devised by the well-named John of Ripa in the fourteenth
century.[50] If someone opens the door and creates a draught, the air
surrounding me will move and I will be surrounded by different air and
a different inner surface, and hence will be in a different place, even
though I have stood stock still. Clearly this puzzle is not after all
genuinely new. It was illustrated above by the example of water being
replaced around a boat, and it is not far from Theophrastus' puzzle
that if my surroundings were removed, I should have lost my place
even though I had not myself changed. Simplicius and Philoponus also
come close to formulating John of Ripa's objection.[51] They describe the
draught of air around me, Simplicius pointing out that the passage of
water would do equally well. If they make it their conclusion not that I
will have *changed* my place, but that in the absence of a suitable
immobile surface, I will have *no* place, this is simply because they read
Aristotle as requiring an *immobile* surface. Simplicius seems to
associate the puzzle with Theophrastus by raising it immediately
before citing Theophrastus' puzzles.

A *partial* line of defence against John of Ripa's version and against
Theophrastus' earlier second puzzle should by now have become clear.
For a mere *rotation* around me of the surrounding air would not put me
in contact with a different surrounding surface. And this supplies a
reason, previously missing, for the line of defence according to which
Aristotle can afford to let places rotate, even though he cannot afford to
let them move in a more thoroughgoing way.

[47] Aristotle *Cael*. 2.2, 285a22-b33.
[48] Themistius *in Phys*. 121,2-4; 8-11, repeated ap. Simplicium *in Phys*. 592,22-4.
[49] Simplicius *in Phys*. 590,27-32; 592,25-7; Philoponus *in Phys*. 565,21-566,7.
[50] Edward Grant, *Much Ado About Nothing*, Cambridge 1981, 125.
[51] Simplicius *in Phys*. 604,3-5; 607,1-2; Philoponus *in Phys*. 564,19-23.

For a fuller defence against John of Ripa and against Simplicius and Philoponus, one might appeal again to Burnyeat's interpretation of Aristotle. For on that interpretation, Aristotle would say that I was always surrounded by an air surface whose immobility, despite local perturbations, was guaranteed by that of a vaster body, and ultimately by that of the heavens. But we have seen that such a defence would require a development that Aristotle does not give it.

Is Theophrastus' puzzle the same as the later ones of Philoponus, Simplicius and John of Ripa? On the face of it, there seems to be a difference, because he appears to envisage a thing's surroundings being *removed altogether*, not merely *replaced*, and its ending up with no place at all. What is supposed to be paradoxical is its coming to have no place, despite not changing itself. But Aristotle could protest, first, that such an experiment could not be performed, and, secondly, that if it were, we ought not to be surprised that the removal of *all* surroundings should suffice to deprive a thing of location. It is likely, therefore, that the experiment Theophrastus intends is the more familiar one of replacing, not removing altogether, a thing's surroundings. As for the paradoxical result of the experiment, this is likely to be that of Simplicius and Philoponus that without changing itself, the thing loses *all* place, because it ceases to have an *immobile* body, as Aristotle axiomatically requires, immediately surrounding it. The alternative version of John of Ripa is that, without changing itself, the thing exchanges its *mobile* place for a *new* one. But this is not so likely to be Theophrastus' point, both because he expresses the paradox in terms of no longer being in a place at all, and because he is said, like Simplicius and Philoponus, to take it as an axiom (*axiôma*) that place is immobile.[52]

The fifth puzzle draws part of its sting from the phrase, 'lose their place, *without their changing themselves at all*'. It thus relates to the recorded views of Aristotle and Theophrastus on another issue. For Aristotle thinks this sort of thing happens when a thing changes its relations, but not when it changes its place. In Book 3 of the *Physics*, he says that there are as many kinds of change (*kinêsis* and *metabolê*) as there are categories. But he clearly includes only four categories, for he also says that changing things (*metaballonta*) always change in respect of substance, quantity, quality, or place.[53] And this is confirmed in Book 5 of *Physics*, where he says that in respect of relatives there is no change, except accidental change (*kata sumbebêkos hê kinêsis*), which he has agreed to dismiss (*apheisthô*, 224b27). There is no change because it is possible that, *without a thing's changing at all* (*mêden metaballon*), a different relational property becomes true of it.[54] The phrase 'without changing at all' had

[52] Simplicius *in Phys.* 606,32-4.
[53] Aristotle *Phys.* 3.1, 200b33-201a9.
[54] ibid. 5.2, 225b11-13 (=*Metaph.* 11.12, 1068a11-13; cf. 5.1, 224a21-2; b26-7); 7.3, 246b11-12; 247b4; *Metaph.* 14.1, 1088a30-5.

been illustrated by Plato who said that Socrates does not change when, through Theaetetus' growth, he comes next year to be shorter than Theaetetus.[55] And it seems to be common ground to Plato, Aristotle and the Stoics that Socrates does not thereby change, except, Aristotle would say, accidentally.[56] But Simplicius disagrees, or is at least worried at *metabolê* being denied in this way,[57] and in two different works he cites in his support Theophrastus' definition of change (*kinêsis*). Theophrastus is like Aristotle in defining change as the actualisation of a potentiality, but unlike him in adding explicitly that this occurs 'in *every* category'.[58] Simplicius rightly or wrongly takes Theophrastus to mean more than the four categories allowed in Aristotle's earlier statement, and he also tells us that Alexander applies Aristotle's definition of change (*kinêsis*) to the category of relatives (*pros ti*).[59] We can now appreciate the extra sting in Theophrastus' puzzle. Theophrastus is saying that loss of place, and John of Ripa that change of place, is just like relational change, in that the thing itself does not change at all. This is the opposite of what Aristotle wanted to say about change of place, since he contrasted it with relational change in this very respect.

The two puzzles on which I have dwelt longest, the second and the fifth, are both about the *immobility* of place. Immobility is needed if the moored boat is to be described as still and the propelled boat as moving. The role of immobility comes out more clearly when we observe the difficulties which subsequent philosophers have felt about finding something immobile, whether a place or another body. Descartes, for example, says:

> We may well end by thinking that no such genuinely unmoving points are to be found in the universe.[60]

And he defines place and motion in terms of a relation not to immobile

---

[55] Plato *Theaetetus* 155B-C.

[56] For the Stoics, see Simplicius *in Cat*. 166,17-29; 172,1-5.

[57] Simplicius *in Phys*. 861,5-28, would rather Aristotle had made a terminological distinction, allowing *metabolê*, while denying *kinêsis*. See Concetta Luna, 'La relation chez Simplicius', in I. Hadot, ed., *Simplicius – sa vie, son oeuvre, sa survie*, Peripatoi vol. 15, Berlin 1987.

[58] Theophrastus ap. Simplicium *in Phys*. 412,31-413,9; 860,19-861,4; 861,19-26; *in Cat*. 435,28-31. At *in Phys*. 413,7-9, Theophrastus is quoted as saying that change occurs for relatives which stand in a relation 'corresponding to capacities (*kata dunamin*)'. But this probably does not depart from Aristotle who, at *Metaph*. 5.15, 1021a14-30 (reference supplied by Pamela Huby), gives as examples of relatives *kata dunamin* the relation of what heats to what is heated and of what makes to what is made, and seems to allow (a20) that the activity of such relatives involves change (*kata kinêsin energeia*).

[59] Alexander ap. Simplicium *in Phys*. 409,24-32. Alexander adds that when one thing is reduced in size, to become half the size of another, the change (*kinêsis*) is in both of them (even though, confusingly, only one is changed – *kekinêtai*), because it is one and the same potentiality that is actualised in the thing that is potentially double and the thing that is potentially half.

[60] Descartes *Principles of Philosophy* 2.13.

bodies, but merely to bodies *regarded* as immobile.[61] How, then, can the relativist view be denied that the moored boat is moving in relation to the water, as much as the water in relation to the boat? Descartes reports a conception according to which a thing is moving only if there is an *activity* in it by which it moves, but this is only the vulgar conception, he says.[62] Leibniz reserves the designation 'true motion' for the body which acts as a *cause* of motion,[63] and Berkeley for the body which has *force applied to it*,[64] but both downgrade true motion by treating it as merely a *species* of relative. Modern physics can distinguish which of two bodies is moving if one is accelerating, while the other is not, but is helpless if bodies move uniformly in relation to each other without acceleration.

What was Theophrastus' role in connexion with the puzzles? He did not invent them all. The third and fourth on the place of the heavens are anticipated by Aristotle himself, and we have seen that the second on immobility was tackled also by another of Aristotle's younger associates, Eudemus of Rhodes. So we cannot be certain that Theophrastus invented any. Furthermore, the five are described by Simplicius as Theophrastus' *puzzles* (*aporein*), not his objections, so he may have hoped for answers. On the other hand, there is no record of him supplying any, as Eudemus clearly did. It would also be wrong to describe him, as he was described in Antiquity, merely as 'touching lightly on those things which he knows to have been said by Aristotle before',[65] or as one who 'followed Aristotle in almost everything'.[66] His importance lies in his having assembled the puzzles together. This in turn will have encouraged the fuller and more formidable formulations and additional objections that we find later in Sextus, Simplicius and Philoponus.[67] I have given only a selection of these developments.

Theophrastus' assembling of puzzles may not only have encouraged further objections to Aristotle, but also opened the way for other theories of place to thrive. Greek philosophers after Theophrastus, including his successor Strato, nearly all rejected Aristotle's definition of place, and I have already commented in Chapter 9 that this may explain why the Stoics appear to feel no need to attack that definition. The wrongness of Aristotle's definition is one of the points on which even those arch-enemies Simplicius and Philoponus agree. Not only did later generations tend to think Aristotle's definition wrong: they were encouraged to accept rival conceptions which the puzzles would not threaten. The commonest idea was that place is an immobile,

[61] ibid. 2.13 and 25.
[62] ibid. 2.24.
[63] Leibniz *5th Letter to Clarke* §53, ed. Alexander, p. 74.
[64] Berkeley *Principles of Human Knowledge* §§113-15.
[65] Boethius *in Int* (2), ed. Meiser, 12,9-12.
[66] Simplicius *in Phys.* 789,4f.
[67] Sextus e.g. *M* 10.30-6; 55-6; *PH* 3.131; Simplicius *in Phys.* 601,25-610,22; Philoponus *in Phys.* 563,26-567,29.

three-dimensional extension.[68] This and Proclus' view that it is an immobile, interpenetrating, three-dimensional body,[69] would be exempt from all the puzzles. Apart from Aristotle and his followers, Simplicius recognises only two exceptions to the view that place is three-dimensionally extended.[70] A striking feature of some of these rival theories of place is that in various ways they come closer to the idea of Newtonian space than Aristotle did, and they might indeed be called theories of *space*. Besides those who thought of space as a three-dimensional extension, a number thought that space could exist vacuously, empty of body.[71] How many did so it is hard to say because of the obscurity of the idea, which is found in Philoponus and ascribed to Strato and to most Platonists,[72] that space could exist without body, 'so far as its own nature was concerned'.[73] The most vivid parallel with Newton's views is found in those theories which allow the whole of physical matter to move in empty space. Cleomedes reports the Stoic argument that we can imagine (*epinoêsai*) the whole physical cosmos

[68] Earlier accounts of space, e.g. in Democritus, Archytas and Plato, are not very explicit on the point, although they treat it as three-dimensional and extended. See David Sedley, 'Two conceptions of vacuum,' *Phronesis* 237, 1982, 175-93. For place as extension in *Epicurus*, see e.g. *Letter to Herodotus* 39-40; Lucretius 1,419-444; Sextus *M* 10.2; Themistius *in Phys.* 113,11; Simplicius *in Phys.* 571,24-5, with David Sedley, op. cit. Epicurus treats space as an intangible substance which may be vacuous, full, or moved through. For the Stoics, e.g. Sextus *M* 10.3 (*SVF* 2.505); *PH* 3.124; Themistius *in Phys.* 113,11 (*SVF* 2.506); Simplicius *in Phys.* 571,24-5 (*SVF* 2.508). For Strato, Simplicius *in Phys.* 601,24; 618,24; Aëtius *Placita* 1.19 in *Dox. Gr.* 317. For Galen, Themistius *in Phys.* 114,7; Simplicius *in Phys.* 573,19-32, Philoponus *in Phys.* 576,13. For 'most Platonists', Simplicius *in Phys.* 601,24; 618,24. For Philoponus, his *Corollary on Place*, at *in Phys.* 557,8-585,4. For a report *without attribution*, Syrianus *in Metaph.* 84,27-86,7. I have not personally been convinced by an alternative interpretation of Epicurus, which should, however, be studied in the skilful formulation of Brad Inwood, according to which place or void should be viewed not as an extension, but as an ideal fluid which flows round the edge of solid atoms. So Brad Inwood, 'The origin of Epicurus' concept of void', *Classical Philology* 76, 1981, 273-85, with further comment in Chapter 5 above.

[69] Proclus, *Commentary on Plato's Republic*, vol. 2, p. 198, lines 8-15, and ap. Simplicium *in Phys.* 611,10-618,25.

[70] Simplicius *in Phys.* 601,16-19.

[71] Of those listed above as making place extension the following allow vacuum to exist, some inside, some outside, the cosmos, and some in microscopic pockets: Epicurus, the Stoics, perhaps Galen, Strato (ap. Simplicium *in Phys.* 693,11-18), and in addition Hero of Alexandria who has a passage corresponding verbally to one of Strato's (Hero *Pneumatics*, introduction). For vacuum in the atomists Leucippus, Democritus and in the fourth century Metrodorus, see Diels-Kranz, *Fragmente der Vorsokvatiker*, index s.v. *kenon*, and David Sedley, op. cit. For the more primitive concept of breathable void in the Pythagoreans, see Aristotle *Phys.* 4.6, 213 6-22-5, and for the Pythagorean Xuthus, Chapter Eight. Plato's denbial of a void into which any body could move at *Timmaeus* 79B1 is complicated by his acceptance of smaller interstices at 58A4-B5; 60E5.

[72] Philoponus e.g. *in Phys.* 579,6-9 (see also the *Corollary on Void*, at *in Phys.* 675,12-695,8). Strato and the Platonists reported by Simplicius *in Phys.* 601,24; 618,24. This is distinct from the attribution to Strato of belief in microscopic pockets of void.

[73] It has been argued that in Philoponus this merely means that its definition must exclude reference to body, not that it could in principle exist without: David Sedley, 'Philoponus' conception of space', in Richard Sorabji, ed., *Philoponus and the Rejection of Aristotelian Science*, London and Ithaca N.Y. 1987.

being shifted,[74] and Epicurus postulates that all atoms fall downwards in the void.[75] The Newtonian view of space as infinite, though not uncommon among the Presocratics,[76] became rarer after Aristotle's attack on more than finite quantities.[77] Renaissance readers of Greek texts found that they had to add the idea of infinity. But even here the Epicureans and Stoics were happy to make space infinite.

The liberation of Antiquity from Aristotle's view is not always recognised. It is set in sharper focus by the contrast with the Middle Ages which reverted to Aristotle. I have already cited the reports which Edward Grant gives of mediaeval efforts to rescue Aristotelian theory from the kind of objections we have been considering.[78] Grant further comments that the anti-Aristotelian conception of a space, void so far as its own nature was concerned, and containing all bodies, gained little support in the Middle Ages.[79] It was known through Averroes to have been Philoponus' conception, and it was with inspiration from Philoponus that Pico della Mirandola[80] reintroduced it in the sixteenth century. Yet we find Descartes still struggling with the Aristotelian legacy, in his concept of extrinsic place as a surface belonging no more to the water than to the boat, and remaining the same, so long as the boat retains its relation to other bodies regarded as unmoving.[81] In contrast with this, Greek theories show far less attachment to Aristotle, and my suggestion has been that this is partly due to the influence of Theophrastus.[82]

[74] Cleomedes, *De Motu Circulari Corporum Caelestium* (=*Theôria Kuklikê*) 1.1.3.

[75] Epicurus *Letter to Herodotus* (in Diogenes Laertius *Lives* 10) §61; Aëtius 1.12.5; Lucretius 2,216-42. Cf. Proclus, who believes that the whole spherical cosmos rotates against a space which is, however, itself physical: Proclus *in Rempublicam* 2.197,17-198,4 (Kroll); ap. Simplicium *in Phys.* 612,32-5.

[76] Anaximander's infinite is construed as a circle by F.M. Cornford, 'The invention of space', in *Essays in honour of Gilbert Murray*, London 1936. But for infinite space in the atomists Leucippus, Democritus and later Metrodorus, and perhaps also in Xenophanes, see Diels-Kranz, *Fragmente der Vorsokratiker*, index s.v. *apeiron*. For Melissus see ibid., frs 3,4,5,6,7, for the Pythagoreans Aristotle *Phys.* 3.4, 203a7; 4.6, 213b22-5, for Archytas Chapter 8, for Anaxagoras Aristotle *Phys.* 3.5, 205b1-3.

[77] Aristotle *Cael.* 1.5-7; *Phys.* 3.4-8.

[78] Edward Grant, 'The medieval doctrine of place'.

[79] Edward Grant, *Much Ado About Nothing*, 19.

[80] Edward Grant, *Ado*, 19; Charles Schmitt, *Gianfrancesco Pico della Mirandola (1469-1533) and his Critique of Aristotle*, The Hague, 1967, ch. 5.

[81] Descartes *Principles of Philosophy* 2.15.

[82] I am most grateful to Myles Burnyeat for discussing his interpretation with me. Other debts in this chapter are to Christopher Rowe, Peter Alexander, Pamela Huby, Eric Lewis and Robert Sharples.

CHAPTER TWELVE

Is space inert or dynamic?
Theophrastus and the Neoplatonists

I now turn to Theophrastus' second fragment, in which he switches from raising difficulties for Aristotle's definition of place to raising difficulties which may be taken to concern his idea of *natural* place, discussed in Chapter 11. The difficulties are still presented only as creating a puzzle (*aporia*), with a 'perhaps' (*mêpote*), so his attitude to them remains to be determined.[1] But the suggestion he entertains, whether he favours it or not (and I shall return to this question below), provides a complete antithesis to Aristotle's theory of natural place. It is to the effect that place (*topos*), that is, natural place, is not a reality in its own right (*kath' hautou ousia tis*), and represents no more than a way of talking about something else, namely, the physical parts of an organism and the arrangement (*taxis, thesis*) of parts dictated by the nature (*phusis*) of the organism. On this view, to say that a dislocated arm is in the wrong place is only to say that its relation to the other parts of the body is contrary to that called for by the nature of the organism. It is not to say that there is such a thing as a place into which the dislocated arm has got itself. And this account can be extended from parts of the *body* to parts of the *cosmos* as a whole.

Theophrastus' idea contradicts Aristotle's view that natural place has power (*dunamis*), and that it helps to explain why the four elements, earth, air, fire and water, move in their different directions. All power is transferred to the *nature* of each organism, including the whole cosmic organism, and the talk of natural place becomes a *façon de parler*. There is a further antithesis with the view which Aristotle expresses in his early dialogue, *On Philosophy*.[2] For he there suggests that the four elements that compose human organisms are robbed of their natural place and forced into an unstable and temporary union. Theophrastus' idea is, not indeed that the elements, but that the

---

[1] I am grateful to David Sedley for pressing this point in his reply to me at the fourth International Conference on Theophrastus. I will leave it to him to explain his own rival interpretation.

[2] Aristotle *On Philosophy* fr. 19b Ross, from Philo *Aet.* 6.28-7.34. The passage was drawn to my attention by Eric Lewis.

organic parts of the human organism have their natural place where
they are.

I have said that Theophrastus' downgrading of talk about place is best
construed as a challenge to *natural* place, not to place in general. For his
suggestion would not cover the ordinary sense in which we say that a
rock rolled along the ground has left its former place. At best it would
only allow us to say that a rock had left its place, when it was thrown
upwards into the air, contrary to the natural arrangement dictated by
form. It is therefore easiest to suppose that Theophrastus believes some
further conception of place is in play when we talk of things changing
their place.

In Chapter 11, it was seen that Aristotle spoke of a left and right, up
and down in the cosmos, but Theophrastus' suggestion goes far beyond
that. Because we are talking about a part's *relation* to other parts, this
has been called a relational account of space (I would rather say of
natural place), the only relational account in Antiquity.[3] To call it
relational is not to say that it automatically implies what I called in
Chapter 11 a relativist account of motion, for it would remain to be seen
whether there were criteria for saying that the arm had moved rather
than the rest of the body. The quotation is again given by Simplicius:[4]

> Theophrastus too appears to have had this conception of place in his
> *Physics* when he continues his account in the form of a puzzle (*hôs en
> aporiâi*), and says: 'Perhaps (*mêpote*) place (*topos*) is not a reality in its
> own right (*kath' hautou ousia tis*), but we speak of it because bodies have
> an order and position (*taxis, thesis*) through their natures and powers
> (*phuseis, dunameis*). And similarly in the case of animals and plants and
> in general of non-homogeneous things, whether animate or inanimate, if
> they have a nature that exhibits form (*emmorphos phusis*). For in these
> too there is an order and position of parts in relation to the whole being
> (*ousia*). And this is why everything is said to be in its own space (*khôra*)
> through having its proper order. For each part of the body too would
> desire and demand its space and position.'

It is very natural for Theophrastus to extend his idea to the parts of
the cosmos: earth, water, air, fire and the heavens. For elsewhere too
he is attracted by the idea of the cosmos as an organism or as like one.[5]

---

[3] S. Sambursky, *The Physical World of Late Antiquity*, London 1962, 13; *The Concept
of Place in Late Neoplatonism*, Jerusalem 1982, 13; Max Jammer, *The Concept of Space*,
London 1954, 21. In making place, so far as it exists at all, relative to bodies,
Theophrastus has nothing in common with modern Relativity Theory. That makes space
relative to frames of reference, but not to bodies. Reference to bodies is often replaced at
the sub-atomic level by the idea of spatially extended fields.

[4] Simplicius *in Phys*. 639,13-22.

[5] Marlein van Raalte, 'The idea of the cosmos as an organic whole in Theophrastus'
Metaphysics', in W. Fortenbaugh and R.W. Sharples, eds, *Theophrastus as Natural
Scientist*, Rutgers Studies in Classical Humanities 3, New Brunswick and London 1988
sees this as the principal theme of his *Metaphysics*. Different assessments are offered by
Glenn Most and John Ellis in the same volume.

In his *Metaphysics* (8a5), he says that the whole heaven has parts and is as well fitted together as a city or an animal. (The word 'heaven' here stands for the whole cosmos, as allowed by Aristotle,[6] not as it usually does for the celestial part of the cosmos.) He further assigns soul to the celestial spheres, and wonders whether the lower elements should not also be assigned soul, and whether they are not *parts* of the heaven (in the wide sense, viz. the cosmos), rather than suffering merely accidental effects from the celestial rotation (5a28-b26). Not only is Theophrastus tempted to think of the cosmos in terms of an organism, but his suggestion would enable him to answer the question he put to Aristotle, how the heaven (in the narrow sense) can have a place. It does so because it has a natural relation to the other parts of the cosmos, a relation dictated by the nature or form of the cosmos. All this makes it plausible that he may have put forward his suggestion not as a puzzle to be refuted, but as something deserving serious consideration. Admittedly, we have seen that it would make Aristotle's talk of natural places only a *façon de parler*. But that Theophrastus should mean to raise that doubt in the *Metaphysics* is not excluded by the report that he himself speaks of natural places, in the Arabic translations of his lost *Meteorology* which Hans Daiber has recently collected in India.[7]

The connexion of place with organic parts was to have wide appeal. It is found not only in the Neoplatonist texts to be discussed below, but even in so anti-teleological an author as Lucretius.[8] Moreover, Theophrastus' whole suggestion is the subject of an unexpected sequel in the Neoplatonist period. Simplicius claims that the Neoplatonist Iamblichus takes the same view of place as Theophrastus[9] and that Theophrastus and Iamblichus in turn support his master Damascius.[10] The point of comparison is that Iamblichus makes a thing's place to be naturally united with it (*sumphues*), rather than external (*exôthen*), just as Theophrastus says that things have a place dictated to them by nature (*phusis*, conn. *sumphues*), which is not merely accidental to them.

I shall return to this comparison, but first I should issue the warning that it overlooks a very big difference: Iamblichus restores to place the independent reality and power of which Theophrastus had drained it.[11]

[6] Aristotle *Cael*. 1.9, 278b9-21.

[7] I am relying on an oral report by Daiber to the Fourth International Conference of Project Theophrastus, held at the Institute of Classical Studies, London, in June 1985. Arabic text and translation, not presented at the conference, are forthcoming in Rutgers University Studies in Classical Humanities 3, as above.

[8] Lucretius 5, 540: the earth stays in its place without weighing down the air beneath, because it is like a head which does not weigh down a neck.

[9] Simplicius *in Phys*. 639,10. Pending the appearance of new translations of the Neoplatonist commentators on Aristotle, the main Neoplatonist texts on place can be found translated in S. Sambursky, *The Concept of Place in Late Neoplatonism*, as above.

[10] Simplicius *in Phys*. 642,17.

[11] Here I must dissent from the very helpful article of Philippe Hoffmann, 'Simplicius: Corollarium de loco', in *L'Astronomie dans L'Antiquité Classique*, Actes du Colloque tenu à l'Université de Toulouse-le-Mirail, Paris 1979, 143-63, p. 158, n. 18.

It is[12] or has[13] power (*dunamis*), a power that acts (*drastêrios dunamis*), for it is not inactive or inert (*adranês, argos*). In particular, it stops bodies from spreading and being dissipated and draws them together (*sunagei*),[14] and it holds up (*anekhei*) bodies that would fall.[15] The idea of holding up is the idea of preventing bodies from descending into prime matter, which is both the lowest thing in the metaphysical hierarchy and the thing which, by being spread out and dissipated, is at furthest remove from the unitary character of the One at the top. By holding up bodies and preventing their dissipation, place gives them their very existence.[16]

Given this *dynamic* conception of place, Iamblichus has to reject the view that place is a bare extension (*diastêma psilôs*),[17] or, as the Stoics said, something which merely supervenes upon bodies (*paruphistasthai*),[18] or, as Aristotle might be taken to mean, the bare boundary of a thing's surroundings (*to peras tou periekhontos psilon*).[19] Iamblichus can accept Aristotle's description 'boundary' (*peras*), if it is taken as implying cause and origin (*aitia, arkhê*). And he can accept his talk of surrounding (*periekhein*), if it is not a merely external surrounding (*exôthen*), but something connected with holding bodies up (*anekhein*).[20] There had been a tradition of giving the idea of surrounding or embracing (*periekhein*) a dynamic connotation ever since the early Presocratics Anaximander and Anaximenes had connected it with governing (*kubernan*)[21] and controlling (*sunkratein*).[22]

Only two sources are named as encouraging Iamblichus to take his dynamic view of place. First, Simplicius emphasises the influence of

[12] Simplicius *in Phys.* 640,3.

[13] Simplicius *in Cat.* 361,15-18, 362,19.

[14] The point is put in several ways. It draws bodies together (*sunagein*), Simplicius *in Phys.* 640,4;11; it does not allow their dissipation (*diastasis*) to proceed indefinitely, Simplicius *in Cat.* 361,18-20, but holds together their dissipation (*sunekhein diastasin*) 362,4; it surrounds or embraces them (*in Phys.* 640,5; 10: *in Cat.* 361,33-362,4; 362,16-18), and gives them a boundary (*aphorizein, sumperainein* 361,15, *horizein in Cat.* 361,20; *in Phys.* 640,11).

[15] It holds them up (*anekhein, in Phys.* 640,13; *in Cat.* 362,23; 25-6); props them up (*diereidein, in Phys.* 640,4); raises those that fall (*anegeirein piptonta, in Phys.* 640,4; *in Cat.* 362,26); fills them (*sumplêroun, in Phys.* 640,5; 9); fills them with a power that raises them (*plêroun anegeirousês dunameôs, in Cat.* 362,25).

[16] Simplicius *in Cat.* 362,1-2.

[17] Simplicius *in Phys.* 639,29-30; *in Cat.* 362,18.

[18] Simplicius *in Cat.* 361,10; 27-8.

[19] Simplicius *in Phys.* 639,29; *in Cat.* 362,10; 24; 27-33.

[20] Simplicius *in Cat.* 361,30-362,4; 362,10; 16-19; 24; 27-33.

[21] Aristotle *Phys.* 3.4, 203b7. The 'Boundless' (*apeiron*) which surrounds and governs might sound like the very opposite of Iamblichus' boundary (*peras*), but the correction according to which the boundless or infinite (*apeiron*) is not well qualified to surround or embrace (*periekhein*) had been supplied in the interim by Aristotle, *Phys.* 3.6, 206b33-207a2; 3.7, 208a3.

[22] Anaximenes fr. 2 Diels-Kranz. Alexander connects it with preserving (*sôzein, Mixt.* 223,9-14). Cf. Empedocles fr. 38.4 Diels-Kranz; Plato *Tim.* 58A; Lucretius 5,467-70; Cleanthes ap. Ciceronem *ND* 1.37; 2.29-30.

the Neopythagorean pseudo-Archytas, although he is not quoted as saying more than that place has a certain priority.[23] Secondly, Iamblichus himself cites the fact that Plato connects space in the *Timaeus* with Creation.[24] But there is another motive for his dynamic theory. Simplicius repeatedly calls it the 'intellective theory' (*noera theôria*) of place. It forms part of Iamblichus' design of defending the harmony of Plato and Aristotle against Plotinus' attack. Plotinus had charged that the Aristotelian scheme of categories could not apply to the world of intelligible things such as Forms.[25] Iamblichus' intellective theory responds that the dynamic conception of place as what preserves a thing by surrounding it does apply, and applies first and foremost, in the intelligible world.[26] Plotinus himself had treated what a thing is *in* as preserving it. Reversing the expected relation, but following Plato in this, Plotinus made the world's body to be in the World Soul in this way, as well as the World Soul in the Intellect and the Intellect in something else.[27] And Iamblichus is credited with just these examples of cosmos in Soul and Soul in Intellect, along with others.[28] It is because the concept of place has so many other applications that a dynamic conception is required to fit all the cases.[29]

So much for the dynamic character of place as Iamblichus conceives it. From its filling bodies with its power Simplicius draws an inference. It is not separate, he conjectures, from what is in place.[30] And elsewhere, without referring to Iamblichus, he takes non-separateness to mean that this sort of place moves with the body when it moves.[31]

> They think that place is immobile (*akinêton*), because they concentrate on this broad, shared (*en platei, koinos*) place. For the unique (*idios*) place of a thing which is essentially united (*sunousiômenos*) with it certainly moves with it (*sunkineitai*), but the shared place which is thought of as broad stays still (*menei*), as being unique to a more universal and embracing (*holikôteron, periektikôteron*) body.

> Now the place which is unique (*idios*) to the position (*thesis*) of the parts of a thing is both inseparable (*akhôristos*) from the thing and moves with it (*sunkineitai*) when it moves. But, as Syrianus himself used to say, the

[23] Simplicius *in Cat.* 361,12-26: place has a certain priority, because it must exist first, in order that other things may exist, move and be moved in it.

[24] Iamblichus ap. Simplicium *in Phys.* 639,23-32.

[25] Plotinus 6.1-3.

[26] P. Hadot, 'L'harmonie des philosophies de Plotin et d'Aristote selon Porphyre dans le commentaire de Dexippe sur les Catégories', in *Plotino e il neoplatonismo in Oriente e in Occidente*, Rome 1974, 31-47, translated in Richard Sorabji, ed., *The Transformation of Aristotle*, London and Ithaca N.Y. 1989.

[27] Plotinus 4.3.20(50); 5.5.9(4-5 and 31-2). I owe the references to S. Sakonji. Cf. Plato *Tim.* 36E.

[28] Simplicius *in Cat.* 362,7-16; 362,33-364,6.

[29] Note the 'so that' (*hôste*) in Simplicius' report at *in Cat.* 362,10.

[30] Simplicius *in Phys.* 640,10.

[31] ibid. 629,8-12; 637,25-30.

broad place which is involved in motion from place to place, is unique to a different body, the more universal one, and is inseparable from that, and moves with it, if it should chance to move, whereas it is both separable from the body contained within it and immobile in relation (reading: *hôs pros*) to it.

By suggesting that a body's place moves with it, Simplicius unites it with that body in a far stronger sense than Theophrastus ever intended. We saw that for Theophrastus the nature of the cosmic organism calls for a rock to stay at ground level, but that none the less the rock can still be tossed into the air. On Simplicius' extrapolation from Iamblichus, by contrast, a body's unique place is united with it in such a sense that the body can never leave that place. It is therefore a tenuous connexion that Simplicius finds between the two philosophers, when he says that both think of a thing's place as naturally united with it.

Dynamic views of place are found in four subsequent Neoplatonists, Syrianus[32] and his pupil Proclus,[33] Damascius and his pupil Simplicius. But I shall move straight on to Damascius, because he is the next to be explicitly associated with Theophrastus. Damascius reaffirms much of what has been said. He agrees with Iamblichus that place has a power of drawing things together (*sunagôgos*).[34] As quoted and elaborated by Simplicius, he also repeats Theophrastus' references to the arrangement of an animal's parts,[35] and compares the parts of the cosmos, while at the same time speaking of a kind of position (*thesis*) that is naturally united (*sumphutos*) with a body's essence:

One kind is naturally united (*sumphutos*) with essence (*ousia*), as for example in my body my head's being up and my feet down. The other is adventitious, as when I have my position sometimes in the house and sometimes in the market.[36]

Simplicius interprets the comparison with an organism as applying to place (*topos*), as well as to position (*thesis*):

So it is because of place, which arranges each of our parts well (*euthetizôn*), that my head has come to be on top of my body and my feet below. And the liver is on the right, the heart in the centre, the eyes by which we see when moving forward are in front, and the back on which we carry loads is behind ... It is also because of place, then, that the parts of the cosmos have their proper good arrangement (*euthetismos*) in the universe ... As the parts of the earth and of the heaven are variously arranged (*tetaktai*) here and there because of place, and some are, it may be, in the North, others in the South, so it is because of place that the

[32] Syrianus ap. Simplicium *in Phys.* 618,25-619,2.
[33] Proclus ap. Simplicium *in Phys.* 613,7-10 and 28.
[34] Simplicius *in Phys.* 625,28.
[35] ibid. 625,15-16; 626,8-12; 627,29; 628,1-2; 629,13-18; 645,12-13.
[36] Damascius ap. Simplicium *in Phys.* 625,14-17.

whole heaven and the whole earth, which are parts of the cosmos, have
the correct measurement (*eumetria*) of their position and have the cosmic
ordering (*diakosmêsis*) which falls to them, with one occupying the
circumference of the universe and the other its centre.[37]

In the same passage, and elsewhere, place is compared with *time*,
which also arranges things, separating the Trojan and the
Peloponnesian war, so that they will not collide.[38] Because of its active
powers, place is not a mere boundary, nor a mere extension.[39]

So far, much of what Damascius says is familiar. But he adds the
new idea that place should be viewed as a measure. He distinguishes
the positioning (*thesis*) of an organism's parts from the place (*topos*),
which he regards as a measure (*metron*) of that positioning,[40] where a
measure is something that prevents indeterminateness.[41] He thinks of
place (although this time Simplicius does not agree) as a measure not
only of positioning, but also of size, and as giving things the right
size.[42] Taking up an idea that is already in Proclus, he compares the
measure to a sort of outline (*proüpographê*) or mould (*tupos*), into
which the organism should fit, if it is to lie properly.[43] The following is
a direct quotation:

> Place is as it were a sort of outline (*proüpographê*) of the whole position
> (*thesis*) and of its parts, and so to say a mould (*tupos*) into which the
> thing must fit, if it is to lie properly and not be diffused, or in an
> unnatural state.[44]

Simplicius explains that the outline or mould is a flexible one, so that it
can allow for a variety of positionings, as it does in the case of the
moving heavens.[45] Since place measures positionings in this way,
Simplicius describes it as an arranger or arranging (*euthetizôn,
euthetismos*).[46] He explains that there are three elements in
measuring: the thing measured, the measure or physical instrument of
measuring, and the unit in the world of Forms by which the measuring
is done.[47] The last is unextended and perhaps this is why Damascius

---

[37] Extracted from Simplicius *in Phys.* 626,8-27.

[38] Simplicius *in Phys.* 625,29; 626,7-9; 13-16; 644,33-4.

[39] ibid. 627,6-9; 12-14. Nor yet is Proclus right that place is a body, for it would then
need another place to arrange its parts: 627,8-11.

[40] Simplicius *in Phys.* 625,28; 627,2; 14-15; 644,14; 23-4; 645,4.

[41] ibid. 640,20-641,7; 773,19-774,23.

[42] ibid. 644,24-5; 26; 645,6-17.

[43] Proclus ap. Simplicium *in Phys.* 613,9; Damascius ap. Simplicium *in Phys.* 645,7-10;
endorsed by Simplicius *in Phys.* 643,12-13; 773,25.

[44] Damascius ap. Simplicium *in Phys.* 645,7-10.

[45] Simplicius *in Phys.* 633,11-18.

[46] ibid. 626,5; 9; 18; 627,29; 644,11; 645,5; 13-14. Similar expressions at 626,3-4; 20;
26; 34; 644,14; 36.

[47] ibid. 634,11-31; 636,7-8; 636,27-637,18.

regards place as unextended.[48] But Simplicius, though agreeing that place is a measure,[49] does not agree with the denial of extension.[50]

In giving his own views, Simplicius exploits Damascius' distinction of adventitious position, and thus enables himself to go beyond the fragment of Theophrastus, and to give an account not only of natural place, but of any place to or from which a body may move. We have seen how Damascius distinguishes the positioning (*thesis*) which is united by nature (*sumphutos*) with a thing's essence and is permanent, like my head's being above, my feet below, from an adventitious position (*epeisaktos thesis*), like my being in the house or in the market place.[51] Simplicius infers that places (*topoi*) as well as positions, can be adventitious or united by nature,[52] and he adds a different, though related, distinction already noted: one kind of place is unique to the individual (*idios*) and moves with it when it moves. Another kind of place can be shared (*koinos*), and abandoned in favour of a new one.[53] The shared kind includes both such adventitious places as *in the market* and such broader places as the earth or some part of it, which may be the proper place of land animals.[54] These distinctions are what facilitate a general account of all place and motion. First, the shared place which I occupy, for example the place of the air, though not unique to me, will be unique to the whole air, and would move with the air if it moved.[55] As for the immediate place within the air which I occupy, that has no permanent existence, but perishes when I leave it. But the place of the whole air retains the ability again to fit me into some part of itself, and when it does so, it will in that part have the same measure as my extension, and will measure my extension again.[56]

In some of his statements Simplicius merely confirms what has already been said. Place draws bodies together,[57] and arranges them,[58] as time arranges events.[59] Theophrastus' appeal to animal parts is

---

[48] ibid. 601,16-19, reporting Damascius' work *On Number, Place and Time*. Elsewhere in his *Dub. et Sol.*, ed. Ruelle, Damascius gives something extended as an example of place in the primary sense (*kuriôs, in Parmenidem* II, 219,18-220,1), namely the sun's being in its extended (*diastatos*) light. This reference to *solar* light is not, I think, an endorsement of Proclus' view that place is a *supra-celestial* light (S. Sambursky, *The Concept of Place in Late Neoplatonism*, Jerusalem 1982, 22; 94). See Chapter 7.

[49] Simplicius *in Phys.* 626,3; 627,14; 630,12; 631,38; 634,2-24; 635,14-27; 638,6-17; 644,14-645,17; 774,6.

[50] ibid. 623,20; 634,21; 30; 636,7-8; 637,8-10; 15-17.

[51] ibid. 625,14-18.

[52] ibid. 627,16-17.

[53] ibid. 620,8-12; 637,25-30. The peculiar place which moves with a thing cannot be merely its volume (S. Sambursky *The Concept of Place in Late Neoplatonism*, Jerusalem 1982, 18, 22, 167). For mere volume has no active arranging power.

[54] Simplicius *in Cat.* 364,26-9; *in Phys.* 627,31-2. An example of parts of the earth would be North and South, *in Phys.* 626,24.

[55] Simplicius *in Phys.* 629,8-12; 637,25-30.

[56] ibid. 632,9-26.

[57] ibid. 631,38; 636,8-13; 637,8; 638,2.

[58] References above.

[59] Simplicius *in Cat.* 364,11-18; *in Phys.* 626,13-16.

repeated,[60] and places are said to be boundaries (*perata*) only in the active sense (*drastikon*) of giving definition (*horistikon*) to the things they belong to.[61] But as he elaborates Damascius' view (and it is often hard to say where Damascius' ideas stop and Simplicius' begin), the resulting concept of place begins to look quite unfamiliar. Going beyond Theophrastus, he is able to say not just that the heaven, but that the whole cosmos, has a place, namely, when its parts have their proper arrangement.[62] Again, although place is three-dimensionally extended, it is unlike the immobile three-dimensional extension which so many post-Aristotelian philosophers identified with place. For, as we have seen, immediate place is subject to perishing, and unique place is subject to motion: immobility is not required.[63] Moreover, Simplicius thinks of place not as an extension (*diastasis*), and hence as something in the category of quantity, but as a substance or essence (*ousia*) which is extended.[64] In fact, Simplicius has the problem that unique place seems to be in competition with the essence or form of a thing.[65] It had been and was in Theophrastus the role of an organism's form or essence to arrange its parts and control its size. Theophrastus left the job to form (*emmorphos phusis*)[66] and downgraded place. If Simplicius does the reverse, and transfers the job to place, what role is left for form? Simplicius says that form is not orderly (*euthetos*) on its own without place, any more than it is perfect without perfection.[67] He admits that on this conception the essence and the place of a thing become hard to distinguish (*dusdiakritos*),[68] even though they are not the same. We could hardly be farther away from Theophrastus' downgrading of natural place.

> In the same way it [the form: *eidos*] also has its position ordered with a good configuration because of place. So let us not seek to separate the form from the participation in the spatial which is essentially united with it and to see it as well arranged in its own right. That would be just like our separating its perfection from it in thought and none the less insisting on regarding it as perfect even without. So this statement says

---

[60] Simplicius *in Cat.* 364,23-6; *in Phys.* 625,15-16; 626,8-12; 627,29; 628,1-2; 629,13-18.

[61] Simplicius *in Cat.* 364,31-5.

[62] Simplicius *in Phys.* 631,7-31. Elsewhere he suggests the alternative that the cosmos has matter as its place, *in Phys.* 467,37. Ps-Archytas' variant is that the place of the cosmos is the boundary (*peras*) of all things, ap. Simplicium *in Cat.* 363,26-7. For the view that the whole can be said to be in the parts because it is inseparable from them, see above references to Eudemus, Themistius, Simplicius, Philoponus.

[63] Simplicius *in Phys.* 633,18-634,3.

[64] ibid. 623,19-20. Iamblichus had said in another context that place has *ousia* (ap. Simplicium *in Cat.* 361,7-30).

[65] Simplicius *in Phys.* 629,13-19; 630,30.

[66] ibid. 639,18-19.

[67] ibid. 630,21-631,6.

[68] ibid. 638,26-7.

nothing in opposition to the one which denies that place is form, even if it says that place is connected with form.[69]

I come now to the last person in the story. For although Simplicius claims that his predecessors are in the tradition of Theophrastus, it is in fact someone else who, for natural places, comes closest to Theophrastus' suggestion: I mean Philoponus. He comes close in two contexts: in one,[70] he is defending the conceivability (though not the actual existence) of a vacuum against Aristotle's charge that a vacuum would not be differentiated in such a way as to explain differentiated natural motions. In the other context,[71] Philoponus is defending his own quite un-Theophrastan conception of place as a three-dimensional extension against a similar charge from the Aristotelian camp: how can this extension, whose definition would allow it to be vacuous, have the power (*dunamis*) by which heavy and light things seek their proper places? In both contexts Philoponus denies any power to place. He explains natural motions by appealing, like Theophrastus, to the parts of animals and comparing the cosmos, since it is a great animal with parts. It is the arrangement of parts in relation to each other that explains natural motion, not some power belonging to different places.[72] Philoponus ridicules the idea that the elements could be striving to reach an Aristotelian place, that is, the surface of their surroundings. His view, in sympathy with Theophrastus' idea, is teleological: it is for their own good and for that of the cosmos that the parts seek their proper arrangement. But he adds his belief in the Creator: it is he who assigned the arrangement. The agreement with Theophrastus concerns only the idea of *natural* place, not of place in general. But he does offer a complete repudiation of the Aristotelian and Neoplatonist ascription of power (*dunamis*) to places.[73] The two passages run as follows:

I reply to the present point that first and foremost there is no need for the place which receives each body to have any power (*dunamis*) or quality. The fact that it is good for each thing to be thus and so does not mean that any power need be stored in the place where it naturally resides. Aristotle himself would not say the fact that the celestial body is arranged with a circumference and that this is good for it means that the

[69] ibid. 630,24-30.
[70] Philoponus *in Phys.* 632,4-634,2.
[71] ibid. 579,27-580,3; 581,18-31.
[72] For a different assessment of Philoponus' position as ad hoc, see Michael Wolff, 'Philoponus and the rise of Preclassical Dynamics', in Richard Sorabji, ed., *Philoponus and the Rejection of Aristotelian Science*, London and Ithaca N.Y. 1987, 96.
[73] Philoponus actually uses the phrase *drastêrios dunamis*, 'power that acts', which Iamblichus had applied to place, but he reserves it for such things as the impetus of impetus theory (Philoponus *in Phys.* 385,7). In discussing the 'much talked-of force (*poluthrulêtos bia*) of vacuum', Philoponus still grants no power to place, on the analysis of David Sedley, 'Philoponus' conception of vacuum', in Richard Sorabji, ed., *Philoponus and the Rejection of Aristotelian Science*.

surface (reading: *tên*) at the circumference has a power and that the
celestial body is arranged with a circumference, because it seeks
(*ephietai*) this surface. After all, he does not even think that the surface
(reading: *tên*) at the circumference is a place. So I deny that the celestial
body occupies this part of the void or (better) of extension, that is, the
extreme part at the circumference, because this part has some power.
Rather it is because it is the nature of the heaven to embrace everything
in its circumference and to contain everything within itself. Because this
is its nature, then, it is to be expected that it occupies the corresponding
part of space (*khôra*). So it is a result that happens not because of a power
especially belonging to extension. The heaven occupies the outermost
part of extension, because this is its nature – I mean, to surround all
things. It is like the case of the generation of animals: each different part
is in contact with a different part of space, whether the latter be the
boundary (*peras*) of the surroundings as Aristotle thinks, or an
extension, as I think and reason shows. For example, the head is in
contact with this part of the surface (*epiphaneia*) of the air, even if the air
is not actually divided, and the hands and feet and so on are each in
contact with a different part. Certainly, no one says that the boundary of
the air in contact with the head, as distinct from that in contact with feet
and hands, has a certain power, and that because of this, in search of
(*ephesis*) that boundary, the head occupies this air-boundary and another
part occupies another air-boundary. Rather, what is said is that the
head's nature is to be on top of the rest of the body and that different
parts of the body are differently arranged because this is good for the
animal. It merely happens as a result of this that the head is in contact
with this boundary and another part with another, although the air
contains no differences in relation to itself. I say this also about the
universe, for the cosmos is a great animal. Because it was good for the
arrangement of the whole cosmos that the bodies which fill it up should
have this sort of relation to each other, and because each body has as its
natural impulse (*hormê*) the search for this sort of relation to the others,
and because it achieves this result when it occupies this part of
extension, it is to be expected that each body seeks this result, not
because that part of extension has some power, but because good accrues
to the universe and its identity and well-being to each part, when it goes
beneath this part and on top of that. So it seeks a spatial extension not to
obtain that extension, but to obtain a certain relationship to the other
parts.[74]

It is quite ridiculous to say that place has a certain power in its own
right. For things move to their proper places not in search of a surface,
but in search of the arrangement (*taxis*) they obtained from the Creator.
So because earth was arranged at the bottom, so as to be beneath
everything else, water in the second position, air and fire in the third and
fourth, it is to be expected that, when something is dislodged from this
arrangement, and what would naturally be on top of another thing is put
by some force beneath it instead of on top, the things will move in search
of the arrangement allotted by the Creator, far enough to be in that
arrangement again. So light things move up, not because they seek

[74] Philoponus *in Phys.* 632,4-634,2.

simply to be in contact with the surface of their surroundings, but in search of the arrangement which the Creator assigned them. For then they most fully have their identity and they have their perfection then. So place does not have any power to make bodies move to their proper places. It is rather that the bodies seek to preserve their arrangement.[75]

Philoponus appears here as a true upholder of Theophrastus. But what separates the two of them from Iamblichus and the Athenian Neoplatonists is only an example of a wider disagreement in Greek Philosophy as to whether place has any power. In particular, there was disagreement on whether it had the power to sort the four elements into regions natural to them. Many would side with Theophrastus' doubts, and indeed Theophrastus' assembling of doubts may well have encouraged his successor Strato to reject Aristotle's idea of natural place altogether. According to Strato, all the elements are heavy, and even fire would fall, if you removed the elements beneath it. What makes fire rise is not a natural place, but mere *ekthlipsis*. This expressive word means its being squeezed upwards by the other elements,[76] an idea which Aristotle had rejected.[77] Theophrastus' contemporary Epicurus, founder of the rival atomist school in Athens, makes his atoms fall in the void,[78] and move upwards only by *ekthlipsis*.[79] Indeed, the idea of *ekthlipsis* had already been employed by the earlier atomists Leucippus and Democritus.[80] Epicurus does not allow that there are such *places* as up, down and centre in the void, because it is never-ending; there is only an up and a down *direction*,[81] and it is denied that atoms, as they fall, will find any permanent resting place.[82]

Of course, these thinkers are unlike Theophrastus, in that they remove organising power from place, not in order to assign it to something else, but in order to replace it with the blind forces of chance and necessity. But there is a much closer sympathy between Theophrastus and the *Stoics*. As pointed out in Chapter 9, the Stoics do not have to address themselves to the view that *place* has any influence on motion, once Theophrastus has questioned the idea and Strato abandoned it. And in fact for them neither filled place (*topos*) nor empty place (*kenon*) can act or be acted on, because this is possible only

[75] ibid. 581,18-31.

[76] Strato, frs 50-52 in Wehrli—Simplicius *in Cael.* 267,29; 269,4; Stobaeus *Eclogae* 1.14.1 (*Dox. Gr.* 311).

[77] Aristotle *Cael.* 1.8, 277b1-9.

[78] Epicurus, *Letter to Herodotus* in Diogenes Laertius *Lives* 10.61; Lucretius 2.216-50; Simplicius *in Cael.* 269,4-5.

[79] Lucretius 2.191-205; Simplicius *in Cael.* 267,30.

[80] See Diels-Kranz, *Die Fragmente der Vorsokratiker*, Leukippos A24, Demokritos A60; A61.

[81] Epicurus *Letter to Herodotus* in Diogenes Laertius *Lives* 10.61; Lucretius 1.1070.

[82] Lucretius 1.1077-8.

for *bodies*.[83] Because of that in turn, non-bodies are downgraded into not having being at all (they are not *onta*),[84] and this is applied to *place*, which is treated as a mere something (*ti*), instead of a real being (*on*).[85] We have already seen in Chapter 9 how Chrysippus makes all bodies tend to the centre of the cosmos because of powers which are lodged in *bodies*, not in *places*. That is, because of a holding power which is identified with an all-pervasive gas or *pneuma* that holds things together by the centripetal motions within it.[86] This might almost be described as a Stoicised version of Theophrastus' idea. The Stoics may be going beyond Theophrastus' guarded remarks, when they make the cosmos a living organism, and when Cleanthes makes its vital heat hold the parts together (*continere*).[87] But Chrysippus sounds extremely like Theophrastus, when he says, in the very context of discussing motion towards the centre, that the parts of the cosmos exist only as parts of the whole, that any part is naturally united (*sumphues*) with the rest, and that the coordinated arrangement (*suntaxis*) of the cosmos is the cause (*aitia*) of the movement towards the centre.[88]

Plato's view of space had been more nuanced. Talking in the *Timaeus* of the prototypes ('vestiges', *ikhnê*) of the four elements during the original period of chaos, he says that his Receptacle, or space (*khôra*), shakes them like a winnowing fan and so distributes them into their four regions.[89] So far his space sounds dynamic, but on the other hand, this power of space is not an active power: space shakes the elements only because it is itself swayed by the qualities it receives. How it can be swayed and shake the elements is a problem that worries commentators, because it is also said to be subject to no change.[90] Insofar as Plato does make the Receptacle sort the elements, he has less influence than might be expected on discussions of space, because a good many ancient authors think that he calls the Receptacle space only as a metaphor or analogy.[91] It is the Neoplatonists who are the most

---

[83] Diogenes Laertius *Lives* 7.55 (*SVF* 2.140); Sextus *M* 8.263 (*SVF* 2.363); Seneca *Ep. ad Lucilium* 106,2ff (*SVF* 3.84); Cicero *Acad. Prior.* 11.39 (*SVF* 1.90); Nemesius *Nat. Hom.* p. 32 Matth. (*SVF* 1.518); Aëtius *Placita* 4.20.2 in *Dox. Gr.* 410,6 (*SVF* 2.287); Plutarch *Com. Not.* 1073E (*SVF* 2.525).

[84] Plutarch *Com. Not.* 1073E (*SVF* 2.525).

[85] Plutarch *Adv. Col.* 1116B-C; Sextus *M* 10,218; cf. *M* 10,3; *PH* 3,124. This point remains true, regardless of whether Plutarch is right (and Plutarch is challenged by David Hahm, *The Origins of Stoic Cosmology*, Ohio State University 1977, 260-5) that Chrysippus wavered on a related issue. According to Plutarch *Sto. Rep.* 1054B-1055C, Chrysippus often denied, but once asserted, that bodies tend to the centre of *space*, as well as to the centre of the cosmos.

[86] References in Chapter 9.

[87] Cicero *ND* 2,29-30.

[88] Plutarch *Sto. Rep.* 1054E-F; 1055B-C (*SVF* 2.550). I am grateful to Victor Caston for connecting this passage with Theophrastus.

[89] Plato *Tim.* 52E-53A.

[90] ibid. 50B, discussed in the commentary of Calcidius §§301; 329; 352.

[91] See Simplicius *in Phys.* 539,8-542,14 (quoting Alexander); Philoponus *in Phys.* 516,5-16; 521,22-5.

whole-hearted in ascribing power to place, and among various other powers, four of them assign it the power discussed by Plato of moving and sorting the four elements into their regions. Syrianus,[92] Proclus,[93] Damascius and Simplicius.

We may ask finally what sympathy we can retain in modern times for the debate on the power of natural place. Of course the whole idea of natural places for the elements lost its justification, when Copernicus' view replaced Aristotle's geocentric conception of the universe. But it has been claimed that Copernicus retained an analogue of natural places,[94] and the idea is still found in Descartes.[95] In some expositions of contemporary physics, the idea is expressed that space itself can have power: it is the geometry of space that deflects bodies in the phenomenon we call gravity, although matter also has power, since it gives space its geometry.[96] But more recent expositions of the General Theory of Relativity[97] tend to treat the causal language as inappropriate, and speak merely of a functional relation between the geometry of space, the distribution of matter and the phenomenon of gravity.

---

[92] Syrianus, as interpreted by Simplicius *in Phys.* 618,25-619,2.

[93] Simplicius *in Phys.* 613,28.

[94] William H. Donahue, *The Dissolution of the Celestial Spheres*, Ph.D. diss, Cambridge 1973, 14-15 (and N.Y. 1981); there is a decreasing tendency of earth, water and air to cling to each other. They also have a natural circular motion around the sun.

[95] Descartes *Principles of Philosophy* 4.26.

[96] S. Sambursky, *The Concept of Place in Late Neoplatonism*, 22; Bertrand Russell, *The ABC of Relativity*, London 1925 (3rd ed. 1969).

[97] I am most grateful to Arthur Fine for information on this subject and to David Sedley for his comments.

Part III
Motion

CHAPTER THIRTEEN

Nature and God: two explanations of motion in Aristotle

Aristotle's rejection of vacuum as a cause of motion was discussed in Chapter 9, and his acceptance of natural place as a cause in Chapter 12. It remains to discuss some of the other causes of motion which he recognises, and their subsequent replacement by impetus theory.

Aristotle had to walk a tightrope in explaining his view that the inner nature of a thing can be the cause of its motion. In *Physics* 2.1, his task is to distinguish natural objects from artificial ones, and he does so by saying that natural objects have an *internal* cause of change, their nature. Contrast an artefact like a bed. It can be subjected to change: you can move it, for example. But in doing so, you have to push it from the *outside*. So it is unlike natural objects which have an *internal* source of change. Or, more precisely, it is only insofar as a bed is made of natural stuffs like wood or bronze, that it has any internal source of motion, in this case of downwards motion. Aristotle concludes that nature is an *internal* (*en*) source and cause (*arkhê* and *aitia*) of motion (*kineisthai*) or rest (*êremein*).[1]

But in *Physics* 8.4, Aristotle is constrained by an opposite consideration. In order to make room for God as that by which the heavens are moved, he has to support Plato's principle that whatever is in motion is moved *by* something. Moreover, he cannot allow that rocks fall downward only because of their inner nature, unassisted by anything else. Or the question would arise why the heavens also should not rotate solely because of their inner nature, unassisted by anything else, and then there would be no role for God.

Plato's principle seems on the face of it independently plausible. Only a few bold spirits denied that what is in motion need be moved *by* anything. Epicurus did so in effect, in the case of his unpredictable swerve,[2] and Galen did so explicitly for quite different reasons.[3] But

---

[1] Aristotle *Phys.* 2.1, 192b13-23.

[2] Lucretius 2.216-93, and see Chrysippus' reply to the Epicureans at Plutarch *Sto. Rep.* 1045B-C.

[3] Galen ap. Alexandrum, *Refutation of Galen's Attack on Aristotle's Doctrine that Everything that Moves is Set in Motion by a Mover*, selective translation in S. Pines, 'Omne quod movetur necesse est ab alio moveri: a refutation of Galen by Alexander of

the Stoics were able to embarrass their opponents by charging them with postulating 'causeless motion'.[4] The modern conception of inertia would allow that a body that had been moving in a straight line at uniform speed without beginning need not be moved *by* any force. But so long as it *acquired* its present speed or direction, a force would have been needed at least originally.

Aristotle needed, then, to find something *by* which a falling rock was moved. And the rock's inner nature would not be sufficiently distinct from the rock to serve his subsequent argument for a prime mover distinct from the heavens. How could he reconcile this constraint with the need to distinguish natural bodies from artefacts as having an *internal* cause of motion?

His solution in *Physics* 8.4 is to find a pair of low-grade causes which do not meet his full requirements for efficient causation. Aristotle repeatedly says that the causes of motion must be in *contact* with what they move.[5] And during his discussion of projectile motion, it becomes clear that he thinks a cause is needed which *maintains* contact with what it moves, throughout the motion.[6] In *Physics* 8.4 this contact requirement is sufficiently met by the inner nature of a falling rock. Aristotle cites its inner nature in response to the question why (*dia ti*) light and heavy things move to their proper places: the reason is that they are constituted by nature for that direction (*aition hoti pephuken poi*).[7] But in answer to the *different* question *by* what (*hupo*) they are moved, Aristotle cites two low-grade causes which do not maintain continuing contact with what they move.[8] To take the example of rising steam, it is moved partly by the man who generates it (*tou gennêsantos*) by boiling the kettle, and partly by the man who takes the lid off the kettle and so removes the obstacle.

Aristotle makes room for these two low-grade causes by explaining that the rock's nature, or internal source of change, is a source not of causing motion, but of passively undergoing it (*paskhein*) at the hands of a further agent.[9] The idea of passivity had been prepared for, because the earlier account of nature in *Physics* 2.1 had described it as a source of *kineisthai*,[10] and the word *kineisthai* stands indifferently for the intransitive *being in motion* and for the passive *being moved*. But having thus introduced his two low-grade causes, Aristotle now

Aphrodisias and the theory of motion', *Isis* 52, 1961, 21-54 (see esp. the translation of Carullah ms. fol. 66b, on pp. 25-6), complete translation in N. Rescher and M. Marmura, *Alexander of Aphrodisias: the refutation of Galen's treatise on the theory of motion*, Islamabad 1969.

[4] e.g. Alexander *de Fato* 15 and 22; Plutarch *Sto. Rep.* 1045B-C; Alexander (?) *Mant.* 174,3-5.

[5] Aristotle *GC* 1.6, 322b22-6; *Phys.* 3.1, 202a6-9; 8.5, 258a20-1; 8.10, 266b28-267a20.

[6] Aristotle *Phys.* 8.10, 267a20.

[7] ibid. 8.4, 255b13-17.

[8] ibid. 256a1-2.

[9] ibid. 255b30-1.

[10] ibid. 2.1, 192b21.

does everything to play them down, or at least to play down the obstacle-remover, as he needs to, if he is to retain the idea that it is forced, not natural, motion which has an external cause. Thanks to its inner nature, the steam is all ready to burst into upward motion, just as soon as the obstacle is removed.[11] As for the obstacle-remover, there is a sense in which he causes the motion and a sense in which he does not. He does so only incidentally (*kata sumbebêkos*), and may be compared with the wall off which something ricochets, without having been moved *by* (*hupo*) the wall.[12] In this way, Aristotle tries to maintain the need for some kind of independent mover, without introducing a full-scale external cause, which would spoil his distinction between natural and artificial motion:

But we must inquire why (*dia ti*) ever light and heavy bodies move to their own places. The reason (*aition*) is that they are naturally constituted for that direction (*pephuken poi*), and that is what it is to be light or heavy, one being defined by up and the other by down. But, as has been said, things are *potentially* light or heavy in more than one way. For while something is water, it is at least potentially light in one way. But equally when it is air, there is a way in which it is still potentially light, since it can be prevented by an obstacle from being aloft. If the obstacle is removed, it is actualised (*energei*), and always gets higher. In a similar way too something with a given quality changes to being actual (*energeiâi*), for a man who possesses knowledge brings it to mind at once, if nothing prevents him, and something of a given size spreads out, if nothing prevents it. As regards the man who moves the obstruction that has been imposed, however, there is a way in which he causes the motion (*kinei*) and a way in which he does not, I mean for example, the man who wrenches away the pillar from beneath or removes the stone from the wineskin in the water. For he causes the motion incidentally (*kata sumbebêkos kinei*), just as a rebounding ball is moved not by the wall, but by the man who throws it. Thus it is clear that none of these things moves itself, but each has a source of motion (*arkhê kinêseôs*), not of causing motion (*kinein*), or of acting (*poiein*), but of passively undergoing it (*paskhein*). If, then, everything in motion is moved either by nature or contrary to nature and forcibly; and if everything in motion forcibly and contrary to nature is moved by (*hupo*) something, and by something else (*allou*); and if again of the things in motion by nature those that are moved by their own selves are moved by something [sc. their souls], and so are those that are not, like light and heavy bodies (for they are moved either by the generator (*gennêsantos*) who made them light or heavy, or by the man who freed the obstacle and obstruction) – in that case everything that is in motion will be moved by something.[13]

The same account of natural rise and fall is given in *de Caelo* 4.3,

---

[11] ibid. 8.4, 253b19-24.
[12] ibid. 255b24-29.
[13] ibid. 255b13-256a3.

with a cross-reference to the *Physics*,[14] although a little earlier in the
de Caelo the generator and obstacle-remover are ignored, and only the
inner inclination (*rhopê*) is stressed.[15] Altogether it is likely that
Aristotle would accept three kinds of explanations for natural rise and
fall, first the inner nature, secondly the obstacle-remover and
generator, and thirdly the natural place of the falling rock. We have
already seen in Chapter 12 that he considers natural place to have a
certain power (*dunamis*),[16] and it is most likely, although he never
says so explicitly, that he would think of natural place as explaining
motion as a final cause or goal. It would not be like the inner nature,
generator, or obstacle-remover, an efficient cause.

Aristotle's treatment of *animal* motion is not wholly similar, but it is
partly so. Animals have an inner nature, but when Aristotle talks of
the inner nature of a living thing, he often means its *soul*.[17] His answer
to the question *by* what an animal is moved, is that it is moved by its
soul.[18] So far there is no analogue of the obstacle-remover. But for
independent reasons, Aristotle does introduce auxiliary efficient
causes here too, to nudge the animal into motion. Even on a sleeping
animal the environment (*to periekhon*) has an effect. For example, the
food which it has absorbed from the outside, on being digested,
prompts it to wake up and search for more.[19]

To return to the case of rising steam and falling rocks, I take it that
Aristotle's strategy in *Physics* 8.4 is to argue by induction. If in all
other cases a thing that is moved is moved by something, the same
must apply to the celestial spheres, and that will enable Aristotle to
introduce his prime mover. The induction has been criticised on the
grounds that the prime mover is of such a very different type from the
generator and obstacle-remover.[20] But the induction is not meant to
show that the celestial spheres must be moved by something of the
same type, but only that they must be moved by *something*.

We might hope that Aristotle would now progress to showing that
that something is his divine unmoved mover. But unfortunately in the
next chapter, *Physics* 8.5, something appears to go wrong. He does
indeed argue that the series of things by which other things are moved
cannot stretch infinitely, but must terminate in something that is
either a self-mover, like the soul in Plato, *or* something unmoved, as he

[14] There is a cross-reference to the *Physics* account of the generator and
obstacle-remover at *Cael*. 4.3, 311a9-12. I take it that earlier in the chapter the internal
origin of 310b24 and 32 is the inner nature, and the one who makes heavy or light at
310a32 is the man who boils the kettle or condenses the steam.

[15] Aristotle *Cael*. 3.2, 301a20-6.

[16] Aristotle *Phys*. 4.1, 208b11.

[17] Aristotle *PA* 1.1, 641a28; b9-10; *GA* 2.4, 740b29-741a2.

[18] Aristotle *Phys*. 8.4, 254b27-33, further explained 8.6, 258a1-b9.

[19] ibid. 8.2, 253a7-20; 259b1-20.

[20] G.A. Seeck, 'Die Theorie des Wurfs, Gleichzeitigkeit und kontinuerliche Bewegung',
in G.A. Seeck, ed., *Die Naturphilosophie des Aristoteles*, Darmstadt 1975; Sarah
Waterlow, *Nature, Change and Agency in Aristotle's Physics*, Oxford 1982, 234, 236, 260.

himself thinks.[21] But first he wants to get Plato's alternative of self-motion out of the way. So he responds that what we call self-motion, that is, animal motion, really involves two distinct elements, something that is moved and an unmoved mover. We might well recall *de Anima* 3.10, 433b13-19, where Aristotle explains that the animal's soul, or the desiring faculty of its soul, is moved by the thing it desires, and that the thing desired is an unmoved mover. If Aristotle had had that in mind he would seem to have arrived in one bound at his eventual conclusion, not expressed until the *Metaphysics*, that God moves the celestial spheres as a final cause, as the object of their desire and love.[22] He need only argue that the ultimate entity by which things are moved must be *unmoved*, and that, as suggested by *de Anima* 3.10, only objects of desire will qualify.

Unfortunately, however, the doctrine he has in mind turns out to be quite different from the one in *de Anima* 3.10, and even appears on the surface to be incompatible with it. It is the doctrine, which is also expressed in the *de Anima*, that the unmoved mover is the soul itself, *not* the object of its desire.[23] This immediately creates a most serious problem: why should not the souls of the celestial spheres then be the prime movers, so that we would not need to appeal beyond them to God? Some modern commentators ascribe to Aristotle a conclusion not so far from this.[24]

The idea of making celestial soul, rather than God, the prime mover was tempting. And this view was actually entertained by Aristotle's successor, Theophrastus.[25] Theophrastus' own successor, Strato, in turn made all motion natural, in order, as Cicero says, to free God from work.[26] And this was how the situation arose (discussed in Chapter 9) which made it unnecessary for the Stoics to launch an attack on Aristotle's now forgotten unmoved mover.

There is an analogous development among Aristotle's interpreters concerning the treatment of *nature*. To the question by what rising steam is moved Alexander answers, in defence of Aristotle: through the natural inclination within. He makes no mention of the obstacle-remover, and in this he is followed by Philoponus, who, in his

---

[21] Aristotle *Phys.* 8.5, 257a25-31.

[22] Aristotle *Metaph.* 12.7, 1072a25; b4.

[23] Aristotle *Phys.* 8.5, 258a7; a19; *DA* 1.3, 406a3; b7-8; 1.4, 408b5-18.

[24] Sarah Waterlow, in the most subtle and interesting of recent discussions, does not believe that Aristotle thought the heavens were ensouled, but the prime mover for which she has Aristotle argue is *analogous* to a soul for them, an idea which she deplores (op. cit. 236-9, 243, 247, 257). G.A. Seeck (op. cit.) thinks that Aristotle commits himself unintentionally to making *nature*, rather than God, the prime mover, when he downgrades the obstacle-remover in *Phys.* 8.4 to a merely accidental cause. He also thinks (controversially) that the *de Caelo* is by and large later than the *Physics*, and that there Aristotle recognises and accepts the earlier *Physics* result, although he subsequently goes back on it in the *Metaphysics*.

[25] Theophrastus *Metaphysics* 7b15-23 and ap. Proclum *in Tim.* 2.122,10-17.

[26] Strato ap. Ciceronem *Acad. Prior.* 2.38.121 (=Wehrli fr. 32).

exposition of Aristotle, explicitly rules out the obstacle-remover.[27] Alexander does, it is true, recognise Aristotle's 'generator' but in an unexpected way. For instead of taking the generator to show that an *external* cause is needed in addition to the natural internal inclination of the steam, he takes it to confirm ('for') that the source of motion from which the steam rises, and through which it rises, is the *internal* inclination itself. The idea is that the generator is not someone who acts externally on the steam, but one who produces its internal constitution, so that his responsibility *confirms* that the cause of rising is the internal inclination itself:

> He likewise made clear in the case of the bodies that move naturally through the inclination within them that their source of motion is from the inclination existing in them by virtue of which they move naturally. For the thing that moves them from heaviness in potency to heaviness in actuality, making them be in a state differing from that they were in, is also the cause of their motion in actuality.[28]

We are now in the tantalising position that Aristotle had his eventual conclusion almost in his grasp, and could have secured it, if he had but held to the view that the only unmoved movers are objects of desire. Now that he has allowed *souls* to be unmoved movers, what is there left that he can do? I think that he will eventually have all the materials he needs, but he does not yet supply them in the *Physics*, nor for that matter in the *de Caelo*. I am thinking rather of the *de Anima*. But I had better explain first why I am virtually ignoring his account of celestial motion in the *de Caelo*.

Others have shown[29] very effectively that it is almost impossible to get a single coherent view out of the *de Caelo*, which I take to be largely earlier than the *Physics*, though with some later additions. In *Cael.* 1.2, Aristotle argues that the rotation of the celestial element is due to its inner nature, without any hint that that nature might be a soul.[30] But in Book 2, we hear that the heaven is ensouled and (sc. for that reason) contains a source of motion.[31] There are even passages in

---

[27] Philoponus *in Phys.* 195,24-32, translated below in Chapter 14.

[28] Alexander *Refutation of Galen* (as above), fol. 67a17-20 of the Carullah ms. in the translation of Rescher-Marmura p. 17; cf. Pines pp. 27 and 42. Sarah Waterlow imagines a view like Alexander's, but rightly says that it would not be accepted by Aristotle (op. cit., 165).

[29] H. von Arnim, *Die Entstehung der Gotteslehre des Aristoteles*, Vienna 1931; W.K.C. Guthrie, 'The development of Aristotle's Theology I', *Classical Quarterly* 27, 1933, 162-71; id., introduction to Loeb edition of Aristotle *On the Heavens*, London and Cambridge Mass. 1960; H. Cherniss, *Aristotle's Criticism of Plato and the Academy*, Baltimore 1944, appendix 19, 581-602.

[30] Aristotle *Cael.* 1.2, 268b14-16; 269a6-18.

[31] ibid. 2.2, 285a29-30; 2.12, 292a18-21; b1-2. Earlier still, in the *de Philosophia*, celestial motion had been described as voluntary, but not natural, fr. 21 Ross=Cicero *ND* 2.16.44.

Books 1 and 2 which have been thought to refer to God as prime mover,[32] but their meaning or date of insertion is disputed, and on the other side it is said in 1.7 that there is nothing greater (*kreitton*) which could move the celestial body.[33] I do not think that the solution to the problems of the *Physics* will emerge from these texts.

What the *de Anima* supplies is a way of reconciling the apparently conflicting statements it contains as to whether the soul is unmoved, or whether it is moved by the object of desire, which alone remains unmoved. Talking of perceptible qualities, Aristotle explains in *de Anima* 2.5 that they move the sense faculty of the soul, or the sensing animal, only in a very special sense. They bring it from a heightened state of potentiality to full actuality. It is like the case of a man who has first reached a heightened state of potentiality by learning geometry, and who is then stirred into full application of his powers. Such a shift to full actuality should not be called a qualitative change (*alloiôsis*, literally: a becoming *other*) at all, or else one must distinguish it as being of a different type.[34] This is applied quite generally to other shifts to full actuality, and it will also cover the case in which the soul, or its faculty of desire, is stirred into fully actual operation by an object of desire. We should say either that the soul is not moved at all, or that it is moved only in the special sense of being brought to full actuality. We can now understand how *de Anima* Book 1 can describe the soul as unmoved, while Book 3 describes it as being moved (that is, stirred to full actuality) by an unmoved object of desire.

In Book 8 of the *Physics*, Aristotle does not explain the view of his later *Metaphysics* that the heavens are alive and that God excites motion in them only as a *final* cause and object of desire. Nor does he explain the view of the *de Anima* 2.5 that such exciting leaves the celestial soul or souls unmoved in a sense. Nor does he add the view of *de Anima* 3.10 that the object of desire remains unmoved in a much fuller sense, and that the soul's faculty of desire can only be excited by such an unmoved object of desire. If he had said this, he would have been able to complete his argument, by maintaining that, although the celestial soul is indeed an unmoved mover in a sense, it needs to be excited by the supreme unmoved mover as the object of its desire. This would have supplied him with a second approach to his conclusion, after the first approach, which seemed so likely to succeed, turned out to be left unused.

It looks as if in the *Physics* he already knew the conclusion he thought right, that God, and not any celestial soul, is the prime mover, but that the tools for securing this conclusion were not available to him until he had written the *de Anima* and *Metaphysics* book 12. What we find instead is an argument which, without mentioning the point, in effect rules out the unmoved mover being a celestial soul, so that we

[32] Aristotle *Cael.* 1.8, 277b9-10; 2.6, 288a27-b7; b22-3.
[33] ibid. 1.9, 279a33-4.
[34] Aristotle *DA* 2.5, 417a31-b16; 418a1-3.

can infer that it must be something else. For in *Physics* 8.10, Aristotle argues that the unmoved mover cannot be a power (*dunamis*) residing in (*en*), or possessed by (*ekhein*), a finite body, since a finite body could not accommodate the infinite power needed for the eternal rotation of the heavens.[35] This argument will be the subject of Chapter 15.

[35] Aristotle *Phys.* 8.10, 266a24-5; b5-6; b25-6; 267b22-3

CHAPTER FOURTEEN

The theory of impetus or impressed force: Philoponus

Thomas Kuhn has called the introduction of impetus theory a scientific revolution, or, in his terminology, a paradigm shift.[1] The credit must go to Philoponus. What he introduced was an alternative to Aristotle's dynamics, and he started with the motion of projectiles.

Aristotle had been puzzled as to what makes a javelin continue to move after it has left the hand. Its 'natural' motion is only downward. Its 'forced' or 'unnatural' motion onwards is produced for the first few feet by the hand of the thrower who grasps it. But after its release, Aristotle thinks, another cause must be sought which, like the hand, is external to, and yet in contact with it. He decides[2] that by pushing the air, the thrower imparts to successive pockets of air behind the javelin the power to push it onwards. They receive this power not only while the thrower's hand is pushing the air, but even after his hand has come to rest. In effect, the pockets of air are unmoved movers, although Aristotle does not put it that way, and he might prefer to say that they are no-longer-moved movers. Philoponus' innovation is to suggest instead that a force (*dunamis, iskhus, hormê, energeia, arkhê*) can be implanted by the thrower directly into the javelin, and need not remain external to it in the air. This force came to be called an impetus and was still a commonplace in the time of Galileo.

Philoponus' impetus theory

Aristotle's theory of projectiles was ripe for replacement. He does not explain sufficiently why air should sometimes help motion, as with projectiles, while at other times[3] it creates resistance to motion and reduces speed. As to why it helps motion, he can only plead that air is light in relation to some bodies, even though it is heavy in relation to

---

[1] Thomas Kuhn, *The Structure of Scientific Revolutions*, Chicago 1962, 2nd ed. 1970, 120.

[2] Aristotle *Phys*. 8.10, 267a2-12; cf. *Cael*. 3.2, 301b23-30.

[3] *Phys*. 4.8, 215a24-216a11. I thank Daniel Law for this point.

others.[4] Philoponus' contemporary Simplicius gets as far as raising the crucial question: why say the motion is impressed on the air, rather than on the missile, so forcing ourselves to make the air a mover?[5] But he is satisfied with Aristotle's explanation of the special propensity of air. The position of air as a no-longer-moved mover seems all the stranger, because Aristotle is otherwise so sparing in explaining motion by reference to unmoved movers. Objects of desire can act as unmoved movers, but otherwise the only acknowledged case of an unmoved mover is the soul,[6] and this is still moved in the sense of being stirred by the object of desire.[7] As seen in Chapter 9, Philoponus has enormous fun ridiculing Aristotle's appeal to pockets of air. If this were the mechanism, direct contact between the thrower and the javelin would be otiose, since the actual work is done by pushing not the projectile, but the air. An army could then perch its projectiles on a thin parapet and set the air behind in motion with 10,000 pairs of bellows. The projectiles should go hurtling towards the enemy, but in fact they would drop idly down, and not move the distance of a cubit. The advantage of the rival impetus theory which has a force transmitted directly to the javelin, is that it explains why the thrower needs to be in contact with the javelin. Philoponus makes his point in the context of Aristotle's denial of motion in a vacuum.

> (641,13) We must first ask those who state this [Aristotle's theory]: when one throws a stone by force (*biâi*), is it by pushing the air behind the stone that he forces (*biazetai*) the stone into its unnatural (*para phusin*) motion? Or does the man who pushes transmit (*endidôsi*) some kinetic power (*dunamis kinêtikê*) to the stone itself? If he transmits no power to the stone, and it is only by pushing the air that he moves the stone, or the bowstring moves the arrow, what advantage was there in the stone's being in contact with the hand, or the bowstring with the notch in the arrow? It would have been possible, without touching these, to stand the arrow, for example, on the tip of a stick which would serve as a thin line, and similarly for the stone, and with ten thousand machines to set a great quantity of air in motion from behind. Clearly, the more air was set in motion, and the greater the force, it ought to push it more and shoot it further. But in fact, even if you were to stand the arrow or stone on a line that really lacked breadth or on a point, and were to set in motion all the air from behind with all your thrust (*rhumê*), the arrow would not move even the distance of a cubit. It is clear that it is not the air pushed by the hand or the bowstring that causes motion in the case of what is thrown or fired either. For why should this happen when the thrower is in contact

[4] *Cael.* 3.2, 301b23-30; used by Themistius *in Phys.* 235,5-6; Simplicius *in Phys.* 1349,30-2.

[5] ibid. 1349,26-9; cited by S. Sambursky, *The Physical World of Late Antiquity*, 1962, 72, and by H. Carteron, *La notion de force dans le système d'Aristote*, Paris 1923, 11-32, translated into English in Jonathan Barnes, Malcolm Schofield, Richard Sorabji, eds, *Articles on Aristotle 1*, London 1965, see p. 170, n. 39.

[6] Aristotle *DA* 1.3, 406a3; *Phys.* 8.5, 258a7; a19.

[7] Aristotle *DA* 3.10, 433b13-19.

with what is thrown more than when he is not?

(641,33) In addition, if what is in contact is continuously connected, the dart with the bowstring and the hand with the stone, and there is nothing in between, what would the air be that is moved from behind? And if it is the air from the sides that is moved, what has that to do with the thing thrown? For the air there falls short of it.

(642,3) From these and many other considerations, then, you can see that things moved by force cannot be moved in this way. Rather, *some incorporeal kinetic power must be transmitted by the thrower to the thing thrown*, and the air that is pushed contributes either nothing at all or very little to this motion. And if things moved by force are moved in this way, it is surely clear that even if one propels an arrow or stone in a vacuum by force and contrary to its nature (*para phusin*), the same thing will happen all the more, and he will not need anything from outside to do the pushing. And surely this account, supported as it is by clear evidence (I mean the account according to which *some incorporeal kinetic process (energeia kinêtikê) is transmitted by the thrower to the thing thrown*, so that the thrower has to be in contact with the thing thrown) – surely it will raise no more problems than Aristotle's opinion that certain processes come to the sight from the thing seen. For from colours too we see certain processes arriving in an incorporeal way and colouring solid bodies in their path, when a sunbeam strikes after passing through colours. This can be seen clearly, when a sunbeam strikes through a coloured glass: wherever the sunbeam falls on a solid body by the way of the glass, it colours that body with the colour it passes through. So it is clear that certain processes pass in an incorporeal way from one thing to another in succession.[8]

There is an important point of translation in the opening sentence of this passage at 641,13-15. When Philoponus asks, 'when one throws a stone by force, is it by pushing the air behind the stone that he forces the stone into its unnatural (*para phusin*) motion?', he is questioning the *mechanism* of projectile motion. But an alternative translation has been suggested according to which Philoponus would be questioning not the mechanism, but the *unnaturalness* of the motion: 'when one throws a stone by force by pushing the air behind the stone, is it an unnatural motion with which he pushes the stone?'[9] This translation seems to me impossible for three reasons. First, it omits the Greek word *houtôs*, 'in this way', which is a device used in the Greek for emphasising that the question is about the way in which the stone is moved, literally: 'by pushing the air behind the stone, is it in *this* way …?' Secondly, the ensuing discussion is devoted to the mechanism, not to the unnaturalness of the motion. Thirdly, Philoponus later confirms that he takes the motion without question to be unnatural, when he says at 642,8 that he is concerned with propelling 'an arrow or a stone

[8] Philoponus *in Phys.* 641,13-642,20.

[9] Michael Wolff, 'Philoponus and the rise of preclassical dynamics', in Richard Sorabji, ed., *Philoponus and the Rejection of Aristotelian Science*, London and Ithaca N.Y. 1987, ch. 4, at p. 98 with n. 47.

in a vacuum by force and contrary to its nature (*para phusin*)'. This will become relevant below.

These ideas appear in Philoponus' *Physics* commentary, of which (though a second edition after 529 has been suggested) the first edition can be dated to A.D. 517. And he applies his impetus theory to another question which had puzzled Aristotle: why the belt of elemental fire, which Aristotle places below the stars, rotates in a circle. Aristotle, we have seen, thinks of the earth as stationary at the centre of the universe, surrounded by belts of water, air and fire and then by the transparent spheres which carry the stars. These last were sharply distinguished from the lower elements (earth, air, fire and water), and assigned a circular, as opposed to a rectilinear, motion. So it puzzles Aristotle why the belt of elemental fire should also move in a circle, as he thinks it does, on the evidence of comets and suchlike, which he takes to be carried along in that belt. His own answer to the question seems to waver,[10] but Philoponus' answer in the *Physics* commentary is again that an impetus is transmitted from the rotating heavens into the fire belt. He exploits Aristotle's claim in *Phys*. 3.3, that there is only one process going on when a teacher teaches and his pupil learns, and in general when an agent acts and a patient is acted on, and that that process resides not in the agent, but in the patient. But he entirely reconstrues Aristotle's claim, to mean that a kinetic power (*kinêtikê dunamis*), a force (*iskhus*) and process (*energeia*) is transmitted into the patient, in this case into the fire belt. He further uses of this force the expression 'power that acts' or 'efficacious power' (*drastêrios dunamis*), which in Chapter 12 we saw associated[11] with Iamblichus' dynamic theory of place:

(384,11) That is what Aristotle says, but someone may ask if the account which says that motion is only in the thing moved and not at all in the mover is true of every mover and thing moved. It is true in cases of growth or change of quality that what is moved is the patient not the agent, for what is whitened is not the whitening agent but the whitened subject, and what grows is not the agent of growth but the growing subject, that is, not the agent, but the patient. But in cases of change of place, in which some movers cause motion by being moved, the account will not be thought to fit. For if the heavens move the belt of fire beneath by being moved themselves, how can I avoid saying that the mover is itself moved precisely in the process of causing motion? In that case, the movements will be two different ones, that of the mover and that of the thing moved, and the movement of the heavens will be different from the movement undergone by things down here, as they are made to change their qualities in this way or that by the movement of the heavens. Thus the actualisations and the movements will be two different ones, differing not only in their definitions, but also in the subject in which they inhere. The fire belt will admittedly share in the motion of the

---

[10] Aristotle *Cael*. 1.2, 269a9-18; a30-b2; *Meteor*. 1.7, 344a11-13.
[11] Iamblichus ap. Simplicium *in Cat*. 361,7-30.

mover, just as my cloak shares in my motion when I move, but as the heavens move round, the neighbouring matter all undergoes not only this motion, but also changes of quality. For just through the sun's changing its position, its change of place makes the things down here change their qualities in this way or that.

(384,29) To this objection I reply in brief that Aristotle does not deny that there are movers which cause motion by being moved. On the contrary, he says clearly that natural things are all such as to be incapable of causing motion except by being moved themselves. Aristotle's claim is rather about the kinetic power (*kinêtikê dunamis*) transmitted (*endidomenê*) by the mover to the thing moved, I mean that force (*iskhus*) and process (*energeia*) by which a potentially moveable thing is moved, and brought to fulfilment (*teleioumenon*): it is one single force, but gets the origin of its coming forth (*arkhê proodou*) from the mover, and has its completion and perfection (*telos, apoteleutêsis*) and, so to speak, persistence (*monê*) in the thing moved. For when it has been generated in what is potential, it does not leave it alone, but persists (*menousa*), so to speak, and brings it to fulfilment, and the bringing forth of the potentiality (*tês dunameôs proagôgê*) is its fulfilment (*teleiôsis*), and that is what motion is. An expert will rehearse his knowledge, even if there is no one to learn from him, but when his rehearsal takes hold in a pupil and works its effect (*dran*) on him, I say that the efficacious power (*drastêrios dunamis*) by which the pupil is affected is one single power, and has its origin in the teacher, but is generated in the pupil, and moves him, and perfects the potential in him, and clearly is in the person moved, not in the mover, for the mover does not change in respect of this power.[12]

The expression 'efficacious force' (*drastêrios dunamis*) is used not only by Iamblichus, but also by Proclus.[13] And it has been well shown by others that this passage of Philoponus is shot through with the concepts of Proclus, who holds that when a higher entity creates an entity lower in the hierarchy, that lower entity must be seen as coming forth (by a *proodos*) from, as returning (by an *epistrophê*) to, and yet as persisting (by a *monê*) in the higher entity. Furthermore, the higher entity itself persists (by a *monê*) undiminished in the process.[14] Yet, as Michael Wolff has further demonstrated, Philoponus transforms Proclus' ideas by insisting that the transmitted power finds its *monê*, or, as he guardedly says, its *monê, so to speak*, not in the original source of motion, but in the thing moved instead. He thereby emphasises the fact of its transmission.[15]

So far, Philoponus' applications of impetus theory are only

[12] Philoponus *in Phys.* 384,11-385,11.
[13] e.g. Proclus *in Parmenidem* (Cousin) 772.
[14] See Proclus *Elements of Theology*, propositions 25-39, and for *monê* 26; 27; 30; 35, and see the discussions in Michael Wolff, *Fallgesetz und Massebegriff*, Berlin 1971, 88-9; id., in *Philoponus and the Rejection of Aristotelian Science*, at 110; Jean Christensen de Groot, 'Philoponus on *de Anima* 2.5, *Physics* 3.3, and the propagation of light', *Phronesis* 28, 1983, 177-96.
[15] Michael Wolff, loci cit.

beginning. Aristotle had split dynamics into unconnected areas. Projectile motion was explained by the external pockets of air. The heavens were thought to be alive and their motion was explained, at least in the *Metaphysics*, in psychological terms. The fall of rocks and rising of flames was explained non-psychologically by reference to an inner nature, while the rotation of elemental fire, we have seen, was a special case. Philoponus' next move has the effect of unifying dynamics. The context is his discussion of Creation in the book of Genesis. This is the subject of his *de Opificio Mundi* which has been dated some thirty or forty years later than the first edition of the *Physics* commentary, either to 546-9 or to 557-60. In a few lines, he extends impetus theory, in one form or another, to all the remaining cases, and he is enabled to do so by the belief, for which he has argued in so many earlier works, that God created the universe. It is God who implants (*entheinai*) a motive force (*kinêtikê dunamis*) in the sun, moon and other heavenly bodies at the time of creation. It is God who then implants the downward inclination (*rhopê*) in earth and the upward inclination in fire. It is God who implants in animals the movements which come from the souls within them.

The impetus which God implants in heavenly bodies seems closely analogous to that which a thrower implants in a javelin, but in the other cases, the analogy is less close. The impetus implanted in the elements, earth, air, fire and water, must be a complex one. For in Philoponus' view, elements lose their weight or lightness, that is, their inclination to move down or up, once they reach their proper places.[16] Earth has weight, he thinks, only when lifted away from its resting position. What God implants in earth, then, is not an inclination to move down, but an inclination to move down, *if* dislodged. Similarly with animals, what God implants when he implants their souls is nothing as simple as the impetus in a javelin, and indeed at this point the analogy begins to be strained. This is how Philoponus puts it:

> Let the champions of Theodorus' view tell us by what divinely inspired text they have been taught that angels move the moon, the sun and each of the heavenly bodies, whether pulling them like yoked animals from in front, or pushing them on from behind like baggage trundlers, or both, or bearing them on their shoulders. What could be more ridiculous than this? Was God who created them unable to implant a motive power (*kinêtikên entheinai dunamin*) in the moon, the sun and the other heavenly bodies, as he implanted inclinations (*rhopai*) in heavy and light ones, and implanted in all animals the movements which come from the souls within them, so that the angels will not have to move them by force (*biâi*). For all that is moved otherwise than by nature has a movement that is forced and contrary to nature (*biaion, para phusin*), and will be

---

[16] Philoponus *contra Aristotelem* ap. Simplicium *in Cael.* 66,8-74,26=frs 37-46 Wildberg.

caused to perish. How then could so many bodies of such a size last if they were forcibly dragged for so long?[17]

I have said that, in so far as Philoponus unified dynamics, it was his belief in a creator God which enabled him to do so. Of course, that was not enough on its own. The Neoplatonists, as has been pointed out in this connexion,[18] also believed in a creator God, and we shall see that Proclus associated his creator God with something not far removed from an impressed force. For the unification of dynamics, it was necessary that Philoponus should also have the idea of applying the concept of impressed force to cases which did not involve God directly, such as the motion of projectiles and of the fire belt. But in fact, he thought of that idea first, and it was only later in a book devoted to the biblical account of Creation that it occurred to him to extend the concept of impressed force to the remaining cases of motion. So it was his reflection on his creator God which in fact completed the unification.

Philoponus did not progress directly to the views of the *de Opificio Mundi*. Earlier, in his *Physics* commentary, he explained the fall of rocks and rising of flames by reference to their inner nature, without any suggestion that that nature was *impressed*. At one point he calls it an innate inclination (*emphutos rhopê*), the word *emphutos*, innate, being connected with *phusis*, nature.[19] At another point he says:

How, then, do natural things differ from things that are not natural? Because those that are natural appear to have within themselves the origin (*arkhê*) of their motion and rest. For when they move, both animals and lifeless things have their mover within themselves, and are not moved by something from the outside. For example, when stones are let go, they are not moved downwards by the person who lets them go, for he merely lets them go. It is rather the natural inclination (*phusikê rhopê*) in them that carries them down. In the same way too, fire that is let go from below is carried upwards by the nature (*phusis*) within it. Then, once they have been carried by the nature within them to join the main body of material to which they belong and to their proper places, they remain at rest.[20]

In these lines, the inner nature that here accounts for the fall of rocks and rising of flames is not yet described as being impressed by God. But although God is not given this role, he none the less enters

---

[17] Philoponus *de Opificio Mundi* 28,20-29,9.

[18] See Michael Wolff, in *Philoponus and the Rejection of Aristotelian Science*, 107-8 and n. 90, who is, however, attacking the different view that belief in a creator God motivates Philoponus' impetus theory, rather than the view proposed here, that it permits the eventual extension of impetus theory to all cases.

[19] Philoponus *in Phys.* 690,21. The word *rhopê* is standardly confined to an *upwards* or *downwards* inclination; see *in DA* 110,12-17; *in GC* 298,10; *in Phys.* 195,24-32; 630,25-631,2; 695,2-8; *Opif.* 29,2-3.

[20] Philoponus *in Phys.* 195,24-32.

into the explanation in a different way. In a passage already discussed in Chapter 12, Philoponus says that it was the Creator who originally assigned to earth, water, air and fire their arrangement (*taxis*) one above the other, and the natural movements of these elements are executed in search of that original arrangement. This teleological explanation in terms of the goal sought is evidently to be *added* to the explanation in terms of inner nature:

> It is quite ridiculous to say that place has a certain power in its own right. For things move to their proper places not in search of a surface, but in search of (*ephiemena*) the arrangement (*taxis*) they obtained from the Creator. So because earth was arranged at the bottom, so as to be beneath everything else, water in the second position, air and fire in the third and fourth, it is to be expected that, when something is dislodged from this arrangement, and what would naturally be on top of another thing is put by force beneath it instead of on top, the things will move in search of the arrangement allotted by the Creator, far enough to be in that arrangement again. So light things move up, not because they seek simply to be in contact with the surface of their surroundings, but in search of the arrangement which the Creator assigned them. For then they most fully have their identity and they have their perfection then. So place does not have any power to make bodies move to their proper places. It is rather that the bodies seek to preserve their arrangement.[21]

In the same earlier work, Philoponus treats the rotation of the heavens as due to an innate power (*emphutos dunamis*), in other words, an inner nature, just like the innate inclination (*emphutos rhopê*) which accounts for the fall of rocks, and once again there is no mention of this power being *impressed*:

> This is the reason, then, why motion takes time, and it will go faster or slower because the innate inclination (*emphutos rhopê*) is different. For if in the other [celestial] bodies all motion takes time, despite there being no medium to cut through, and the difference between faster and slower is due to nothing other than the powers innate (*emphutoi dunameis*) in them, it should readily follow for bodies which are subject to generation and destruction that, even if they should move through a vacuum, their movement would take time, and would admit of faster and slower speeds.[22]

In order to be clear how the earlier treatment of celestial motion and of the fall of rocks differs from impetus theory, it is important to be clear what is meant by impetus. Modern historians and the thinkers they are commenting on use the term in two different ways. Impetus in the strict sense is equated with *vis impressa* – an internal force impressed from without. But there is also a looser use of the term to apply to a force internal to the moving body, whether impressed from

---

[21] ibid. 581,19-31.
[22] ibid. 690,20-7.

without or not.[23] I believe that this ambiguity has led to confusion. If we drop the requirement of impression, we shall be inundated with examples of impetus. There will be no need to wait for Philoponus' *de Opificio Mundi*. Already in the *Physics* commentary he would ascribe the fall of rocks and the rotation of stars to internal forces. And he would have had many forerunners. Aristotle and the Stoics both ascribe many cases of motion to internal forces, and both have been cited as forerunners, as has the Aristotelian Alexander of Aphrodisias, and the author of the pseudo-Aristotelian *Mechanica*.[24] But these thinkers do not treat their internal forces as *impressed*.[25] In addition, Alexander still has the force that moves a projectile implanted in the *air*, not internally in the projectile, and others have warned that the same may well be true of Hipparchus of Nicaea in the second century B.C.,[26] although he has been accepted by Galileo and many others[27] as a

[23] Walter Böhm, *Johannes Philoponos, ausgewählte Schriften*, Munich, Paderborn, Vienna 1967, 369, even suggests that the concept of impetus changed eventually from that of an impressed force to that of an inner soul-like principle.

[24] On Aristotle, see H. Carteron, *La notion de force dans le système d'Aristote*, 11-32, translated into English in Jonathan Barnes, Malcolm Schofield, Richard Sorabji, eds, *Articles on Aristotle 1*, 170, n. 40; G.A. Seeck, 'Die Theorie des Wurfs', in G.A. Seeck, ed., *Die Naturphilosophie des Aristoteles*, Proceedings of the 4th Symposium Aristotelicum, Darmstadt 1975, 386. On the Stoics, see Walter Böhm, op. cit., 346; 369; G.E.R. Lloyd, *Greek Science after Aristotle*, Cambridge 1973, 158ff; Michael Frede, 'The original notion of cause', in Malcolm Schofield, Myles Burnyeat, Jonathan Barnes, eds, *Doubt and Dogmatism*, Oxford 1980, 249. On Alexander of Aphrodisias, see the report on Girolamo Cardano in E. Wohlwill, 'Die Entdeckung des Beharrungsgesetzes' I, *Zeitschrift für Völkerpsychologie und Sprachwissenschaft* 14, 1883, 387. See also S. Pines, 'Omne quod movetur necesse est ab alio moveri': a refutation of Galen by Alexander of Aphrodisias and the theory of motion', *Isis* 52, 1961, for the view that Alexander has as good a claim as Philoponus to have provided the framework for impetus theory. On ps-Aristotle *Mechanica*, see H. Carteron, loc. cit.; F. Krafft, 'Zielgerichtetheit und Zielsetzung in Wissenschaft und Natur', *Bericht zur Wissenschaftsgeschichte* 5, 1982, 53-74, at 60.

Stoic influence has indeed been found by J.E. McGuire (1985) in Philoponus' account of nature in the *Physics* commentary. But when Philoponus turns nature into a *vis impressa* many years later in the *de Opificio Mundi*, he is departing from the Stoics.

[25] The most that Aristotle's God implants in the heavens is *love* (*Metaph.* 12.7, 1072b3), and, on a certain interpretation of his causal theory (*Phys.* 3.3), the activity of inspiring it. But neither of these is like an impetus. In inspiring love, God acts as a final, not an efficient, cause, and his activity of inspiring it would not be considered numerically distinct from the heavens' activity of being inspired (*Phys.* loc. cit.)

[26] Alexander ap. Simplicium *in Phys.* 1346,29-1348,5, and Alexander *Refutation of Galen's Attack on Aristotle's Doctrine that Everything that Moves is Set in Motion by a Mover*, cited by S. Pines, op. cit. (1961), p. 30, and translated by Rescher-Marmura, p. 26. For Hipparchus, see Walter Böhm, op. cit., 16-18; Michael Wolff, op. cit. (1987), 102.

[27] Galileo *On Motion*, translated by I.E. Drabkin, *Galileo Galilei, On Motion and On Mechanics*, Madison 1960, at 89-90; E. Wohlwill, 'Ein Vorgänger Galileis im 6. Jahrhundert', *Verhandlungen der Gesellschaft deutscher Naturforscher und Ärzte* 77, part 2, reprinted in *Physikalische Zeitschrift* 7, 1906, 23-32, at 26-7; A.E. Haas, 'Über die Originalität der physikalischen Lehren des Johannes Philoponus', *Bibliotheca Mathematica*, series 3, vol. 6, 337-42; W. Hartner and M. Schramm, 'La notion de "l'inertie" chez Hipparque et Galilée', in *Collection des Travaux de l'Academie Internationale d'Histoire des Sciences*, vol. 11, Florence 1958, 126-32; F. Krafft, op. cit., 60; S. Sambursky, 'John Philoponus', in C.C. Gillispie, ed., *Dictionary of Scientific*

proponent of impetus theory for projectiles. The most likely fore-runner of Philoponus as an advocate of impressed force is Proclus. But we have already encountered earlier in the chapter Michael Wolff's warning that despite Proclus' influence, Philoponus needed to alter Proclus' concept of *monê*, before it would fit impetus theory. Proclus' best claim to have anticipated the idea of impressed force is one which I shall only be able to assess at the end of the next chapter. But I should mention here, to get it out of the way, a use of the idea of implanting a power (*endidonai dunamin*), which turns out not to be relevant.

In *Elements of Theology*, Proposition 81, Proclus runs into the problem that the soul is meant to be immune from being acted upon, and yet is supposed to act upon the body. He thinks this one-sided action will only be possible, if the soul acts through an intermediary, a power (*dunamis*) which it implants (*endidonai*). I am grateful to A.C. Lloyd for pointing out that this is fully explained in another work, the *de decem dubitationibus circa providentiam*, ch. 31.[28] It turns out that Proclus' *dunamis* is quite unlike the force implanted in a projectile, for it is the mortal part of the soul, as opposed to the immortal part. Even when the *Elements of Theology* seeks to generalise the need for an impressed power to other cases, the generalisation does not appear to apply to projectiles, or to anything very much other than Soul and Intellect. For an impressed power is needed only when a thing is, as he puts it, participated in and yet remains separate, in the way that the soul is participated in by body (for body is ensouled), and yet remains separate from the body.

To return to Philoponus, the fact that his impetus is an *impressed* force bears not only on the question of antecedents, but also on that of his influence. Shlomo Pines made a major contribution when he detected impetus theory in Avicenna, Barakât, and in all subsequent theories of the Islamic East.[29] Subsequently he found it in an earlier source, which draws directly from Philoponus,[30] and which has been dated as early as the ninth century.[31] But Pines was modest about his own important discovery, saying that there was no direct evidence for Philoponus, but that 'perhaps' the intrinsic movement he postulated in the case of light and heavy bodies 'facilitated a framework for' impetus

Biography, New York 1970, 134-9, at 136; id., *The Physical World of Late Antiquity*, London 1962, 73.

[28] Proclus, pp. 91-3 of the Budé edition of *de decem dubitationibus*, ed. D. Isaac, Paris 1977.

[29] S. Pines, 'Études sur Awḥad al-Zamân Abu' l-Barakât al-Baghdâdî, *Revue des Études Juives* n.s. 3 and 4, 1938, 3-64 and 1-33, reprinted in S. Pines, *Studies in Abu' l-Barakât al-Baghdâdî Physics and Metaphysics*, Collected Works of Shlomo Pines, vol. 1, Jerusalem and Leiden 1979; id., 'Les précurseurs musulmans de la théorie de l'impetus', *Archeion*, 1938, 21, 298-306.

[30] S. Pines, 'Un précurseur bagdadien de la théorie de l'impetus', *Isis* 44, 1953, 246-51.

[31] Fritz Zimmermann, 'Philoponus' impetus theory in the Arabic tradition', in Richard Sorabji, ed., *Philoponus and the Rejection of Aristotelian Science*, ch. 5, n. 19.

theory.[32] At most he pointed out that Philoponus went beyond Alexander in extending the idea of internal force to projectiles,[33] but he did not draw attention to the fact that in Philoponus, unlike Alexander, the force was *impressed*.[34] Had he done so, I believe he might have felt free to see Philoponus as taking the crucial step in relation to Islamic theories of projectile motion. For in the Islamic accounts, the impetus in projectiles is also an impressed force. The idea of impression is applied, however, only to forced motion, not to the fall of rocks or the rotation of the heavens, and this is what is what we should expect. For Islamic writers had no interest in Philoponus' Christian tract, the *de Opificio Mundi*, where impression is extended to these other contexts.

The sources of medieval Latin treatments of impetus have been a matter of controversy, but Buridan and his pupil Oresme seem astonishingly close to Philoponus. Not only did Buridan make Philoponus' point that violent beating of the air will not move a stone;[35] he also shares the idea of Philoponus' *de Opificio Mundi* that stellar movement is due to an impressed impetus, impressed by God at the time of the Creation.[36] This idea is not likely to have come via Islamic sources, nor is there a known Latin translation of the *de Opificio Mundi*. But scholars have probably been too cautious in following Anneliese Maier's view that impetus theory was not transmitted to the Latin West at all, but was an independent development there.[37] The minimal idea of impetus as an internal force (*mayl*) would have been transmitted to the Latin West, as Zimmermann has shown,[38] when Ghazâlî's summary of Avicenna was translated into Latin in the second half of the twelfth century. The idea of impetus as a force not only internal, but also impressed, features in Avicenna's work, and although that was not itself available in Latin, Zimmermann has pointed out that the Latin of Ghazâlî's summary of Avicenna contains the idea of an impetus that is *forced* (*inclinacio motionis violens*),[39] and this would naturally have been taken to mean an internal impetus impressed by the thrower. By the time of Galileo, impetus theory is such a commonplace that he does not name any

[32] S. Pines, op. cit. (1961), 54; Michael Wolff was also hesitant in his *Geschichte der Impetustheorie*, Frankfurt 1978, 168.

[33] S. Pines, op. cit. (1961), 49, 51.

[34] ibid., 53

[35] See *Quaestiones Super Libros IV de Caelo et Mundo*, lib. 2, q. 12-13; lib. 3, q. 2; *Quaestiones Super Libros VIII Physicorum*, lib. 8, q. 12.

[36] e.g. *Quaestiones Super Libros IV de Caelo et Mundo*, lib. 2, q. 12, Latin quoted, with German translation, by Michael Wolff, *Geschichte der Impetustheorie* 226. See his ch. 7, 212-46 for Buridan and Oresme.

[37] A. Maier, *Zwei Grundprobleme der scholastischen Naturphilosophie*[2], Rome 1951, 127-33, who influences S. Pines, op. cit., 1961.

[38] Fritz Zimmermann, op. cit., 1987.

[39] ibid. n. 12, citing Ghazâlî, *Intentions of the Philosophers*, Arabic ed. S. Dunya, Cairo 1960, 264, 18f = J.T. Muckle, *Algazel's Metaphysics: a medieval translation*, Toronto 1933, 100,12-15.

authorities for the view. On the other hand, in the strict sense according to which impetus is an *impressed* force, Galileo does not go as far as Philoponus or Buridan.[40] Michael Wolff has drawn my attention to the discussion in the *Dialogue Concerning the Two World Systems*,[41] in which Galileo makes Salviati express ignorance of whether the causes of fall and of celestial motion are external or internal. Galileo himself applies the notion of impressed force only to projectile motion and to a limited range of other cases. Philoponus' application of the notion was very much wider than that.

Impetus theory was not finally replaced until a certain conception of inertia came to be accepted. It is a matter of controversy whether the relevant conception of inertia is to be found in Galileo, or in Descartes, in Newton, or in Newton's eighteenth-century followers.[42] On one view, Newton's third definition shows him to be in the tradition of Philoponus, treating inertia as itself a type of inner force (*vis insita*), like the innate inclination (*emphutos rhopê*) of Philoponus' *Physics* commentary, though not impressed like the implanted inclination of his *de Opificio Mundi*.[43] On another view, Newton's theory of inertia makes a force necessary only for limited purposes, for example, to *start* a javelin moving. It is needed to *divert* any body from its state of rest or of uniform rectilinear motion. But so long as a body continues to move at uniform speed in a straight line, this persistence requires no force at all. Instead, the body merely conforms to Newton's first law, which says that a body will persist in a state of rest, or of uniform rectilinear motion, unless a force acts on it to the contrary. In particular, the impressed force which brings a body into a state of uniform rectilinear motion does *not* persist in the body thereafter,[44] and in this it differs from the older conception of impressed impetus. Certainly, this conception of inertia came to prevail eventually.

The idea of inertia was unificatory, because it showed how rotation related to rectilinear motion. It was no different in principle from accelerated or decelerated rectilinear motion. All these equally required a force, since they were equally deviations from the motion now viewed as basic: uniform rectilinear motion.[45] The problem of how to accommodate in one system two such different motions as the rectilinear and rotatory had troubled Greek philosophers such as

[40] Walter Böhm is taking impetus theory in the loose sense, when he ascribes it quite generally to Galileo, op. cit. (1967), 365, 369-70.

[41] Translated into English, S. Drake, 2nd ed. 1967, 234.

[42] On the claims of Galileo and Descartes see Richard S. Westfall, 'Circular motion in seventeenth-century mechanics', *Isis* 63, 1972, 184-9.

[43] Newton, *Opera Omnia*, vol. 2, London 1979, p. 2, definition 3, cited by J.E. McGuire, comment on I.B. Cohen, in R. Palter, ed., *The Annus Mirabilis of Sir Isaac Newton*, 1666-1966, Cambridge, Mass. 1970, 186-91, Michael Wolff, *Geschichte der Impetustheorie* (1978), and Walter Böhm, op. cit. (1967), 371.

[44] 44. Newton op. cit., p. 2f, definition 4, cited and discussed by Michael Wolff, op. cit. (1978), 315-16, and Richard Westfall, op. cit., 189.

[45] See Westfall, op. cit.

Democritus from Presocratic times, as David Furley has shown.[46] Philoponus leaves us with the diverse intentions of the Creator in implanting this or that impetus, and in this respect his unification of dynamics does not go as far as Newton's.

Philoponus on the distinction between natural and unnatural motion

I have now described the contribution which I believe Philoponus made through his introduction of impetus theory and unification of dynamics. But certain further contributions have been suggested by the pioneers in this subject, and to put his work in perspective, I must explain why I disagree. It has been suggested that Philoponus' introduction of impetus theory had the effect of demolishing Aristotle's division of motion into natural and unnatural (or forced), and that it was intended to do so.[47] It has been maintained, secondly, that as part of the demolition process, Philoponus rejected Aristotle's auxiliary causes of natural motion. On the most natural interpretation that would be the generator of steam and the remover of the kettle lid. But, thirdly, on another interpretation, the one that has actually been pressed,[48] the eliminated cause is taken to be the *final* cause or *goal* of natural motion.

As regards the first point, the distinction between natural and unnatural motion, it is true that impetus theory violates Aristotle's criterion for classifying projectile motion as *unnatural*, for it leaves no external cause which (like Aristotle's pockets of air) maintains active contact with the projectile. And Aristotle defines unnatural motion as produced by such an external cause. But Philoponus would evidently wish to find an *alternative* criterion for distinguishing unnatural motion, for he views the distinction between natural and unnatural motion as valid even after the introduction of impetus theory. He does so in the *Physics* commentary, when applying impetus theory to projectiles.[49] Again in the *contra Aristotelem*[50] and the *de Opificio Mundi*[51] he speaks as if the only alternatives were that motion should

---

[46] David Furley, 'The Greek theory of the infinite universe', *Journal of the History of Ideas* 42, 1981, 571-85.

[47] This suggestion was introduced by Pierre Duhem, *Études sur Léonard de Vinci*, 3 vols, Paris 1906, 1909, 1913; *Système du Monde*, vols 1-5, Paris 1913-17. It has been defended by Michael Wolff, *Fallgesetz und Massebegriff* (1971) 45-52; id., *Geschichte der Impetustheorie*, 168; id., in *Philoponus and the Rejection of Aristotelian Science*, 96-8; Walter Böhm, op. cit. (1967), 18 and 339.

[48] Michael Wolff, loc. cit. If I am led to disagree at a number of points with Michael Wolff in this chapter, it is only after learning from what I regard as the most thoroughgoing investigation to date of Philoponus' impetus theory.

[49] Philoponus *in Phys.* 641,13-642,20, translated above.

[50] Philoponus *contra Aristotelem* ap. Simplicium *in Cael.* 34,9. Compare also the statement in a work from the same period in Philoponus' career, *in Meteor.* 97,6, though this is recording a Platonist view.

[51] Philoponus *Opif.* 29,6.

be natural or forced and unnatural. Moreover, the context of this claim in the *de Opificio Mundi* is the very one in which he extends impetus theory to the maximum. Throughout his works, he consistently describes as natural the fall of rocks and rise of flames,[52] and the movement of the heavens.[53] The view that the passage in the *Physics* commentary consciously violates the distinction between natural and unnatural motion is based on the idea, which I have argued to be mistaken, that it is questioning the unnaturalness of projectile motion, rather than simply questioning the mechanism by which it occurs.[54] All this makes it clear that Philoponus does not see himself as overturning the division of motion into natural and unnatural. He merely needs a new criterion for classing projectile motion as unnatural, and this need not be an insuperable difficulty: unnatural motion might, for example, be motion *not* due to God's impetus.[55]

I know of only one group of contexts in which Philoponus is dissatisfied with the distinction between natural and unnatural motion, and this has nothing to do with impetus theory. It is most marked in his early *de Anima* commentary which precedes the introduction of impetus theory, and it has already disappeared by the time of his late writings when impetus theory is given its fullest extension. The main problem concerns the rotation of the fire belt, and the problem arises because Aristotle appears to contradict himself on whether this relation is unnatural.[56] Faced with this puzzle, Philoponus, in the *de Anima* commentary, combines the idea that the rotation of fire is a forced motion (*biaios*), produced by the celestial rotation above it, with the idea that (contrary to the normal implication) it is none the less natural (*kata phusin*). It is natural because, in rotating, fire does not leave its proper place at the periphery of the sublunary world.[57] This creates for him a secondary problem about the rise of flames, which can hardly be natural as well, or fire would have two different natural motions, one upwards and one rotatory. Philoponus was later to accept this,[58] but here his solution is that the upward movement of fire is not natural, except in the sense in which convalescence is natural. Rather it is a move *in the direction of*

[52] Philoponus *in Phys.* 195,24-32; 690,21, both translated above; *contra Aristotelem* ap. Simplicium *in Cael.* 34,23-4; 30-2; 35,14-16.

[53] Philoponus *in DA* 66,11; *in Phys.* 690,24-5, translated above; *contra Proclum* 278,21-8; 492,20-493,5 (but this last passage describes the motion as above the nature of the *body*, although not the *soul*, of the heaven); *contra Aristotelem* ap. Simplicium *in Cael.* 49,19.

[54] See my discussion earlier in this chapter of Philoponus *in Phys.* 641,13-15.

[55] Philoponus would then be exploiting the fact that God is not the one who impresses the impetus into a javelin, although he might have difficulty with his view that God *is* the one who implants the impetus in animals.

[56] Aristotle *Meteor.* 1.7, 344a11-13 recognises the rotation of fire; *Cael.* 1.7,269a9-18 argues that it could not rotate either naturally or unnaturally; *Cael.* 1.7, 269a30-b2 speaks as if rotation would be unnatural for it.

[57] Philoponus *in DA* 66,1-4.

[58] Philoponus *contra Aristotelem* ap. Simplicium *in Cael.* 35,14-16.

its natural state.[59] Not content with this first treatment, Philoponus goes on in his *Physics* commentary and his *contra Proclum* to try out another view which is embraced by Ammonius' other pupils, Damascius and Simplicius, and by one of Ammonius' successors, Olympiodorus. According to this view, the rotation of the fire belt is *neither* in accordance with, *nor* contrary to, its nature, but is instead *above* its nature (*huper phusin*).[60] The concept of what is *above* nature or *supernatural* is freely used by Philoponus in a number of further applications. For example, he suggests that the motion of the heaven is above the nature of its body, but not of its soul.[61] And in Chapter 15 we shall see him applying the concept of the supernatural not to the motion, but to the unending preservation, of the cosmos, which is postulated not by himself, but by Proclus. The use of the concept, however, does not represent an attempt to *replace* Aristotle's distinction between natural and unnatural motion, but simply to elaborate it. Finally, in two works more or less simultaneous with each other, the *Meteorology* commentary and the *contra Aristotelem*, Philoponus abandons even this variation on the distinction between natural and unnatural motion, and plumps straightforwardly for the view that the rotation of fire is *natural*.[62] I know of no further departure by him from Aristotle's dichotomy between natural and unnatural motion.

Philoponus on the efficient causes of natural motion

The next suggestion is that Philoponus eliminates Aristotle's auxiliary causes of natural motion. I shall start with the more natural interpretation, even if it is not the intended one, according to which the causes to be eliminated are the obstacle-remover and the generator. It is in fact quite true that in a passage of the *Physics* commentary already quoted, Philoponus dispenses with Aristotle's obstacle-remover, or letter-go:

> For example, when stones are let go, they are not moved downwards by the person who lets them go, for he merely let them go. It is rather the natural inclination (*phusikê rhopê*) in them that carries them down.[63]

This rejection of the obstacle-remover as a cause is, admittedly, an important difference from Aristotle. For, as was seen in Chapter 13, if the fall of rocks calls for no such causes, this will affect the strategy for

---

[59] Philoponus *in DA* 66,4-17.

[60] Philoponus *in Phys.* 198,12-19; 378,21-31; *contra Proclum* 240,28-241,10; 259,27-260,2; 278,21-8; 492,20-6. For Damascius, see Philoponus *in Meteor.* 97,20-1; for Simplicius *in Cael.* 21,1-25; 35,13; 51,22-6; 80,23; for Olympiodorus, his *in Meteor.* 2,19-33; 7,21-30.

[61] Philoponus *contra Proclum* 492,26-493,5.

[62] Philoponus *in Meteor.* 97,20-1; *contra Aristotelem* ap. Simplicium *in Cael.* 34,8; 35,2-8; 35,14-20; 35,28-30.

[63] Philoponus *in Phys.* 195,27-9.

arguing that celestial motion requires God as an external final cause. None the less, Aristotle himself had already played down the obstacle-remover, when he introduced him in *Phys*. 8.4 as a merely *accidental* cause (*kata sumbebêkos*) of motion.[64] So Philoponus need not be conscious of departing from Aristotle. Indeed, the suggestion has been made[65] that Philoponus is simply following Alexander's interpretation of Aristotle, as described in Chapter 13. If that is so, Philoponus will accept Aristotle's *other* efficient cause, the generator who makes the steam by boiling the kettle,[66] but will follow Alexander in interpreting him as an *internal* cause, on the grounds that he is responsible for the internal constitution of the steam, rather than being an agent who works on the steam from the outside.[67]

When Philoponus moves on to the *de Opificio Mundi*, he brings Aristotle's 'generator' still more into his own, even though that is the very moment when he finally diverges from Aristotle by having the natural inclination in rocks and flames *impressed* by the Creator. For by an unconscious irony, his introduction of the un-Aristotelian impetus theory has the side-effect of enshrining Aristotle's 'generator', but now transformed from the humble boiler of a kettle into God who created the light elements in the first place and impressed their natural inclinations into them.

Despite this, it is tempting to infer that Philoponus sought to eliminate Aristotle's generator as well as his obstacle-remover from among the efficient causes of natural motion. Evidence might be drawn from a valuable recent study of Philoponus' treatment of *nature*.[68] Whereas Aristotle viewed a rock's inner nature as a cause of its being *passively* moved (*paskhein*)[69] by auxiliary agencies, Philoponus insists that its inner nature has an *activity* (*energeia*) and *acts* (*poiei*). Still more significantly, Philoponus uses a Stoic term, when he calls the inner nature of a thing a *cohesive* (*sunektikê*) cause of its being, a cause that holds it together.[70] And it has been argued that he will have taken over not only the Stoic term, but also the Stoic idea that cohesive causes are *self-sufficient* (*autotelês*), which means that they can produce effects self-sufficiently by themselves (*autarkôs di' hautôn*

---

[64] Aristotle *Phys*. 8.4, 255b27.

[65] S. Pines, 'Omne quod movetur necesse est ab alio moveri: a refutation of Galen by Alexander of Aphrodisias and the theory of motion', *Isis*, 52, 1961, 21-54, at 48.

[66] *gennêsantos*, Aristotle *Phys*. 8.4, 256a1.

[67] Alexander, *Refutation of Galen's Attack on Aristotle's Doctrine that Everything that Moves is Set in Motion by a Mover*, fol. 67a17-20 of the Carullah ms., translated by Rescher-Marmura, p.17, and by Pines, pp. 27 and 42.

[68] J.E. McGuire, 'Philoponus on *Physics* 2.1: *physis, dunamis* the motion of the simple bodies', *Ancient Philosophy* 5, 1985, 241-67, commenting on Philoponus *in Phys*. 197,30-198,8, translated below.

[69] *arkhê tou paskhein*, Aristotle *Phys*. 8.4, 255b30-1.

[70] This cause is the *pneuma* which pervades it: Galen *On Cohesive Causes*, ch.1, translated from the Arabic by M. Lyons, in *Corpus Medicorum Graecorum, supplementum orientale* 2, Berlin 1969, discussed in Chapter 6 above.

poiêtika).[71] A probable example[72] of such a cause is the internal state of a cylinder and of its *pneuma* after it has been pushed. This inner state is sufficient on its own, without the aid of further causes, to make the cylinder roll. From this someone might infer (although the study cited does not do so) that if Philoponus makes nature an *active* and *self-sufficient* cause, he is dispensing with Aristotle's auxiliary causes.

Such a conclusion would not follow. The Stoic connexion is important and well established, but the reference to *action* is not of itself a challenge to Aristotle's view that a thing's nature needs to be *moved* to action by something else. Moreover, even if Philoponus is taking over the Stoic view that a thing's inner nature is a self-sufficient cause of a thing's natural motion (and this is less than certain), nothing will follow about the elimination of Aristotle's auxiliary causes. For the Stoics themselves see a prior push of the cylinder as a necessary prerequisite for putting it in the right internal state for rolling: the Stoics' pusher is as much a prerequisite as Aristotle's obstacle-remover. The passage, then, establishes nothing germane, but it is of sufficient interest to deserve quotation:

> This, then is the definition of nature. But it is worth recognising that this definition does not signify what nature is, but rather its activity (*energeia*). For by learning that nature is the source of motion or rest, we learnt not what it is, but what it does (*poiei*). So in order to give the definition of what it is as well, let us say that nature is life (*zôê*) or power (*dunamis*) diffused through bodies, moulding them and managing them, as the source of motion or rest in the thing in which it belongs primarily, and in it in virtue of itself and not coincidentally. It is clear that nature manages not only animate things, but also inanimate. For everything has a natural power (*dunamis*) cohesive of its being (*sunektikê tou einai*), since it would perish and change into non-being if there were nothing to hold it together (*sunekhein*). But it is clear that as it is in animate things that form is more evident, so too the providence of nature is more evident in them. And so from this too it is clear that the definition will cover the nature of animate things as well, which is their soul. For the life of animate things is nothing other than soul.[73]

Another passage in Philoponus has been taken as launching a conscious attack on Aristotle's account of nature as a passive (*paskhein*) cause subject to other influences, and hence an attack on the need for auxiliary causes of motion. For he describes the weight of a rock not as a passive, but as a productive (*poiêtikê, poiêtikon*) cause of its falling.[74] But his contrast here is not with *passive* causes of motion, but with the resistance of the medium as an *impeding* (*empodistikon*)

[71] Clement of Alexandria *Stromateis* 8.9.
[72] See Cicero *de Fato* 41-4.
[73] Philoponus *in Phys.* 197,30-198,8.
[74] Michael Wolff, in *Philoponus and the Rejection of Aristotelian Science*, 96-7, citing Philoponus *in Phys.* 678,19-29; 679,29.

cause.[75] In general, the notion of a *poiêtikê* cause takes on different meanings, according to the different contrasts with which it is associated. Elsewhere, Philoponus calls the internal inclination in a rock a *poiêtikê* cause of motion, meaning that it is an efficient as opposed to a *final (telikon)* cause.[76] Neither use of *poiêtikê* is intended to deny that a rock's weight or nature is a *passive* cause, requiring auxiliary causation. Indeed, Simplicius thinks that Aristotle himself would want to call a thing's inner nature a *poiêtikon* cause, just as Philoponus calls its weight *poiêtikon*. Admittedly, Simplicius is puzzled, and wonders how Aristotle can combine this with calling inner nature *passive*.[77] But he thinks the question can be answered, and one of his answers is that the inner nature is *poiêtikê* because it still *contributes* (*suntelei*) to the achievement of the goal or final cause. This point can be reinforced in the case of natural rise or fall, for Aristotle does everything to make clear that the inner nature of a falling rock makes the *major* contribution. It makes the rock all ready to burst into motion once released from obstacles, and the obstacle-remover moves it only 'in a way' and 'accidentally', like the wall which accounts for a ricochet.[78]

Further evidence for Philoponus' rejection of auxiliary causes has been sought in his insistence that weight does not belong to things in virtue of '*something else*'.[79] But again there is no reason to think that this is an attack on the obstacle-remover or generator by which heavy bodies are passively moved. The kind of idea Philoponus has been rejecting in the surrounding context is that weight is due to the ability to cut through resistant air, or to reduced air-resistance.[80] It is these factors concerning resistance which he is denying to be causes of weight; he is not excluding the 'generator'.

A final passage that has been cited in evidence comes from the earlier *de Anima* commentary. Philoponus is discussing Aristotle's remark that, if the soul is thought of as a self-mover, it cannot, except accidentally, be moved by something *else*. Philoponus comments that equally if motion is due to nature, it cannot need a *further* mover, or the work of nature would be done in vain.[81] But the further mover which Philoponus here declares unnecessary is clearly some rather substantial kind of mover. He is not ruling out the need for an obstacle-remover or generator.

Philoponus on the final causes of natural motion

I turn now to the *final* causes which Philoponus has been thought to

[75] Philoponus *in Phys*. 678,19.
[76] ibid. 630,17-18; 26.
[77] Simplicius *in Phys*. 288,8-11; 17-32; 1218,13-19.
[78] Aristotle *Phys*. 8.4, 255b19-29.
[79] Philoponus *in Phys*. 678,22-4; 679,23-9.
[80] ibid. 679,14-18; 21-2.
[81] Philoponus *in DA* 109,24-8.

eliminate from his account of natural motion. The cause by which Aristotle's naturally falling rock is passively moved is, according to one interpretation,[82] an interpretation from which I dissented in Chapter 13,[83] not the obstacle-remover or the generator but the rock's natural place acting as a *final* cause. And it is this final cause that Philoponus has been alleged to eliminate in the passages just discussed. It has further been held, though again I am not able to agree,[84] that Aristotle's final cause is much more of a cause than the obstacle-remover or generator, since it acts as an external cause which continues to maintain contact with the falling rock. It is said, finally, that Philoponus systematically removes Aristotle's final causation not only from the motion of rocks, but also from that of animals and of the heavens, thereby again modifying Aristotle's concept of *natural motion*.

In connexion with falling rocks, I think that this interpretation must be mistaken. We have already seen how Philoponus insists in his *Physics* commentary that rocks fall in search of (*ephiemena*) the arrangement (*taxis*) which God originally gave them.[85] All he does is to substitute that God-given arrangement as the final cause in preference to Aristotle's natural places. And he has no need to retract this claim of final causality, when he finally applies impetus theory to heavenly bodies in the *de Opificio Mundi*. As regards the motion of animals and men, the view has been ascribed to Philoponus that evil action has no final cause, but I have not found that view expressed in the passage of the *de Opificio Mundi* cited.[86] In the *de Anima* commentary, Philoponus does no more than reveal, quite accurately, the extent to which Aristotle himself had already qualified the claim of final causes

---

[82] Michael Wolff, *Fallgesetz und Massebegriff* (1971) 48, 71-9; and in *Philoponus and the Rejection of Aristotelian Science*, 96-7, 111-13, 117.

[83] I argued there that *Phys.* 8.4, 256a1-2, offers the obstacle-remover and the generator as the *only* justification for saying that heavy bodies are moved by (*hupo*) something. Admittedly, an answer to the *different* question why (*dia ti*) they move to their own place (*ho hautôn topos*) had earlier been that they are naturally constituted for that direction (*pephuken poi*, 255b15). But that answer draws attention to the directed inner nature, not just to the direction, and is in any case not offered as the answer to the question by (*hupo*) what they are moved.

[84] As regards the claim of *contact*, Aristotle denies in *GC* 1.6, 323a20-33, that, in the ordinary sense of 'touch', unmoved movers *touch* what they move. Nor in any case could the natural place of a falling rock, or the breakfast of an approaching human, be described easily as in *contact* with them. As regards the claim of *externality*, that notion is not applied to that by which a thing is moved in *Phys.* 8.4, 255b13-256a3. We have already noted a difference of opinion as to whether the generator there mentioned is external. Aristotle is better known for the claim, in *Phys.* 2.1 and elsewhere, that external causes are the mark of *unnatural*, not of natural, motion. I do not think that he makes all final causes external: if a purely imaginary breakfast stirs someone to movement, it would be better described as an *internal* final cause.

[85] Philoponus *in Phys.* 581,19-31.

[86] *Opif.* 301,11-303,19 says only that our deliberate choice (*proairesis*) gives to evil its derivative existence. I do not think that that excludes a purpose or final cause for evil action.

to act as *movers*. As observed in Chapter 13, Aristotle thinks that the soul is in a certain sense unmoved, and if it is also described as moved by final causes, this is only in the weak sense (*de Anima* 2.5) that the soul's capacity for desire is *aroused* by final causes out of a state of heightened potentiality into full actuality. This transition to full actuality is not motion or change in the strict sense and this is what Philoponus explains, though in connexion with objects of perception rather than of desire: they do not move the soul, except in the sense of *arousing* it, in the manner of *de Anima* 2.5.[87] None of this corrects Aristotle on final causes; it merely expounds him. Admittedly, one thing is missing from the extant first half of Philoponus' *Physics* commentary. For he does not there follow Aristotle in the attempt of *Physics* 8.2 and 8.6 to attribute animal motion partly to external influences in the environment.[88] But we do not know whether he did so in his missing discussion of *Physics* Book 8, and in any case, these external influences are efficient, not final, causes, if my account was right in Chapter 13.

The big difference concerning final causes comes when in the *de Opificio Mundi* Philoponus turns God into an efficient cause impressing motion into the heavens, instead of a final cause who inspires the heavens to motion by arousing their love. Indeed, he denies that the heavens are animated by souls at all. But I would not agree that this change is anticipated in his earlier writings. It has been suggested that after the early commentary on the *de Anima*, he eliminates the souls of the heavens, and thereby eliminates the final cause which inspires those souls with love. But in fact celestial motion seems to be explained by celestial souls (as well as by inner nature) in the *contra Proclum* of 529 and still later in the *contra Aristotelem*.[89] And the reference to a greater power (*kreittôn dunamis*) in the *contra Proclum*, if it is not just a repeated reference to celestial souls, may even be an explicit reference to the unmoved mover who fills those souls with love.[90]

Coherence and continuity in Greek science

I began this chapter with a reference to Thomas Kuhn, and am now

[87] Philoponus *in DA* 109,29-111,15.

[88] Aristotle *Phys.* 8.2, 253a7-20; 8.6, 259b1-20.

[89] This is clear in Philoponus *contra Aristotelem* ap. Simplicium *in Cael.* 78,19-79,14; 80,16-17; 91,17-18. Here, incidentally, at 79,8-13, soul and nature seem to be treated as two independent causes of celestial motion. In the *contra Proclum*, at 492,18-20, Philoponus reminds us of the two views that celestial rotation is due to the *body* of the heaven and that it is due to the *soul* in it. He then says, 492,26-493,1, that its rotation is above (*huper*) the nature of its *body*, implying that it comes from the *soul*. In that context, he is likely to think of soul not as independent of nature, but as one *kind* of nature, which is a view he shares with Aristotle (*PA* 1.1, 641a28; b9-10; *GA* 2.4, 740b29-741a2), and expresses in a passage already translated (*in Phys.* 197,34; 198,2; 6-8).

[90] Philoponus *contra Proclum* 492,26-7.

in a position to refer to another of his ideas. In recent work, he has emphasised the coherence of much of Aristotle's physics, a point with which I entirely agree.[91] it is often hard to unpick one piece of Aristotle's physical theory without unravelling much of the rest. If one rejects his two-dimensional conception of place and reverts to the more familiar conception of space as a three-dimensional expanse, it becomes harder to deny the existence of extra-cosmic space (Chapter 8), and to avoid the conclusion that such space is both vacuous and infinite in extent, so that void and infinity exist. Philoponus was among those who reverted to the three-dimensional conception of space, and even allowed that space could be vacuous, 'so far as depended on it' (Chapter 11). But his theory too needed to be coherent, and he could not easily draw the obvious conclusion that there was infinite vacuous space outside the cosmos, because if he once allowed infinite space, he might have to allow infinite time,[92] and that would wreck his project of arguing for the Christian belief in God as the creator of time a finite number of years ago. We therefore see him struggling to agree with Aristotle that there is no extra-cosmic space (*in Phys.* 582,19-583,12, Chapter 8), and indeed that space can never be empty in fact.[93] The Stoics took the other route and allowed empty extra-cosmic space. But by the familiar unravelling process, that disagreement with Aristotle led on to another, and they had to give air and fire a centripetal motion, to balance the centrifugal one and save them from dispersing in the surrounding void.

This chapter has supplied further examples of coherence in Aristotle, for we have seen his distinction between natural and unnatural motion threatened by Philoponus' anti-Aristotelian idea that all motion alike is due to an internal force impressed from without. Philoponus evidently wanted to retain the Aristotelian distinction, but it could well be that his effort to do so would need to appeal once again to his anti-Aristotelian belief in a Creator, natural motion being motion whose impetus was impressed by the Creator. One disagreement with Aristotle depends again on another, and in both contexts the Creation proves crucial to Philoponus' theory, as it earlier proved crucial in enabling him to extend impetus theory to all cases of motion,[94] so that his theory too acquires a certain coherence of its own.

Ancient science, however, differs from modern in the frequency with which coherent alternatives exist side by side, instead of replacing

---

[91] Thomas Kuhn, Shearman Lectures, University College London, 1987

[92] Unless he could argue, contrary to Aristotle, that the existence of infinite space is less objectionable than the existence of infinite past time, on the grounds that infinite space, unlike infinite past time, does not have to be *traversed*.

[93] For his (not wholly convincing) rationale, see David Sedley, 'Philoponus' conception of space', in Richard Sorabji, ed., *Philoponus and the Rejection of Aristotelian Science*.

[94] For further examples of its centrality, see Richard Sorabji, 'John Philoponus', in *Philoponus and the Rejection of Aristotelian Science*, at p. 30.

each other sequentially. A system as opposed to Aristotle's as that of the atomists could exist side by side with his and with intermediate systems, and all in dialogue with each other. So the preservation of a thoroughgoing alternative to Aristotle did not have to wait until a later date. The coexistence of coherent alternatives gives Greek science a certain continuity, in that one system does not so much replace another as develop in competition with it.

One task which I have left over is to consider whether Proclus might have anticipated Philoponus in introducing the idea of an *impressed force*, or whether at least he inspired this idea in Philoponus. One of the contexts in which Proclus appears to postulate an impressed force is that of an argument about infinite power, an argument which was started by Aristotle. This is the argument mentioned in Chapter 13 as ruling out the idea that the unmoved mover could be a celestial soul. The history of the argument is of considerable interest in its own right, and it will be the subject of the final chapter.

CHAPTER FIFTEEN

Infinite power impressed: the Neoplatonist transformation of Aristotle

The infinite power arguments in antiquity

This chapter concerns a line of thought which began in Aristotle as an argument in dynamics, and was converted by his commentators into an argument about Creation although it retains implications for both subjects. The argument was eventually used, we shall see, to urge that Aristotle's God was not only the mover of the heavens, but also their creator or sustainer. Such a view of Aristotle is hard to believe,[1] for the main role of his God in the world appears to be merely that of a *mover*. He moves the heavens, but does not seem to give them *existence*. Admittedly, he has a certain indirect responsibility for the existence of compound bodies down below the heavens. For he unconsciously inspires[2] the sun's motion, which by the obliquity of its angle in turn produces seasonal disturbances, and so makes earth, air, fire and water turn into each other, relocate themselves and mingle to form new compounds.[3] But the existence of the main masses of these four elements and of the heavens themselves appears to be quite independent of him. At least this is so for the four elements, even if for the case of the heavens we accept the argument which will be encountered below, that to maintain them in motion is to preserve their essence and sustain them.

The Aristotelian argument with which I am concerned comes from *Physics* 8.10, and will be familiar from Chapters 9 and 13 above. Aristotle maintains that the unmoved mover which is required to account for the motion in the cosmos is not a spatially extended entity. In his words, it lacks size (*megethos*), and is indivisible and without parts (*adiaireton, ameres*).[4] Among his premises he needs to include

---

[1] There is an extended essay on Aristotle's position by R. Jolivet, 'Aristote et Saint Thomas ou la notion de création', in his *Essai sur les rapports entre la pensée grecque et la pensée Chrétienne*, Paris 1955, 3-79.

[2] Aristotle *Metaph.* 12.5, 1072b3.

[3] Aristotle *GC* 2.10.

[4] Aristotle *Phys.* 8.10, 266a10-11; 267b25-6.

two points that he had sought to establish earlier. The motion in the cosmos lasts for an infinite time (as in *Phys.* 8.1),[5] but on the other hand nothing can extend over an infinite space (as in *Phys.* 3.5).[6] The contrast in his treatment of infinite time and infinite space is therefore not fortuitous, but essential to the argument.

He first maintains[7] that the cause of an infinitely long motion cannot be finite. This should already be enough to yield his conclusion that the cause, being infinite, will not be a spatially extended entity. For all spatially extended entities are finite in extent. But Aristotle also seeks to rule out the suggestion that the cause might be an infinite power *housed in and belonging to* a finitely extended entity, that it might, for example, be some infinite power belonging to the heavens. And this would incidentally rule out the cause being the soul or souls which animate the heavens. He therefore argues not only that it cannot be finite, but also more generally (*holôs*, 266a24) that it cannot be an infinite power of a finite entity: an infinite power, he objects, cannot reside in, and belong to something of finite extent.[8] His overall conclusion might be re-expressed in modern terms by saying that perpetual motion, produced by physical means, would impose the impossible requirement of an infinite fuel tank.

If the unmoved mover neither is, nor belongs to, a spatially extended entity, this opens the way to the view later spelt out in *Metaph.* 12.7, where the argument is repeated, that the unmoved mover is God.[9] But in the Neoplatonist period, Aristotle's argument became a locus of controversy. First, Syrianus (died c. A.D. 437) protested that once Aristotle had admitted the source of infinite kinetic power could not be the heavens themselves, he should have admitted that the same source which lends the heavens its infinite kinetic power would also be the source of *being* which produces them. He would then have been in agreement with Plato, whose God is a *creator*, producing the orderly cosmos. Plato actually makes a connexion, according to Syrianus, between the source of motion and the source of being. For at *Phaedrus* 245E he says that without the world soul, which causes motion, the

---

[5] ibid. 267b24-5.

[6] ibid. 267b22-4.

[7] ibid. 266a12-24; 267b23-4.

[8] ibid. 266a24-b6; 266b25-7; 267b22-4. There are cross-references to his argument at *Cael.* 1.7, 257b21-3; 2.12, 293a10-11, and it is repeated in summary at *Metaph.* 12.7, 1073a5-11.

[9] Infinity is associated with divinity both here at *Metaph.* 12.7, 1073a5-11 and at *Cael.* 1.9, 279a27. The association is found in pagan thought well before Plotinus (cf. Xenophanes, Anaxagoras), but see Hilary Armstrong, 'Plotinus' doctrine of the infinite and its significance for Christian thought', *Downside Review* 73, 1955, 47-58; David L. Balas, 'A Thomist view on divine infinity', *Proceedings of the American Catholic Philosophical Association* 55, 1981, 91-8.

heavens and all generation would crash into motionless ruin.[10]

Syrianus' reference to the *Phaedrus* suggests a *direct* connexion between preserving the *motion* and preserving the *existence* of the heavens. The connecting link is not explained, but one might be found in Aristotle's view that circular motion is the *defining* characteristic of the celestial body, so that *in* preserving its circular motion, God is *thereby* preserving its existence. The point is emphasised by Aristotle's successor Theophrastus, when he asks in his *Metaphysics* (6a5-13; 10a10-15) whether the heaven would stop existing as a heaven, if it lost its rotation. We shall find a connexion between motion and existence like that suggested by Syrianus recurring arguably in Simplicius and in Avicenna, Aquinas and Abravanel. But an alternative is supplied by Syrianus' pupil Proclus (c. A.D. 411-485), and this is the one of more direct relevance, because it is a re-application of the infinite power argument.

The same infinite power argument, says Proclus, which excludes an infinite power of *moving* from a finite body will also exclude from it an infinite power of *existing*. So this power (*dunamis*) of existing must *equally* be derived by the cosmos from elsewhere, most plausibly from the same divine source.[11] Proclus adds that Aristotle's argument also prevents the infinite power of existing from being received *as a whole* by the cosmos. It must rather be given in finite dollops as large as the cosmos can take,[12] and this justifies Plato's claim in *Timaeus* 27D-28A that the cosmos is always in process of being brought into being, but never actually has being. Proclus' postulation of dollops may remind us of Leibniz's caricature of Newton's views: God has repeatedly to rewind the cosmic clock.[13] Proclus puts his argument as follows (key parts in italics).

> The Peripatetics say that there is something separate from matter, but it is not an efficient cause (*poiêtikon*), only a final. And this is why they also removed the paradigms and set at the head of all things an Intellect without multiplicity. Plato, however, and the Pythagoreans hymned the demiurge of the universe as something separate from matter, far removed, the creator (*hupostatês*) of everything and providence of all.

[10] Syrianus *in Metaph.* 117,25-118,11. I am indebted for my account of Syrianus and Proclus especially to Carlos Steel, who was kind enough to show me an advance copy of his article, 'Proclus et Aristote sur la causalité efficiente de l'Intellect divin', in J. Pépin and H.D. Saffrey, eds., *Proclus – lecteur et interprète des Anciens*, Actes du Colloque Proclus, C.N.R.S. Paris 1987.

[11] Proclus *in Tim.* (Diehl) 1,267,16-268,6, and in a work entitled *Examination of Aristotle's Objections to Plato's Timaeus*, ap. Philoponum *contra Proclum*, 238,3-240,9; 297,21-300,2; 626,1-627,20.

[12] Michael Wolff, *Fallgesetz und Massebegriff*, Berlin 1971, 94-9, holds that Proclus is committed to the *opposite* view, that the power impressed, as effect, must be equal to its cause, and so infinite, and he takes it that Philoponus' reply turns on denying this. But in fact Proclus and Philoponus *agree* that God can, Proclus would say *must*, implant only a finite force at any one time.

[13] *The Leibniz-Clarke Correspondence*, ed. G.H. Alexander, Manchester 1956, starting with Leibniz's first letter §4.

(267,4) This is the most reasonable view. For if the cosmos loves the Intellect, as Aristotle himself says, and moves in relation to it, from where does it get this desire? Since the cosmos is not the first of things, it must get this desire from a cause which moves it to love. Aristotle himself says that the object of desire moves the thing that desires. If that is true, and the cosmos by its very being and nature desires the Intellect, clearly it gets all its being from the same source as its desiring.

(267,12) But from where does it get its movement *ad infinitum* seeing that it is finite? For as Aristotle says, all body has finite power. From where, then, does the universe get this infinite power of being, if it is not in Epicurus' way from chance?

(267,16) To speak generally, if the Intellect is the cause of the infinite, uninterrupted and unitary *motion*, there is some efficient cause (*poiētikon*) of the eternity. And if so, what prevents the cosmos also being eternal and derived from a cause – the Father? It obtains (*lambanei*) the power by which it *moves ad infinitum* from the thing it desires. *In just the same way, everything shows that it will obtain its infinite power (dunamis) of existing from there, because of the argument which says that an infinite power never exists within a finite body.* Either, then, it does not have a power that keeps it together at all (but how could that be? For anything divisible has something indivisible that holds it together, as Aristotle himself says somewhere. And the universe is a living animal, since at least he himself says that God is an eternal living animal, and every animal is held together by the life in it.) Or as an alternative the cosmos has a power that holds it together, but only a finite one. But that is impossible, for the power would run out, if it is finite. Or it has a power that is infinite, and again it would get this not from itself. *Something else, then, will give (didonai) it the power of existing, and will give it not all at once, since it would not be capable of receiving it (dektikon) all at once. It will give it, then, in the amounts it can take (lambanein), in a stream that flows and ever flows on to it (epirreon). No wonder the cosmos is for ever coming into being and never has being.*[14]

Proclus frequently reiterates his theory, although he does not always describe what is received as a *power* (*dunamis*) of moving or existing.[15] He appeals in addition to yet other passages in Plato for the view that the world's imperishability is *derivative*. This, he says, is expressed in the *Statesman* by saying that the cosmos receives immortality through a process of repair (*athanasia episkeuastê*) by the creator, and in the *Timaeus* by saying that such lesser gods as the celestial bodies are kept in being only by the bond (*desmos*) of the Creator's good will, since, being composite, they would otherwise dissolve in course of time.[16] Plato has the Creator explain this point to the lesser gods:

---

[14] Proclus *in Tim.* 1.266,28-268,6.

[15] Proclus *in Tim.* 1.260,14-15; 1.267,16-268,6; 1.278,20-1; 1.279,11; 1.295,3-12; 1.473,25-7 (=schol.); 2.100,18-29; 2.131,3; 3.220,1-3; and *Examination of Aristotle's Objections*, as above.

[16] *Statesman* 270A; *Tim.* 41A-B; cited by Proclus *in Tim.* 1.260,14-15; 1.278,20-1; 2.100,25-6; 3.220,1-3.

Gods, among the gods of which I am the Creator and the works of which I am the father, those (*ta* for *ha*) created by me are indissoluble, at least so long as I do not will their dissolution. Now although all that has been bound together can be unfastened, still it would be an evil agent who wanted to unfasten what is well fitted together and in a good state. For that reason, although you are not immortal nor indissoluble altogether – you have been created – none the less you will not be dissolved, nor will death fall to your lot, since you have been allotted a bond, in the form of my will, which is greater and more powerful than those with which you were bound together when you were created.[17]

The idea that the heavens are composite, and so, but for God, subject to dissolution, eventually took on a special meaning, as applied to Aristotle. According to the usual, but not universal, interpretation, Aristotle's heavens are a composite of *matter* and *form*. We should expect the prime matter to *survive* the dissolution, but some Islamic discussions (substituting atoms for prime matter as the ultimate substratum) urge that even the ultimate substratum would not survive.[18] On the other side, Averroes denies, as we shall see, that the heavens are any kind of composite, precisely in order to avoid the threat of dissolution.

Syrianus and Proclus saw that Aristotle made God responsible only for the *motion*, not like Plato for the *existence* of the heavens. But with Proclus' pupil Ammonius (c. A.D. 435/45-517/26), the harmonisation of Aristotle with Plato was taken a stage further. He represented Proclus' argument for God's preservative role as one that Aristotle did not overlook, but actually *intended*. Ammonius wrote a whole book to show that Aristotle, like Plato, considered God to be the efficient or productive cause (*poiêtikon aition*) of the cosmos. Simplicius, who gives us this information, cites a series of arguments from Ammonius' book,[19] of which the last is the infinite power argument, slightly reformulated. What the celestial body is said to receive from God is not the *power* of eternal motion and being, but simply eternal motion and being themselves.[20] Or more exactly, it is said to receive its being as a *body*. One would have expected Ammonius to say: its being as a *celestial* body, or as a *cosmic* body. For if it has prime matter, that ought to survive, although it might then constitute not an orderly body, but a chaotic body, like that described in Plato's *Timaeus*.

[17] Plato *Tim.* 41A-B.

[18] H.A. Davidson, 'John Philoponus as a source of medieval Islamic and Jewish proofs of creation', *Journal of the American Oriental Society* 89, 1969, 357-91; at sec. VI, 388-91.

[19] Simplicius *in Phys.* 1361,11-1363,12 translated below. The book is referred to again at Simplicius *in Cael.* 271,18-21. The most helpful account of Ammonius' theology I know is that of Koenraad Verrycken in his Ph.D. dissertation, *God en Wereld in de Wijsbegeerte van Ioannes Philoponus*, Louvain 1985. A summary of his views on Ammonius will appear in Richard Sorabji, ed., *The Transformation of Aristotle*.

[20] Simplicius *in Phys.* 1353,4-8; similar formulation at Simplicius *in Cael.* 369,32, London 1989.

And if, according to Aristotle, the power of any finite body is itself finite, clearly whether it be a power of moving or a power that produces being, then, just as it must get its eternal *motion* from the unmoved cause, so it must receive its eternal *being* as a body from the non-bodily cause. My teacher Ammonius wrote a whole book offering many proofs that Aristotle thought God was also an efficient cause of the whole cosmos, a book from which I have here taken over some items sufficient for present purposes. More complete instruction on the subject can be got from there.[21]

Ammonius' interpretation of Aristotle's infinite power argument was widely accepted. Asclepius' repetition of it[22] does not admittedly constitute an independent endorsement, since his commentary on Aristotle's *Metaphysics* is taken 'from the voice of' Ammonius, with little editorial assessment. In other words, it is an edition of Ammonius' own lectures.[23] But Simplicius construes the infinite power argument in the same way, in the passage just mentioned and elsewhere, as does Olympiodorus.[24] Moreover, Ammonius' general conclusion, that Aristotle's God is the efficient cause of the cosmos, is, we shall see, still more widely endorsed. As regards infinite power, Simplicius and Philoponus take the harmonisation of Plato and Aristotle further still, for they think that Plato already knew and exploited the argument in Aristotle that an infinite body can contain only finite force.[25]

Ammonius' pupil Philoponus (c. A.D. 490-570) was a Christian. In an early work, his commentary on the *Categories* (50,28-51,12), he was content merely to expound Aristotle's version of the infinite power argument. But later he tried to rewrite the argument, to support the Christian belief that God created the universe in the sense of giving it a *beginning*, a belief that is actually incompatible with earlier versions of the argument. To achieve this result, he made the argument much more complex and his rewriting, which is repeated in several works, has been very variously interpreted.[26] But I believe it can be seen to fall into four stages.

First, he argues on behalf of Christianity that since the world cannot

[21] Simplicius *in Phys.* 1363,4-12.

[22] Asclepius *in Metaph.* 186,2-3.

[23] M. Richard, '*Apo phônês*', *Byzantion* 20, 1950, 191-222.

[24] Simplicius *in Phys.* 1361,11-1363,12; *in Cael.* 143,9-144,4; 271,16-18; 301,4-7 (translated below); Olympiodorus *in Phaedonem* lecture 13, §2, lines 37-40 Westerink (=76N).

[25] Simplicius *in Cael.* 143,10-29; Philoponus *contra Proclum* 235,4-19.

[26] Compare Michael Wolff (as above), H.A. Wolfson, H.A. Davidson, Carlos Steel and Lindsay Judson (as below). Philoponus' earlier exposition is picked out by Koenraad Verrycken, who argues that it represents a stage of Philoponus' career in which he is merely editing the lectures of his pagan teacher Ammonius ('The development of Philoponus' thought and its chronology', in Richard Sorabji, ed., *The Transformation of Aristotle*, London 1989).

contain infinite power, it has a beginning and end.[27] But this is only an opening move, for he knows that Proclus and Ammonius would have a reply on behalf of the pagans, and this reply constitutes the second stage of his argument. The belief that the world is of endless duration remains perfectly possible, so long as the infinite power responsible for its eternity is something distinct from it. For that will avoid infinite power being housed within the finite world. Philoponus preserves fragments of Proclus which argue for this alternative conclusion.[28] He also ascribes it to Plato[29] and others, though not explicitly to Aristotle, while propounding it a number of times without ascription.[30]

In the third stage of his argument, Philoponus replies by re-importing into the discussion a subtlety earlier introduced by Alexander. The question is whether the world is thought by his pagan opponents to be imperishable 'in accordance with the *logos* of its own *nature*', or rather because of God's intervention. He asks his opponents to concede that on their view the world is imperishable only in the latter way. It is imperishable because God intervenes to *override* its nature. It is, however, *perishable* in accordance with the *logos* of its own nature, because it is a composite entity and composites can in principle be dismantled. That already enables him in the *contra Proclum* to refute Proclus' interpretation of Plato. Proclus had argued that, since Plato thinks the world imperishable, he must think it ungenerable.[31] But in fact Plato thought the world composite and perishable in accordance with the *logos* of its own nature.[32] Therefore Plato's world should also be *generable* in accordance with the *logos* of its own nature, contrary to Proclus' claim.[33]

But Philoponus has not yet reached his final conclusion. In the fourth stage of his argument, he ascribes to Plato, and himself upholds, the view that the world is generable not just by nature, but without qualification, and even that it was actually generated and had a

---

[27] Philoponus *contra Proclum* 1.18-2,11; *contra Aristotelem* ap. Simplicium *in Cael.* 142,22-5; in an Arabic summary, of a later lost work, translated into English by S. Pines, 'An Arabic summary of a lost work of John Philoponus', *Israel Oriental Studies* 2, 1972, 320-52, at 323-4 (reprinted in his *Collected Works*, vol. 2, Jerusalem and Leiden 1986), and into French by G. Troupeau, 'Un épitomé arabe du "contingentia mundi" de Jean Philopon', in *Memorial A.J. Festugière = Cahiers d'Orientalisme* 10, Geneva 1984, 77-88, at 84; and in a work recorded by Simplicius *in Phys.* 1327,14-16; 1329,17-19.

[28] Philoponus *contra Proclum* 238,3-240,9; 297,21-300,2; 626,1-627,20, quoting Proclus' *Examination of Aristotle's Objections to Plato's Timaeus*.

[29] Philoponus *contra Proclum* 235,4-19.

[30] Philoponus *contra Proclum* 2,11-14; Arabic summary translated by Pines, 324, Troupeau, 84.

[31] Philoponus *contra Proclum* 119,14-120,14; cf. 549,7-550,24.

[32] Plato *Tim.* 41A-B.

[33] Philoponus *contra Proclum* 225,14-226,19; 240,23-6; 241,26-7; 242,11-22; 304,4-9.

beginning in time.[34] Here, as Lindsay Judson has shown,[35] there appears to be a gap in the reasoning. Philoponus finds arguments in Plato to show that the world is generable,[36] but if that generability is merely natural, he will need to explain why God cannot *override* the natural generability and give the world a beginningless existence after all. Judson thinks that Philoponus would not consider natural generability on a par with natural perishability and overrideable in the same way. And this seems to be confirmed in a later work of Philoponus which is lost, but which is summarised in an Arabic version to be quoted below in Pines' translation. There Philoponus is said to hold that 'it is impossible for what is by nature created in time to be eternal *a parte ante*' and he concludes that the world had a beginning. But, as Judson says, Philoponus does not have any very convincing case for treating natural generability and natural perishability differently in respect of God's power of overriding.

The four stages stand out particularly clearly in the Arabic summary, and I will mark them (i) to (iv):

Then he said: (i) if the world is a finite body, as has been demonstrated by Aristotle in the first treatise of his book on the Heaven and (if) the forces of every finite body are finite, as has been likewise demonstrated by Aristotle at the end of the eighth treatise of the book of Physics, (then) because of what we have, and of what Aristotle has demonstrated, the world must have been created in time (and) have come into existence after not having existed.

Supposing, however, that someone says: (ii) As Aristotle has explained and demonstrated, the body of the world is finite and it is impossible that a finite body should have an infinite force; however the force which has ensured the preservation of the essence of the world in the eternity *a parte ante*, [and] which will ensure its permanent and perpetual preservation in the future, is the force of the Creator, may He be blessed, who causes the heaven to move in perpetual motion; we should answer:

(iii) The disagreement between us and him does not (concern) the (portion) of the Creator, may He be blessed and exalted, which is imparted to the world. The disagreement solely (concerns) your saying that the world is by nature eternal *a parte ante*. For if, as you say, it is eternal *a parte ante*, it does not need a force ensuring the preservation of its essence to be imparted to it. For it is a characteristic of a thing which is eternal *a parte ante* by nature that the force which ensures the preservation of its essence should be a force natural to it and not drawn by it from something else. [If however this force is drawn by it from something else] and received by it from a thing other than its essence (the thing in question) is not eternal *a parte ante* by nature, for it is not

[34] Philoponus *contra Proclum* 237,7-15.

[35] See the excellent analysis, which I am here following, of stages two to four in the *contra Proclum*, in Lindsay Judson, 'God or nature? Philoponus on generability and perishability', in Richard Sorabji, ed., *Philoponus and the Rejection of Aristotelian Science*, London and Ithaca N.Y. 1987.

[36] e.g. Philoponus *contra Proclum* 201,22-204,28.

then ... It is (106b) eternal *a parte ante* only because of a force (belonging) to some other (thing).

(iv) Its eternity *a parte ante* would have been abolished if there had not been this thing which gives it permanent (being) and ensures the preservation of its essence preventing the latter from perishing. Since the world is not by nature eternal *a parte ante* as is asserted by the Eternalists, it must be by nature created in time, as is asserted by us. And it is impossible for what is by nature created in time to be eternal *a parte ante*. In this there is a refutation of the doctrine concerning the eternity *a parte ante* of the world adopted by the Eternalists.

The main points of the first treatise of the discourse of John the Grammarian are finished. Glory be to God always, for ever, eternally.[37]

I suspect that Philoponus used the four stages of his argument again elsewhere, but that Simplicius has obscured the fact. The evidence concerns both Philoponus' *contra Aristotelem* and a later discussion of infinite power which Simplicius records, a discussion which may or may not be distinct from the first part of the work summarised in Arabic. Simplicius ascribes to Philoponus only the opening move, according to which the world, lacking infinite power, must be perishable.[38] And he counters with Proclus' alternative view that existence could be meted out to the world by a distinct infinite power in finite dollops.[39] But in fact that reply would have been perfectly familiar to Philoponus, and there are signs that here, as elsewhere, he replied to it, using the third stage of his argument, which protests that the world is at least by *nature* generable, and perhaps a further stage arguing that it is generable *tout court*. Thus Philoponus is said to aim at proving that the heavens are generable (*genêtos*) as well as perishable.[40] And further, he repeatedly refers to *nature*. For he says that the heavens are perishable so far as their own *nature* is concerned,[41] and he refers to God's power to override nature, when he concedes for the sake of argument that God will prevent this potentiality for perishing from being actualised.[42] Simplicius also feels

[37] 'An Arabic summary of a lost work of John Philoponus', translated by Pines, 323-4; French translation by Troupeau, 84-5.

[38] Philoponus *contra Aristotelem*, ap. Simplicium *in Cael.* 142,22-5, and in the later work ap. Simplicium *in Phys.* 1326,38-1336,34; 1358,26-1359,4, whose strategy is summarised at 1327,11-19 and 1329,17-19. The fragments of the *contra Aristotelem* are translated by Christian Wildberg in *Philoponus against Aristotle on the Eternity of the World*, London and Ithaca N.Y. 1987, vol. 1 of Richard Sorabji, ed., *The Ancient Commentators on Aristotle*.

[39] Simplicius *in Phys.* 1327,21-1318,35, partly repeated at 1358,26-1359,4.

[40] Simplicius *in Phys.* 1331,27.

[41] Philoponus' third argument in the later treatise in support of Aristotle's principle that a finite body lacks infinite power, ap. Simplicium *in Phys.* 1331,9; 14; 24; 28-30; 1333,19; 23-32. Similar wording in the second argument at 1329,36-7; 1330,12-17.

[42] Simplicius *in Phys.* 1333,28-30. H.A. Davidson gives a valuable summary of Philoponus' argument and notes Philoponus' appeals to nature in his 'John Philoponus as a source of medieval Islamic and Jewish proofs of creation', *Journal of the American Oriental Society* 89, 1969, 357-91 at 362-5. On the other hand, in this article and another,

forced to address himself to the argument about nature. Like Proclus before him,[43] he replies that, even if the imperishability of the heavens is derived from God, it can still be perfectly *natural*. And in giving his reason, he uses one of his favourite ideas, that the heavens are fitted (*epitêdeios*) by their *nature* to receive the imperishability which God gives them.[44] Simplicius' failure to articulate the third stage of Philoponus' argument, even when replying to it, leads me to take a critical view of his reliability as a reporter of Philoponus.[45]

Philoponus' introduction of the idea of *nature* into the argument is not an innovation. He is simply reversing what had been said much earlier by the Aristotelian Alexander of Aphrodisias. Alexander argued in two passages which have been translated by Robert Sharples,[46] that if the orderly cosmos had a beginning, as Plato claimed in the *Timaeus*,[46a] then it would be perishable by its own *nature*, and in that case not even God could override its perishing. It is a reversal of this when Philoponus expounds the pagan Neoplatonist view as being that, although the world is perishable in its own nature, it is *possible* for God to override its perishability.

Simplicius, writing after 529, attacks not only Philoponus, but also

---

he follows an interpretation of H.A. Wolfson, according to which Philoponus does not, as I suggest, concede for argument's sake that God might prevent the potentiality for perishing from being actualised. On the contrary, Philoponus, on this view, protests, like Averroes after him, that such a potentiality could not remain unactualised for ever: H.A. Wolfson, 'The Kalam arguments for creation in Saadia, Averroes, Maimonides and St. Thomas', *Saadia Anniversary Volume, American Academy for Jewish Research, Texts and Studies* 2, New York 1943, 197-245, at 202-3 and 240; *The Philosophy of Kalam*, Cambridge Mass. 1976, 373-82, at 377-8 and 381; H.A. Davidson, as above, at p. 362, and in 'The principle that a finite body can contain only finite power', in S. Stein and R. Loewe, eds, *Studies in Jewish Religious and Intellectual History Presented to Alexander Altmann*, Alabama 1979, 75-92, at 80 and n. 37. In the last note, however, Davidson recognises that the interpretation fits badly with other things that Philoponus says.

[43] Proclus ap. Philoponum *contra Proclum* 242,4-5.

[44] Simplicius' reply to Philoponus' *contra Aristotelem* is at Simplicius *in Cael.* 143,17-29, and his reply to Philoponus' later treatise at Simplicius *in Phys.* 1331,30-3; cf. 1358,29-30. Andrew Smith has drawn my attention to analogous rationales in Proclus *in Tim.* 1,267,10, translated above, and 3,212,21-2, where he adds that the heavens get their nature from God.

[45] Cf. the comparatively friendly remarks of Christian Wildberg, *John Philoponus' Criticism of Aristotle's Theory of Ether*, Ph.D. diss., Cambridge 1984, 126-7; Philippe Hoffmann, 'Simplicius' polemics', in Richard Sorabji, ed., *Philoponus and the Rejection of Aristotelian Science*, 63-4. Michael Wolff, ibid., p. 106, accepts the impression created by Simplicius that Philoponus deployed only his opening move. He rebukes Simplicius for failing to see that Philoponus would have thought the counter move unsound. But he does not consider the possibility canvassed here that Philoponus gave an answer to the countermove, and that Simplicius failed to bring this out except by inadvertent hints.

[46] Alexander *Quaest.* 1.18,30,25-32,19; and fragment of the lost *in Cael.* preserved at Simplicius *in Cael.* 358,27-360,3, both translated by R.W. Sharples, in 'Alexander of Aphrodisias: problems about possibility 2', *Bulletin of the Institute of Classical Studies* 30, 1983, at 99-103, and discussed by him there and in 'The unmoved mover and the motion of the heavens in Alexander of Aphrodisias', *Apeiron* 17, 1983, 62-6.

[46a] But most Platonists, to meet Aristotle's criticisms, took this beginning as only a metaphor.

Alexander, for both take the view that if God keeps the world in being, that must represent an *overriding* of the world's nature. Simplicius' response once again is that the world is not perishable by its own nature, for it is by *nature* fitted (*epitêdeios*) to receive eternity from God:

> Just as [the world] is created and finite in itself, so it is at the same time fitted (*epitêdeios*) for eternity and perpetual motion. So it is not true without qualification even to say that it is perishable by its own nature. For its nature in its entirety is conceived together with its fitness (*epitêdeiotês*) in relation to the creator (*dêmiourgos*), through whom it shares in eternal benefits.[47]

Simplicius objects, finally, to another aspect of Alexander's argument, according to which, if the world escapes destruction, it must do so 'from itself', and the cause cannot be something external like God, but must be 'in itself'. This is said by Alexander to hold whether the world is created or uncreated. In reply, Simplicius insists that Aristotle *himself* allows God to prevent the world's finite power running out, and makes God the cause of its eternity. Here is one of his answers:

> How does [Alexander] fail to realise that Aristotle himself says the heaven is uncreated, but says none the less that by its own nature it has only finite power, and that Aristotle, like Plato, referred the cause of its eternal motion, which is to say its eternity, to God.[48]

When Simplicius speaks here of the heaven's eternal motion, '*which is to say (t'auton de eipein) its eternity*', one interpretation has made him in effect a follower of Syrianus. His idea will then be that in eternally preserving the circular motion which defines the celestial body, God *thereby* keeps that body in eternal being.[49] But an alternative is that Simplicius is rather following Proclus and Ammonius. He will then be re-applying the infinite power argument, which he endorses elsewhere, and saying that it shows God is needed as much for the world's eternal existence as for its eternal motion.

Influence of Philoponus' infinite power arguments

What happened to Philoponus' arguments? The detective work of modern Arabists has established that his *contra Proclum*, his *contra Aristotelem* and the summary of the third treatise were all available in Arabic, and that various stages of Philoponus' argument recur. The Jewish philosopher Saadia (died 942) discusses the *contra Aristotelem*

---

[47] Simplicius *in Cael*. 361,10-15; cf. 360,23-4.
[48] Simplicius *in Cael*. 301,4-7, replying to 'if uncreated, the cause is in itself' (301,3). Cf. 360,4-19, replying to 'if created' (358,29), the escape is achieved 'from itself' (359,14).
[49] I believe that some such interpretation is suggested by Robert Sharples in *Apeiron* 17, 1983, at 63. For Syrianus, see *in Metaph*. 117,25-118,11.

version of the argument. He does not mention Philoponus, but he endorses what I have called Philoponus' opening move, the inference from finite power to the universe having a beginning and end.[50] Scholarship has further shown that this move is correctly ascribed to Philoponus by Farabi's pupil Yaḥya Ibn 'Adî (892-973),[51] a Christian, and in turn by Ibn 'Adî's Christian pupil Ibn Suwâr (born 942).[52] Ibn Suwâr further ascribes to Philoponus the more complex argument from the world's *nature* to its having a beginning. But what has not to my knowledge been discussed, nor translated, is a portion of a fragmentary commentary by Avicenna (c.980-1038) on Aristotle's version of the infinite power argument, as it appears in his *Metaphysics*. I am very grateful to Fritz Zimmermann for supplying me with an abstract. I believe that Avicenna here does four things. First, he endorses, without ascription, the view of Ammonius that Aristotle's God is the sustainer, as well as the mover, of the heavens (sec. 6). Secondly, he distinguishes the type of argument which we found in Syrianus which moves directly from God's preservation of the heavens' motion to his preservation of their *existence* (beginning of 2b). Thirdly, he re-applies the infinite power argument, to make Aristotle's God the sustainer of the heavens, very much in the way we saw Proclus doing and Philoponus recording in the second stage of his argument (end of 2b). Finally, he adopts a position like that which Philoponus forces on Proclus in the third stage of his argument: the existence of the heavens is necessary (they are imperishable) because of God, not because of their own nature (3,4). Averroes actually describes Avicenna as following Philoponus' point here, although he misidentifies it as Alexander's. Avicenna does not himself put his point in terms of God versus *nature*, perhaps because there was no mention of *nature* in his opponent Abu Bishr. But Averroes

---

[50] See (i) H.A. Wolfson, 'The Kalam arguments for creation in Saadia, Averroes, Maimonides and St. Thomas', 197-245, at 199-203; (ii) id., *The Philosophy of the Kalam*, 373-82; (iii) H.A. Davidson, 'John Philoponus as a source of medieval Islamic and Jewish proofs of creation', 357-91; (iv) Seymour Feldman, 'The end of the universe in medieval Jewish philosophy', *Association for Jewish Studies Review* 1986, 53-77, at 56-8. The fullest discussion, with translated excerpts from Saadia, is by Davidson in (iii). The relevant passage is translated into English by S. Rosenblatt, in Saadia Gaon, *Book of Beliefs and Opinions*, book 1, ch. 1, New Haven 1948, 41-2.

[51] S. Pines, 'A tenth century philosophical correspondence', *Proceedings of the American Academy for Jewish Research* 24, 1955, 114-15, and 'An Arabic summary of a lost work of John Philoponus', 320-52, at 345.

[52] Ibn Suwâr, *Treatise on the Fact that the Proof by John the Grammarian of the Contingency of the World is More Acceptable than That of the Theologians*, ed. Badawi, *Neoplatonici apud Arabes*, Cairo 1955, 244-8 at 246, lines 11-18, translated into French by B. Lewin, 'La notion de muḥdat dans le kalâm et dans la philosophie', *Orientalia Suevana (Donum Natalicum H.C. Nyberg Oblatum)* 3, Uppsala 1954, 84-93, at 91, §8, lines 25-33. For discussion of these two authors see, besides Pines above: Joel L. Kraemer, 'A lost passage from Philoponus' *contra Aristotelem* in Arabic translation', *Journal of the American Oriental Society* 85, 1965, 318-27, at 320-1; H.A. Davidson, 'John Philoponus as a source of medieval Islamic and Jewish proofs of creation', 360-1; H.A. Wolfson, *The Philosophy of the Kalam*, 374.

does put Avicenna's point that way, and so, later still, does Abravanel.[53]

Avicenna's argument is hard to follow, because the commentary is fragmentary and probably a reconstruction, whether by Avicenna or by someone else, of part of a larger production in twenty volumes, which he lost in the sack of Isfahan some three years before his death.[54] It may be significant that, in announcing his hope of reconstructing the work, Avicenna mentions that he is studying not only Alexander and Themistius, but also Philoponus. It will also become relevant that he cites Farabi, whom he puts in a class above all others.[55]

I need to show in more detail how the line of thought I have sketched emerges from the passage. Avicenna is attacking (sec. 6) various commentators on Aristotle, especially (sec. 4) the Nestorian Christian Abu Bishr Matta of Baghdad (died c.940), who had translated Aristotle, Alexander and Themistius into Arabic. Abu Bishr had said (2a) that the *existence* of the celestial sphere was necessary in itself, so that (4) God's role was confined to producing *motion* and to making motion perpetual. Avicenna objects (3,4) that nothing other than God is intrinsically necessary; all other necessity is owed to God, including that of the celestial sphere. This is the view that Averroes later associates with 'Alexander', that is, with the third stage of Philoponus' argument. Avicenna goes on (6) to accept the interpretation of Aristotle provided by Ammonius and his followers including Philoponus, according to which Aristotle's God is the *sustainer* of the universe. On the other hand he thinks (1) that neither the commentators he is attacking nor Aristotle himself supply an actual *argument* for this conclusion. In particular they do not use the argument which we found in Syrianus (beginning of 2b). The only argument they supply for God concerns the *motion*, not the *existence*, of the universe. Avicenna then refers to an argument which would do the trick (end of 2b), and would show God to be the sustainer of the universe. It is the infinite power argument used by Proclus and recorded *inter alios* by Philoponus. According to the misguided interpretation of certain commentators, the celestial sphere will be receptive (a concept found also in the Greek

[53] Averroes *Middle in Cael.* 293v, a, G-H. Avicenna is here described as the *companion* (*comes* in the Latin) of 'Alexander', and the terminology of *nature* is associated with Avicenna here and later by Abravanel in *Mif'alot Elohim (The Deeds of God)* 2.3, p. 126, translated from the Hebrew by H.A. Wolfson, *Crescas' Critique of Aristotle*, Cambridge Mass. 1929, 597 and 682 (italics added): 'For Plato says that the heavens were generated from that eternal matter which had been in a state of disorderly motion for an infinite time, but at the time of creation was invested with order. Consequently by their own *nature* the heavens are corruptible just as they were generated, and it is God who implants in them eternity, as it is written in the *Timaeus*. It is from this view that Avicenna has inferred that the celestial sphere is composed of matter and form and is corruptible and possible by its own *nature*, but necessary and eternal by virtue of its cause.'

[54] S. Pines, 'La "philosophie orientale" d'Avicenne et sa polémique contre les Bagdadiens', *Archives d'histoire doctrinale et littéraire du moyen âge* 19, 1952, 5-37.

[55] Avicenna, *Letter to al-Kiyâ* ed. M. Badawi, *Aristû 'ind al-'Arab*, Cairo 1947 (repr. Kuwait 1978), 120-2, second part translated into French by S. Pines, op. cit., 6-9, at 9.

texts) of the motion imposed by God, and even *perpetually* receptive of it, despite the absurdity that no provision has been made for its *existence* to be perpetual. For its own power is finite, so that it has no *internal* source of perpetual existence. Its perpetual existence cannot then be assured, unless these commentators are wrong, and the divine source of its motion, which acts as a *final* cause filling it with desire, also acts as an *efficient* cause, that is, I take it, as an efficient cause of its *existence*. ('It will have to be receptive and perpetually in motion despite the fact that its power is finite – unless there is another cause and emanation on account of which the desired end becomes an *effective* principle'.) The following are the relevant sections of Avicenna (key parts in italics):

Comments on *Metaph*. 12.6, 1071b5-31:

(1) *He [Avicenna] criticises Aristotle and his commentators as follows. It is improper to argue to God from the fact of motion, by saying that God is the principle of motion. For in that way they cannot show that the true One is the principle of all existence.* If the first principle is a principle of the *motion* of the sphere, it need not therefore be a principle of the *substance* of the sphere.

(2) He [Avicenna] says: Let us examine their belief that the motion of the sphere is necessary and without beginning or end. They do not show: (a) that the existence of the sphere is necessary in itself; (b) *that for it to exist is to be in motion, so that if it ceased to move it would cease to exist.* Rather, they argue as follows: the sphere exists; it is in motion; therefore its motion must be beginningless. But that is a *non sequitur*. For suppose we knew it existed but not whether it had ever moved. From the fact of its existence we could not tell whether it moved, perpetually or otherwise. Neither could we proceed to tell that or how it, its matter and form, had been brought into being [by a creator] *In a word, how is it supposed to be desirous, receptive, eternal? It will have to be receptive and perpetually in motion despite the fact that its power is finite – unless there is another cause and emanation on account of which the desired end becomes an effective principle.* ...

Comments on *Metaph*. 12.7, 1072b3:

(3) If it sets things in motion by dint of being intrinsically desirable, and because the thing moved is fit so to be affected as to get moving, that motion is brought about jointly by conditions in both mover and moved. If so, act and effect follow necessarily, if only in faculties close to reason (Greek: *logos*). Act and effect will thus be necessary, and will constitute a sublime necessity which has a noble existence, since the order of the whole universe flows from it. We do not mean to say that that necessity is one of compulsion or ineluctability. It is one where things could not conceivably be different. This passage (1072b3) does not mean that the heavenly motion is intrinsically necessary, in the sense that it could not be otherwise. It is necessary in the sense explained earlier. Nothing exists of necessity when considered in isolation from the First. Indeed, if

the tie to the First could be broken, everything would reduce to nothing. In respect of itself, everything is vain and perishes, except the face of the First Truth.[56]

(4) Some, failing to distinguish between 'conditionally necessary' and 'strictly necessary', have mistaken the necessity here at issue for an intrinsic necessity, arguing as follows. [lacuna].[57] *Now I ask Abu Bishr: on the assumption that it is of necessity that things are as they are, what part is there for the first cause to play in relation to them? He answers: [to maintain them in] perpetual motion. But that is absurd. The part played by the First is that it is on its account that there is necessity, with no necessity accruing to anything else on account of itself.* A sign of the folly of that [Baghdadian] position is that it makes necessity *intrinsic* to things, perpetuity *extrinsic*: that intrinsic necessity of theirs is not supposed to entail perpetuity; perpetuity is to be furnished from outside. But if perpetuity is bestowed on motion, does that not also make it necessary? If it were necessary by itself, would not that make it independent of its mover?

(5) The truth is that motion, its existence, the necessity of its existence while it exists, and the perpetuity of its existence, all depend on the causes of motion. God is too lofty to be made a cause only of motion. It is He that furnishes the existence of every substance. It is through Him that everything else must be and acquires the necessity consequent upon the relation there must be between Him and it.

Comments on *Metaph.* 12.7, 1072b13-14:

(6) *'Thus upon such a principle does the heaven depend'.* The principle in question is primary, single, simple, intrinsically intelligible, absolutely good, the object of desire for the whole universe. It is from it that necessity emanates to things. Besides it, no importance can be attached to non-existence [as a possible source of existence]. It either excludes non-existence altogether or extends a power to exist to things in the measure in which they are fit to be affected.

'Depends': Aristotle means to say that the universe itself is sustained by that principle. It is wrong to confine its blessings to the lowly gift of motion, as do those commentators.[58]

Avicenna does not at the beginning of (2b) endorse the argument which we found in Syrianus and possibly in Simplicius which moves directly from the heavens' motion to their existence. He merely says his opponents did not use it. It may be wondered whether he could have learnt of the argument from Simplicius. There is a further hint that Avicenna might be aware of Simplicius, and guarding against his claim[59] that the heavens are fitted (*epitêdeios*) by their nature to

[56] cf. Koran 28:88 'All things perish, except his Face' (transl. Arberry).

[57] The missing passage will have contained a statement in the name of Abu Bishr Matta.

[58] Avicenna, Notes on *Metaphysics* 12, from his *Kitâb al-Inṣâf* ed. Badawi, *Arisṭû 'ind al-'Arab*, pp. 23-6 (whole text: 22-33).

[59] Simplicius *in Cael.* 361,10-15, translated above.

receive the eternal rotation and existence which God gives them. At any rate, Avicenna says (3) that the *fitness* of the heavens, in this case their fitness to be moved, does not prove their motion to be intrinsically necessary. On the other hand, this second argument is also sketched by Proclus, and in any case there is no testimony that Simplicius' *de Caelo* commentary (or for that matter his *Physics* commentary) was translated into Arabic.[60]

That the infinite power arguments recur in Averroes (c.1126-c.1198), has been well brought out in some very interesting work by modern Arabists. What I think can now be clarified is the relationship of Averroes' texts to the preceding ones of Avicenna, Philoponus and others. Perhaps the most important point is that the first three stages of Philoponus' argument recur in Averroes, but modern scholarship has not been aware of this, because Averroes was not aware himself. He correctly identifies the first stage as coming from Philoponus, but by a signal error ascribes the second and third to *Alexander*. His ascription is taken at face value by Thomas Aquinas and more recent commentators.[61]

It is clear that the second-stage argument for God as sustainer, which is taken from Proclus, should not be ascribed to Alexander. For we have already seen Alexander arguing that, if the world escapes destruction, this cannot be due to an external cause, such as God.[62] And a further passage of Simplicius to be translated below informs us that Alexander went only as far as making his Aristotelian God an efficient, as well as a final, cause of the world's *motion*, not a cause in any way of its existence.[63] It is equally wrong to ascribe to Alexander the idea from the third stage of Philoponus' argument, that the world is destructible by nature, but preserved by God. That, as we have seen, is another thing that Alexander argued to be impossible.[64]

Averroes addresses the subject in five of his commentaries on Aristotle and in his *de substantia orbis*.[65] The three most informative

[60] H. Gätje, 'Simplikios in der arabischen Überlieferung', *Der Islam* 59, 1982, 6-31. So also H.A. Davidson, 'John Philoponus as a source of medieval Islamic and Jewish proofs of creation', 357-91, at 359, drawing on the earlier work of M. Steinschneider, *Die arabischen Übersetzungen aus dem Griechischen*, repr. Graz 1960.

[61] For Thomas Aquinas see below, and see H.A. Wolfson, *The Philosophy of the Kalam*, 381; H.A. Davidson, 'John Philoponus as a source of medieval Islamic and Jewish proofs of creation', 361, n. 41; S. Feldman, 'The end of the universe in medieval Jewish philosophy', 76.

[62] Simplicius *in Cael.* 301,3; 359,14.

[63] Simplicius *in Phys.* 1361,31-3; 1362,11-15. On this particular point I diverge from C. Genequand's valuable discussion, *Ibn Rushd's Metaphysics*, Leiden 1984, 36.

[64] Alexander *Quaest.* 1.18, 30,25-32,19; fragment of lost *in Cael.* ap. Simplicium *in Cael.* 358,27-360,3.

[65] The three commentaries I shall use here are: (i) Averroes *Long Commentary on Physics* 8, comm. 79 (available in the thirteenth-century Latin translation by Michael Scot, in the Juntine edition of 1562-1574, vol. 4, reprinted Frankfurt 1962, p. 425v, col. a,

commentaries are the *Long Commentary on Physics* 8 (text (1)), the *Middle Commentary on de Caelo* (texts (2A) and (2B)) and the *Long Commentary on Metaphysics* 12 (text (3)). Averroes reports, and correctly ascribes to Philoponus, the opening move according to which, since the heaven has finite power, it is subject to destruction.[66] He further records the stage-two reply, that the heaven, despite its finite power, could receive eternity from its divine mover. We can recognise this as the view which Philoponus quotes against himself from *Proclus*, but this is one of the views which Averroes ascribes wrongly to Alexander. He does so in his *Long Physics Commentary*, and he mentions the view again in his *Long Commentary on Metaphysics* 12, where Genequand translates it as if it were Philoponus' very own view. But again it will make better historical sense if the quotation from Philoponus is taken as ending earlier, leaving the stage-two view recorded, but not ascribed.[67]

Finally, Averroes reports the elaboration which we know indeed to be that of Philoponus: by its own *nature* the celestial sphere has the possibility of being destroyed; it is free of destruction because of something *else*. Once again, however, Averroes ascribes the view to Alexander,[68] adding, as we have seen, that Avicenna was his follower (*comes*, in the Latin of the *Middle Commentary on de Caelo*).[69] It is plausible enough to present Avicenna as endorsing this view; the only mistake is to associate Avicenna on that account with Alexander instead of with Philoponus.

Averroes' reply reveals his relation to a number of further thinkers. It includes a point which modern scholars have understood as coming from Philoponus, and indeed as forming the nub of Philoponus' argument, but in fact, I believe that Philoponus does not use it.[70] The

---

sec. H – 427r, a, B); (ii) *Middle Commentary on de Caelo* translated from Hebrew version into Latin by Paul Israelita (available as 'paraphrasis', Juntine edition, vol. 5, p. 293v, a, G – 293v, b, K; 294r, b, D – 295r, a, B), also translated in part, directly from the Hebrew version by H.A. Wolfson, *Crescas' Critique of Aristotle*, 596-7; 681-2; (iii) *Long Commentary on Metaphysics* 12, comm. to text 41, 1626-38, translated from Arabic by C. Genequand, op. cit., pp. 162-8.

Other works are: (iv) *Long Commentary on de Caelo* 2, comm. 71, Latin translation by Michael Scot, in Juntine edition, vol. 5; (v) *de substantia orbis* 3; and (vi) a work inaccessible to me, the Hebrew ms. version of the *Middle Commentary on Metaphysics* 12 (Casanatense Heb. ms. no. 3083, 140(141)b-141(142)a). This reference is given, along with the others, by H.A. Davidson 'The principle that a finite body can contain only finite power', in Stein and Loewe, eds, *Studies in Jewish Religious and Intellectual History*, 75-92, at nn. 53, 58, 61, 64.

[66] Averroes *Long in Metaph.* 12, 1628; more briefly *Long in Phys.* 8, p. 426v, b, K-L; *Middle in Cael.* p. 293v, a, I.

[67] Averroes *Long in Phys.* 8, 426v, b, K.

[68] Averroes *Middle in Cael.* 293v, a, G-H.

[69] ibid.

[70] I am persuaded of this by Lindsay Judson, op. cit. I do not think that Philoponus is represented as using the point at Simplicius *in Phys.* 1333,28-30; 1334,37-9. This passage rather concedes for the sake of argument the position of Plato and Proclus that the potentiality for destruction will be eternally unactualised. The contrary

point is that the pseudo-Alexandrian view violates Aristotle's conclusion in *Cael.* 1.12, that an eternal thing cannot have a potentially for destruction.[71] Averroes concludes that the view must be rejected. He prefers to return to the idea which we found in Abu Bishr, that the heavens' existence is necessary *of itself*. One reason why the celestial body endures by its own essence is that, as we saw in Chapter 3, Averroes thinks that it is not a composite of matter and form.[72] Another reason is that it lacks a contrary and so has no potentiality for destruction. Motion differs, according to Averroes, because it does have a contrary, rest, and so it gets its perpetual motion, unlike its necessary existence, from elsewhere.[73] In thus separating motion from existence, Averroes distances himself from the type of argument initiated by Syrianus, according to which preserving the celestial body's motion is the *same thing* as preserving its existence. For this departure he was later rebuked by Abravanel.[74]

Had he but known it, Averroes should have claimed Alexander as an ally, not an opponent. For the true Alexander would have agreed that the heavens' existence is necessary of itself. I shall quote from three of Averroes' commentaries, marking the stages of Philoponus' argument from (i) to (iii) and the relevant part of Averroes' reply with (r):

Text (1)

H   Now in the proposition here assumed, according to which the power of
      any body is finite, it can be doubted whether it applies to the celestial
I   body or not. If it does, then the power of the celestial body will be finite,
      but what has finite power is destructible ...
K   This first question at least is very difficult and full of snags. (ii) *And
      Alexander answers in some of his treatises, and says that the celestial
      body receives (adeptus fuisse) eternity from its mover, who is not
      enmattered.*

---

interpretation is the one cited above as initiated by H.A. Wolfson and followed with some misgivings by H.A. Davidson: H.A. Wolfson, 'The Kalam arguments for creation in Saadia, Averroes, Maimonides and St. Thomas', 197-245, at 202 and 240; id., *The Philosophy of the Kalam*, 373-82, at 377-8 and 381; H.A. Davidson, 'John Philoponus as a source of medieval Islamic and Jewish proofs of creation', 357-91, at 362; id., 'The principle that a finite body can contain only finite power', in Stein and Loewe, eds, *Studies in Jewish Religious and Intellectual History*, 75-92, at 80 and n. 37. Something closer to what Wolfson seeks is found in a *different* context at Philoponus *contra Proclum* 131,26-132,28: God's eternal power of dissolving the cosmos cannot remain for ever unactualised.

[71] Averroes *Long in Phys.* 8, 426v, b, K; *Middle in Cael.* 293v, a, I; 294r, a, F; *Long in Metaph.* 12, 1628, but consistency with other texts suggests that again we should move Genequand's inverted commas, so as not to present this as part of Philoponus' views: it represents Averroes' reply.

[72] Averroes *Middle in Cael.* 294r, b, D – 295r, a, B. Its non-compositeness does not seem to be used for the same purpose at *Long in Phys.* 8, 426v, b, M – 427r, a, B, nor (*pace* Genequand, p. 46) at *Long in Metaph.* 12, 1634.

[73] Averroes *Long in Metaph.* 12, 1631; *Middle in Cael.* 294v, b, M.

[74] Isaac Abravanel, *The Deeds of God* 9.9, cited by S. Feldman, 'The end of the universe in medieval Jewish philosophy', 53-77, at 76.

(r) *But this would make it something that can be destroyed, yet never will be destroyed. This is Plato's opinion too, that is, that there is* something eternal which can be destroyed. But Aristotle proved at the end *of the first book of the de Caelo et Mundo [1.12] that there could not be anything eternal that contained a potentiality for destruction.*

(i) *Now John the Grammarian maintained his own opinion* (reading: *opinionem,* with the 1489 edition) *against the Peripatetics, in that he thinks that the world is destructible and generable. And this is the* L *strongest of all the doubts that can befall us on this, especially because Aristotle says expressly in the second book of the de Caelo et Mundo [2.12, 293a10-11] that the power of the heaven is finite,* when he gives the reason why it does not contain more stars than it does. If it did, he says, it would be exhausted.

Now Avicenna heard these words of Aristotle after already hearing the words of Alexander, and he thought there were two ways of being necessary, that is, necessary in dependence on another while merely possible of itself (*necessarium ex alio contingens et possibile ex seipso*), M and necessary of itself (*necessarium ex se*) – necessary through another like the heaven, necessary of itself like the movers of the heaven (*necessarium ex alio ut caelum, necessarium ex se ut motores caeli*). Now we say [75]

Text (2A)

There is room here for the following great doubt. It has been shown that G nothing eternal has the possibility of being corrupted nor can there be in it a potentiality for corruption. But it has also been shown in this treatise that a body which is finite in magnitude cannot but have a finite force. Now, since the celestial sphere is finite in magnitude, the force within it must necessarily be finite. (iii) *The inference must therefore be that while the sphere by its own nature has the possibility of being corrupted, it must* H *be free of corruption on account of the infinite immaterial force, outside the sphere, which causes its motion.* That this is so is maintained by Alexander in a treatise of his, and he is followed by Avicenna, who says that to have necessary existence may mean either of two things. First, to have necessary existence by one's own nature. Second, to have only possible existence by one's own nature, but necessary existence by reason of something else. (And he thought that this latter was the opinion of Aristotle when he said in the second book of the *de Caelo et Mundo,* 'if any of the planetary spheres contained extra stars from the higher sphere which imparts diverse motions to the lower ones, its power would be exhausted, since we have often taken it that every finite body has I finite power. And it was for this very reason that each of the planetary spheres received only one star'. This text from his discourse, then, shows that the power residing in the spheres is finite.

If his view were correct, there would be a potentiality for corruption in an eternal thing. But the subsequent discussion, as you know, denies that it has a potentiality for corruption, whereas he affirms that this can

---

[75] Averroes *Long in Phys.* 8, comm. 79 translated from Latin of the Juntine ed., 1562-1574, vol. 4, p. 426v, a, H – 427r, a, B.

happen. John the Grammarian did not neglect this doubt either, and at
its dictate concluded that the world was generated).

K   Our own answer to this difficulty, however, is that a body may be said
to have a finite force in two senses. First, in the sense that its motion is
finite in intensity and speed. Second, in the sense that its motion is finite
in time.[76]

Text (2B)

(r) On this account, i.e. by virtue of its being simple, the celestial body
has no substratum and no contrary. Hence Aristotle maintains that it is
ungenerated and incorruptible, seeing that it has no subject and no
contrary. It is thus stated by him at the end of the first book of *de Caelo*.
It is no surprise that this was overlooked by Avicenna, but what
surprises us is that it should have been overlooked by Alexander, despite
his admission that the celestial body is simple and not composed of
matter and form, as is evident from a passage in his commentary on Book
Lambda. I believe that there is no difference of opinion among the
commentators on this point, for it is very clear from Themistius'
commentary on *de Caelo et Mundo* that the celestial body has no
substratum. A similar view was expressed by Alfarabi in the name of
Aristotle, i.e. that such was his own view.[77]

Text (3)

(i) *John the Grammarian [Philoponus] raised strong objections against
the Peripatetics concerning this problem. He says: 'if every body has a
finite power and the heaven is a body, then it will have a finite power; but
everything finite is necessarily corruptible, so that the heaven is
corruptible.'*[78] (ii) *If it is said that it acquires incorruptibility from the
eternal separate power, there will be something destructible but eternal.*
(r) *But this has been shown to be impossible at the end of the first book of
the de Caelo et Mundo.*

*There must not be in the celestial body the power to corrupt because it
has no contrary.* It endures by its essence and its substance, not by a
quality (*ma'nâ*) inherent in it. Motion cannot endure by its essence
because it has a contrary which is rest. To account for the permanence of
motion we have to postulate a concept permanent in itself, unlike
permanence in a substance ...

Therefore it is not correct to say that there is something contingent by
itself and eternal and necessary by something else as Ibn Sinâ
[Avicenna] says that the necessary is partly necessary by itself and

[76] Averroes *Middle in Cael.* 1, Juntine ed. 1562-1574, vol. 5, p. 293v, a, G – 293r, b, K,
sections in rounded brackets translated from Latin version of the Hebrew, remainder
translated by H.A. Wolfson directly from Hebrew version, in his *Crescas' Critique of
Aristotle*, 681-2.

[77] Averroes, *Middle in Cael.* 1, translated from Hebrew version by H.A. Wolfson, op.
cit., 596-7, corresponding to the Latin of the Juntine edition, vol. 5, 294v, b, M – 295r, a,
A.

[78] I propose to terminate the quotation from Philoponus here, instead of below after
the reference to the *de Caelo*.

partly necessary by something else, except for the motion of the heaven only. It is not possible that there should be something contingent by its essence but necessary on account of something else, because the same thing cannot have a contingent existence on account of its essence and receive a necessary existence from something else, unless it were possible for its nature to be completely reversed. But motion can be necessary by something else and contingent by itself, the reason being that its existence comes from something else, namely the mover; if motion is eternal, it must be so on account of an immovable mover, either by essence or by accident, so that motion possesses permanence on account of something else, but substance on account of itself. Therefore there cannot be a substance contingent by itself but necessary by something else, but this is possible in the case of motion.[79]

Averroes' treatment of the subject is important because of its extensive influence on subsequent Christian and Jewish discussions. Of the three commentaries just quoted, two were translated into Latin from Arabic in the thirteenth century by Michael Scot, namely the long commentaries on the *Physics* and *Metaphysics*. Consequently, Averroes' discussion was known to such thirteenth-century scholars as Robert Grosseteste and Roger Bacon in Oxford and Bonaventure and Thomas Aquinas in Paris, to name them in rough chronological order.[79a] The most thoughtful discussion was the last, that of Thomas Aquinas, to be found for example in his *Commentary on Aristotle's Physics*.[80] Thomas refers to Averroes' long commentaries on *Physics* and *Metaphysics* 12, and to his *de substantia orbis*, and he reports the controversy in Averroes' own terms as being between Averroes and *Alexander*. In giving his own opinion, Thomas does not support Philoponus' opening move, because unlike Philoponus, he accepts that although the world had a beginning, it will have no end.[81] Instead, therefore, of arguing from finite power to destructibility, he adopts the Proclan position outlined in stage two of Philoponus' discussion, but here wrongly ascribed to Alexander. He does so, even though 'Alexander' goes beyond him by treating the celestial body as not only endless, but also beginningless in time. The celestial body receives its eternity, Thomas agrees, not from any infinite power of its own, but from the infinite power of the divine mover. Thomas defends this familiar position from Averroes' attack by making an addition, which he presents as conforming to 'Alexander's' intention. The celestial body

---

[79] Averroes, *Long in Metaph.* 12, comm. to text 41, 1628; 1631; 1632, translated from Arabic by C. Genequand, *Ibn Rushd's Metaphysics*, 163-5.

[79a] Richard Dales tells me he finds arguments against infinite power also in the relatively early Vatican ms Latin 185.

[80] Thomas Aquinas, *Commentary on Aristotle's Physics* 8, lectio 21, 1147, 1152, 1154. There is an English translation of the *Commentary* by Richard J. Blackwell, Richard J. Spath, W. Edmund Thirlkel, London 1963. See also the *Commentary on Lombard's Sentences* 2, d. 1, a. 1, q. 5, contra 8 and *Summa contra Gentiles* 1.20, obj. 3.

[81] Thomas Aquinas, *Commentary on Aristotle's Physics* 8, lectio 21, 1147 and *Summa Theologiae* 1, 104, a. 4, ad 1 and 2.

does not have a potentiality for non-being, but is indestructible. So 'Alexander' is immune to the objection that he violates the ban in Aristotle *Cael.* 1.12 on an eternal thing being destructible. This is not to deny that the celestial body receives its perpetual existence from God, since necessary things can perfectly well have a cause of their necessity. It has been found puzzling that the celestial body can be necessary both by its own nature and because of God. But we may compare the argument of Proclus that it gets its nature from God, or that of Simplicius that it is by nature fitted (*epitêdeios*) to receive the indestructibility God gives it.[82]

The introduction of Alexander into the discussion takes on added significance in Thomas, for his belief that this was Alexander's view will have confirmed him in the further belief, which he expresses both in the commentary on *Physics* 8 and in many other places, that Aristotle too made God responsible for the beginningless and endless existence of the world.[83] The following is Thomas' main discussion, with his answer to the relevant part of Averroes italicised:

(1147) [Problem 4] Again, the conclusion seems to be false. For the greater the power of a body, the longer it can remain in being. So if no body had infinite power, none could last infinitely. And that is evidently false, both in Aristotle's opinion and in the view of the Christian faith, which posits that the substance of the world will last infinitely long ...

(1152) But Alexander solves the fourth problem, according to Averroes here in his commentary, by saying that the celestial body acquires both eternity and perpetual motion from a separate mover of infinite power. So just as it is not by any infinity of the celestial body that it has perpetual motion, so it is not by any such infinity that it lasts perpetually. Rather each is due to the infinity of a separate mover. Averroes tries to disprove this answer in this commentary, and in that on *Metaphysics* 12, by saying that it is impossible for anything to acquire perpetual being from another. Otherwise it would follow that something destructible in itself (*in se*) became eternal. Perpetual *motion*, however, can be acquired from another, because motion is the actualisation by the mover of the thing subject to motion. So he says that insofar as the

---

[82] References in n. 44. I am grateful to Andrew Smith for linking up the Proclus–Simplicius position. The puzzle re Aquinas is expressed by Carlos Steel on p. 223 of ' "Omnis corporis potentia est finita" ', in Jan P. Beckmann, Ludger Honnefelder, Gangolf Schrimpf, Georg Wieland, eds, *Philosophie im Mittelalter*, Hamburg 1987, 213-24).

[83] Thomas Aquinas, *Commentary on Aristotle Physics* 8, lectio 2, 974-5; lectio 3, 996; *Commentary on Aristotle Cael.* in 1, lectio 8, 91; lectio 21, 216; *Commentary on Aristotle Metaph.* in 2, lectio 2, 295; in 6, lectio 1, 1164; in 12, lectio 12, 2614; *de potentia Dei* q. 3, a. 5, in c; *Commentary on Lombard's Sentences* 2, d. 1, a. q. 5, ad 1 in contr.; *Expositio primae decretalis* n. 1163; *de substantiis separatis* n. 100; *Summa Theologiae* 1, q. 2, a. 3, Respondeo; *contra Gentiles* 1, ch. 13. A contrary interpretation of Thomas given by É. Gilson is discussed by Luca Bianchi, *L'errore di Aristotele, La polemica contro l'eternità del mondo nel XIII secolo*, Florence 1984, 36; the 'was not made' of *Aquinas de articulis fidei* n. 601 should be understood to mean that for Aristotle the world was not made *in time*, but beginninglessly. On Thomas' position, see further R. Jolivet, 'Aristote et Saint Thomas ou la notion de création', as above.

celestial body has being of itself (*de se*), it does not contain any potentiality for non-being, because its substance has no contrary. But it does contain a potentiality for being stationary, since stationariness is the contrary of its motion. That is why it does not need to acquire perpetual being from another, but does need to acquire from another perpetual motion.

(1154) So let us see whether Averroes successfully impugns Alexander's solution, which is that the celestial body acquires eternity from another. His disproof would indeed be successful, if Alexander had posited that the celestial body had of itself (*de se*) potentiality for being and for non-being and that it acquired being for all time from another. (And in saying this, I grant his intention that we should not exclude God's omnipotence by which 'this corruptible can put on incorruption' [cf. 1 Corinthians 15:53]. It is not pertinent to the present point to discuss this now).

Even with his intention granted, however, *Averroes cannot draw a conclusion against Alexander. For the latter did not posit that the celestial body acquires eternity from another on the understanding that of itself (de se) it has potentiality for being and non-being, but on the understanding that its being does not come from itself (ex se).* For everything which is not identical with its own being partakes of being from the first cause, which is its own being. Hence Averroes himself acknowledges in his book *de substantia orbis* that for the heaven God is a cause not only as regards its motion, but also as regards its substance. And that is only because it has its being from him. *Now the only being it has from him is perpetual being. So it has its perpetuity from another. And Aristotle's words are also consonant with this, when he says in Metaphysics 5 and above at the beginning of this eighth book*[84] *that some necessary things have a cause of their necessity.* With this granted, then, the solution intended by Alexander is clear: just as the celestial body has its motion, so it has its being, from another. Hence just as perpetual motion demonstrates the infinite power of the mover, but not of the thing subject to motion, so its lasting perpetually demonstrates the infinite power of the cause from which it has its being.[85]

Thomas tries out yet other approaches to the issue elsewhere.[86] He also rules out the type of move found in Syrianus, according to which the heavens would have a potentiality for acquiring existence from their motion.[87]

Of the other thirteenth-century scholars, Grosseteste has a

[84] Aristotle *Metaph*. 5.5, 1015b9-10; *Phys*. 8.1, 252a32.

[85] Thomas Aquinas *Commentary on Aristotle's Physics* 8, 1147, 1152, 1154.

[86] He denies that the heavens have a finite potentiality for being, on the grounds that their only potentiality is for motion, not for being (*Commentary on Lombard's Sentences* 2, d. 1, a. 1, q. 5, ad contra 8). But that solution clashes with *Physics Commentary* 1153 and *Summa contra Gentiles* 1.20, obj. 3, which allow a potentiality for being. Consequently the latter passage offers a different solution. The heaven has the power of changeless being, but that need not be an *infinite* power, because changeless being has no quantitative extension and is not touched by time, whether the heaven endures for an instant, or for infinite time.

[87] Thomas Aquinas *Commentary on Lombard's Sentences* 2, d. 1, a. 1, q. 5, ad contra 8.

discussion of the infinite power argument which is very close in wording to Averroes' commentary on *Metaphysics* 12. Although he does not mention Averroes by name, he agrees with him that there is no possibility of destruction in the celestial body, because it has no contrary, and is therefore permanent in itself and through its essence.[88] Bacon contents himself with the argument that constitutes Philoponus' opening move: the finiteness of the world's power proves that it must have begun.[89] Bonaventure refers simultaneously to Averroes' commentaries on Aristotle and to the common idea that Aristotle said nothing against the Christian faith. On this view, Aristotle denied not a beginning of the world, but only a beginning in accordance with *nature*.[90] Bonaventure does not know if this is correct as an interpretation of Aristotle (later he was to deny it), but he accepts it, like Thomas after him, as an expression of the actual facts. He thus aligns himself with Averroes and with the genuine Alexander, in agreeing that a beginning of the world is *unnatural*. But his Christian faith makes him diverge from them and insist that the world none the less did have a beginning, through God's *overriding* its nature. He thus completely reverses the view of Philoponus, who urged on Proclus that God must override the world's nature if he is to *prevent* it ending.

Averroes' discussion influenced not only thirteenth-century Christian writers but also such later Jewish thinkers as Gersonides (1288-1344), Crescas (1340-1410) and Abravanel (1437-1509). Abravanel agrees with Philoponus' use of the infinite power argument in his opening move. The other two attack the foundations of the infinite power arguments, including the Aristotelian physics that lies behind them. Gersonides allowed a *finite* force to produce unending motion in certain conditions, and we have already encountered in Chapter 9 Crescas' denial that infinite force would (absurdly) produce motion

[88] Robert Grosseteste *de motu supercaelestium*, in L. Baur, ed., *Die philosophischen Werke Grossetestes*, p. 96, line 29-99.4.

[89] Roger Bacon *Opus Maius*, part 7, Moral Philosophy part 4, ch.1, ed. J.H. Bridges, vol. 2, Oxford 1897.

[90] Bonaventure *Commentary on Lombard's Sentences* 2, d. 1, par. 1, a. 1, q. 2, conclusion (translated by Paul Byrne, in Cyril Vollert, Lottie Kendzierski, Paul Byrne, *St Thomas Aquinas, Siger of Brabant, St Bonaventure, On the Eternity of the World*, Milwaukee Wisconsin 1964, 105-13, at 109-10; and *in Hexaemeron* Collatio 6, n. 5 and Collatio 7, n. 2. He decided that Aristotle denied a beginning by the time of his *de decem praeceptis* of 1267, *Opera Omnia*, ed. Quarrachi, vol. 5, 515.

For Albert's version of the idea see *in Sent.* 2, d. 1, a. 1 (Borgnet vol. 27, p. 29); and *Summa Theologiae* 2, tract. 1, q. 4, a. 5 (Borgnet vol. 32, p. 99a).

Thomas Aquinas *in Phys*. 8, lectio 2, 987 repeats that Aristotle's arguments prove only that a beginning cannot have been in accordance with *nature*, but he thinks it wrong (986) to conclude that that was all Aristotle intended to prove. Cf. *Summa Theologiae* 1, q. 46, a. 1, Reply to Objection 3.

that took no time: the motion would merely need no *extra* time above the minimum.[91]

Ammonius and Aristotle's God as Creator

I have tried to trace how the first three stages of Philoponus' argument were transmitted to the Latin West, and how the second and third stage, which make God a sustainer overriding the world's tendency to disintegrate, came to be associated with the greatest exponent of Aristotelianism, Alexander of Aphrodisias. But how did the idea of a Sustainer or Creator come to be associated with Aristotle *himself*?

Part of the story is well known. A number of texts which do indeed postulate a Sustainer or Creator came to be ascribed to Aristotle in error. The earliest is the *de Mundo* (chs 2, 5, 6), which dates from around the beginning of our era. One of the most important was the *Theology of Aristotle*, which was compiled in Arabic in the ninth century in the circle of Kindi. Fritz Zimmermann has identified a number of passages in it which postulate creation out of nothing.[92] Another relevant treatise is the *Liber de Causis*, which Zimmermann believes originally to have been part of the *Theology*, and whose inauthenticity was first exposed by Thomas Aquinas. Then there was a purported letter from Aristotle to his pupil Alexander the Great, sometimes known as the *Secretum Secretorum*, and first encountered in Arabic in the ninth century.[93] All of these except the *Theology* were freely available in Latin translation in the thirteenth century.

The references in the *Theology of Aristotle* to creation out of nothing do not necessarily imply that the world had a beginning. For in the pagan Neoplatonist sources from which the *Theology* draws, the claim that the world is created not out of anything, not even out of matter, means only that the world and its matter both owe their beginningless existence to God. On the other hand, in the mouth of a Christian like Philoponus, creation out of nothing *would* mean creation with a beginning. And one passage in the *Theology*, a passage wrongly

[91] On Crescas, see *The Light of the Lord*, translated in part by H.A. Wolfson, *Crescas' Critique of Aristotle*, proposition 12, part 2, p. 271, with Wolfson's notes; and H.A. Davidson, 'The principle that a finite body can contain only finite power', in Stein and Loewe, eds, *Studies in Jewish Religious and Intellectual History*, 75-92, at 85-9. For Gersonides *Wars of the Lord* 5.3.6; 6.1.3.; 6.1.14, I am relying on the account of C. Touati, *La pensée philosophique et théologique de Gersonide*, Paris 1973, 308-15; and for Abravanel *Deeds of God* 9.9, on S. Feldman, 'The end of the universe in Jewish medieval philosophy', 53-77, at 74-6.

[92] See his magisterial account, 'The origins of the so-called *Theology of Aristotle*', in Jill Kraye, W.F. Ryan, C.B. Schmitt, eds, *Pseudo-Aristotle in the Middle Ages*, London 1986, 110-240, at 174-5; 199-200.

[93] In Roger Bacon's rearranged Latin edition, published as fascicle 5 of Roger Bacon *Opera hactenus inedita*, Oxford 1905-1960, the relevant passage appears in Latin at part 3, ch. 6, p. 127. There is an English translation from Arabic, in discourse 4, sec. B, *de exitu rerum*, at p. 228.

associated again with Alexander,[94] looks as if its ultimate source could well be Philoponus. For the view it expresses, that Nature creates form, but God creates both form and matter out of nothing, is precisely the view that Philoponus propounds in his own person, not as a view of Aristotle's.[95] It could well have suited the compilers of the *Theology* to draw on Philoponus and to understand creation out of nothing in his way. We shall see that Farabi seems to understand the *Theology* as proposing a beginning of the universe, and the *Secretum* certainly appears to imply creation with a beginning.

Another well known reason for ascribing to Aristotle a creator God was the pressure felt by some thirteenth-century thinkers to present Aristotle as agreeing with the Christian faith. This pressure, already noticed in connexion with Bonaventure, is found in Albert, Bacon and many others,[96] although Thomas Aquinas warned against it.[97]

Those who were motivated, whether Christian or Islamic, had other stray texts besides the spurious ones to appeal to. We have already seen that Avicenna and Averroes appealed to Aristotle's theological treatise, *Metaphysics* Book 12, where he says that the heavens depend (*êrtêtai*) on a certain principle. And before them, Dexippus (perhaps following Porphyry), Proclus, Ammonius, David and Farabi made use of the same work.[98] Farabi, Maimonides and many thirteenth-century authors used a passage from Aristotle's *Topics* which was thought to imply uncertainty or undemonstrability, since Aristotle says that it is hard (not, however, impossible) to prove whether the world is eternal or not.[99] Thomas Aquinas, though conversant with the passage[100] is well aware that Aristotle does think the world eternal.[101] But for

---

[94] Excerpted from ps-Alexander, *On Coming-to-be*, in a work called *Extracts by Alexander from Aristotle's Theology*. The excerpt from ms. Istanbul, Carullah 1279, fols 64v-65r, is translated by Zimmermann at p. 174. It explicitly ascribes belief in creation out of nothing to 'The Philosopher' (Aristotle). See also Richard Sorabji, *Time, Creation and the Continuum*, 248, 315.

[95] Philoponus *contra Aristotelem* ap. Simplicium *in Cael.* 1141,5-30; 1142,1-28.

[96] See Delorme's introduction, p. xxxvi, to Roger Bacon *Opera hactenus inedita*, fasc. 13; and Luca Bianchi, *L'errore di Aristotele: la polemica contro l'eternità del mondo nel XIII secolo*, 20, 23-4, 26-7, 37-8.

[97] Thomas Aquinas *Commentary on Aristotle Physics* 8, lectio 2, n. 986.

[98] The references of Dexippus (who will be discussed below), of Proclus (*Elements of Theology* prop. 12 (Dodds p. 14, lines 22-3)), Avicenna and Averroes, are to *Metaph.* 12.7, 1072b13-14 (*êrtêtai*). Farabi makes a general reference to *Metaph.* 12, as well as to *Cael.* 2.8, 270a29ff and *Phys.* 2.6, 198a10, in his *Harmony of Plato and Aristotle*, ed. Dieterici, Leiden 1892, from which extracts are translated below. Ammonius and David cite *Metaph.* 12.10, 1075a11-25, on the general as the source of order in an army, as supporting the view that things down here come from (*ginesthai ek*) Platonic Forms existing as creative *logoi* in the Demiurge (Ammonius ap. Asclepium *in Metaph.* 44,35-7; David *in Isag.* 115,27-32).

[99] Aristotle *Topics* 1.11, 104b12-16, cited by Farabi loc. cit.; Maimonides *Guide* 2.15. See for example L. Bianchi, op. cit., 19-39; R. Jolivet, op. cit., 6-17 for thirteenth-century controversy.

[100] Thomas Aquinas *Summa Theologiae* 1, q. 46, a. 1, Respondeo.

[101] Thomas Aquinas *Commentary on Aristotle Physics* 8, lectio 2, 986.

showing that Aristotle none the less thinks it owes its beginningless existence to God, his favourite passage is a text in Aristotle's *Metaphysics*, book 2.[102]

What I want to add to these influences is that of *Ammonius*, who first used the infinite power argument to show that Aristotle's God was a Creator. But the 'infinite power' argument was only one in a series. The report of Simplicius deserves to be quoted in full:[103]

As for Aristotle, no one disputes that he calls God or the prime mover a final cause (*telikon*). But that he also calls God an efficient cause (*poiêtikon*) is sufficiently shown, I think, by his designating as efficient cause that whence comes the origin of change in his distinction of causes in the second book of the *Physics*: 'again that whence comes the first origin of change or stability, as the adviser is a cause, or the father of the child and in general the maker of what is made.'[104] What could be said clearer than this, to show that the prime mover is an efficient cause?

(1361,18) And in the first book of *On the Heavens* Aristotle said clearly that neither God nor nature does anything in vain.[105]

(1361,19) In the same book he said 'Eternity (*aiôn*) got its name from its being always (*aei einai*), immortal and divine, so that on it depend (*exêrtêtai*) the being and life of other things, of some in fuller detail, of others more obscurely.'[106] And it is clear that just as everything gets its good through the final cause, so it gets its being and life through the demiurgic cause.

(1361,24) He showed in the first book of *On Generation and Corruption* that the prime mover is also an efficient cause, when he inquired into the reasons for continuous generation and said: 'One cause is that whence we say the origin of change comes' – he clearly means by this the efficient cause – 'and one is matter. It is the latter sort that we must discuss, because we have discussed the former already in the treatise *On Motion*, and said that one thing is unmoved throughout all time and another is always being moved.'[107] Thus he too says that the efficient cause is twofold: that which is unmoved is the cause of everything, while the heavens are the cause of sublunary things. Alexander explains these words as follows: 'Aristotle says that the prime mover is the efficient cause of the movement of the divine body, a body which is itself ungenerated.'

(1361,33) In Book Alpha Major of the *Metaphysics*[108] Aristotle praises Anaxagoras, and Hermotimus before him, because they not only gave material causes of the universe, but also saw Intellect as an efficient and final cause. He writes: 'When someone said that Intellect was present in nature, just as in animals, and was the cause of the cosmos and of all

---

[102] Aristotle *Metaph.* 2.1, 993b19-31, cited by Thomas Aquinas *in Phys.* 8, lectio 2, 974; lectio 3, 996; *in Cael.* 1, lectio 21, 216; *in Metaph.* 2, lectio 2, 295; *Summa Theologiae* 1, q. 2, a. 3, Respondeo; *contra Gentiles* 1, ch. 13.
[103] Simplicius *in Phys.* 1361,11-1363,12.
[104] Aristotle *Phys.* 2.3, 194b29
[105] Aristotle *Cael.* 1.4, 271a33.
[106] ibid. 1.9, 279a27 (cf. *Metaph.* 12.7, 1072b14)
[107] Aristotle *GC* 1.3, 318a1.
[108] Aristotle *Metaph.* 1.4, 984b15.

order, he seemed like a sober man beside the random talk of his predecessors.' So after saying that Anaxagoras and before him Hermotimus touched on these explanations, he adds: 'Those who made these suppositions postulated that there is a principle of things which is at once the cause of goodness and the sort of cause from which change comes to things.' Thus he praises those who posited Intellect as both final and efficient cause, just as a little earlier he praised Anaxagoras because he called Intellect the origin of movement and kept it impassive and unmixed.

(1362,11) Alexander and certain other Peripatetics believe that Aristotle thinks there is a final cause of the heavens and one which moves it, but not an efficient cause of it, as was shown by the text cited a little earlier from Alexander who says: 'The prime mover is the efficient cause of the movement of the divine body, a body which is itself ungenerated.' So come let us show that Aristotle thinks Intellect is the efficient cause of the heavens themselves.

(1362,16) This fact alone would be enough, that he defines as efficient cause that whence comes the origin of change, and that he calls Intellect, or the unmoved cause, that whence comes the origin of the motion that follows celestial motion. (For it is the unmoved that is the origin, via the motion of the heavens and that of sublunary things.)

(1362,20) But besides in the second book of the *Physics*[109] he says that luck and haphazard are accidental causes which supervene on Intellect and nature, the causes that are in themselves efficient. 'Both', he says, meaning luck and haphazard, 'belong to the mode of causation called whence-comes-the-origin-of-change.' And he adds this: 'Haphazard and luck are the causes of things which might have been caused by intellect or nature, but which have in fact resulted from some accidental cause arising. But nothing that is accidentally such is prior to what is in itself such. So haphazard and luck are posterior to intellect and nature. In that case, however much haphazard might be the cause of the heavens above, intellect and nature must be prior causes of our universe and of many things in it besides.'

(1362,30) Against this some disputatious person may perhaps take refuge in saying that Aristotle is not showing by this that Intellect and nature are causes of the heavens, but only that he who says haphazard and luck are efficient causes of the heavens will be forced to make Intellect and nature prior causes. The objector should reflect that what is moved by something else must also get its existence from elsewhere, if being is superior to moving.

(1363,4) And if, according to Aristotle,[110] the power of any finite body is itself finite, clearly whether it be a power of moving or a power that produces being, then, just as it must get its eternal *motion* from the unmoved cause, so it must receive its eternal *being* as a body from the non-bodily cause.

(1363,8) My teacher Ammonius wrote a whole book offering many proofs that Aristotle thought God was also an efficient cause of the whole cosmos, a book from which I have here taken over some items sufficient

---

[109] Aristotle *Phys.* 2.6, 198a5.
[110] ibid. 8.10, 266a10-b27; 267b17-26.

for present purposes. More complete instruction on the subject can be got from there.

It was Ammonius in the sixth century A.D. who, by writing a whole book, put this interpretation of Aristotle's God on the map, and his sponsorship of the idea is documented not only here, but in a number of other places, from which we learn of other arguments.[111] But this is not to deny that there had been earlier hints of such an interpretation, notably in the fourth century in Iamblichus, Dexippus, and St Ambrose. Iamblichus denies that Aristotle contradicts Plato's belief in Ideas.[112] This issue is connected, because Plato was often understood to hold that the Ideas, residing in the divine Intellect, acted as efficient causes for producing the cosmos.[113]

Iamblichus' pupil, Dexippus, refers to Aristotle's *Metaphysics* 12, and to the expression there used according to which the heaven and nature 'depend on such an origin' (*ek toiautês arkhês êrtêtai*). Aristotle believes, according to Dexippus, that substances owe their unity to this origin, the One, and that some at least owe their life and form to the movement which the unmoved mover induces. Such a way of preserving Aristotle's credentials as a Platonist had been hinted at by Plotinus, and dismissed,[114] but it is possible that, before Dexippus, Porphyry had defended it in his longer commentary, now lost, on the *Categories*. Here are Dexippus' words:

> For Plotinus puts substance forward as a single genus among the intelligibles, because it provides being universally to incorporeal forms and implants being in all perceptibles and enmattered forms. If that is how things are, the origin of substance also stretches through all things, and remains the same while it takes a first, a second and a third rank, and provides being according to these ranks, to some things in a primary way and to other things in a different way. So if everything goes back to this origin (*arkhê*) because it depends on it (*ap' autês êrtêmena*),[115] Aristotle's description of substance can imply (*emphainein*) also the first origin (*arkhê*) of substance from which substance has fallen to the lowest of the subordinate levels ...

[111] Ammonius ap. Simplicium *in Phys.* 1361,11-1363,12; *in Cael.* 271,18-21; ap. Farabium *Harmony of Plato and Aristotle*, ed. Dieterici, Leiden 1892, ed. Nader, ch. 11, Beirut 1968, translated below. There is a German translation by F. Dieterici, in *Alfarabi's philosophische Abhandlungen*, Leiden 1892, 39-40. Similarly, Asclepius *in Metaph.*, from the voice of Ammonius, 28,20-29,2; 44,35-7 (which adds an appeal to *Metaph.* 12.10, 1075a11-25); 69,17-21; 103,3-4; 148,10-13; 151,15-32.

[112] Iamblichus ap. Elian *in Cat.* 123,1-3.

[113] This development circumvents Aristotle's objection that Plato's Ideas are not suited to play the role of efficient causes, *Metaph.* 1.9, 991b3-9; *GC* 2.9, 335b9-24, discussed by Julia Annas, 'Aristotle on inefficient causes', *Philosophical Quarterly* 32, 1982, 311-26. The objection will now apply only to Ideas conceived as existing outside the divine Intellect. See also p.278, n.118.

[114] Plotinus 6.1.3(1-5).

[115] Aristotle *Metaph.* 12.7, 1072b14; cf. *Cael.* 1.9, 279a27, used by Ammonius.

For this enquiry, then, I shall use what is said in the *Metaphysics*.[116] For Aristotle there are two kinds of substance, intelligible and perceptible, and between them natural substance. It is composite substance which is perceptible; that divided between form and matter [sc. the heavenly spheres] is natural; the substance above these is intellective and incorporeal. Aristotle often calls it an unmoved mover, because it is the cause of movement which gives form and life. That is what Aristotle demonstrates about these substances in *Metaphysics* 12, and with that he collects the many substances into the total substance. He put them together under a single arrangement, and referred them back to a single origin (*pros mian arkhên anêgage*). For how will anything else participate in the One, if substance itself which has its being in the One is deprived of the cohesion which refers back to the One?[117]

The remaining figure from the fourth century is a Christian writing in Latin, St Ambrose. He refers to Aristotle's efficient cause as the *operatorium* for whom it was enough to bring something effectively about (*competenter efficere*) that he should merely have thought it was to be undertaken (*adoriendum putasset*).[118] But Ambrose will have been influenced by an entirely different work, now fragmentary, Aristotle's early treatise, the *de Philosophia*, written when he was still in Plato's Academy. Further, Ambrose may have been misreading the work, for if fr. 19C (Ross) speaks of God as a Creator, that may be merely because the treatise incorporates Plato's views in dialogue form. Meanwhile fr. 13 refers to a Creator only in the context of how *other* men come to believe in divinity.

There were other anticipations by Arius Didymus in the first century B.C. and by Hierocles in the fifth A.D.[119] But it was left to Ammonius to develop the interpretation, and even to read some such idea back into Syrianus. Syrianus himself says that, whether or not Aristotle agrees with Plato, he is compelled to voice the same things in a way, because his own principles show that the world needs an efficient cause.[120] But Ammonius, or the commentary from the voice of Ammonius compiled

[116] Aristotle *Metaph*. 12.1, 1069a30.

[117] Dexippus *in Cat*. 40,28-41,3; 41,7-18.

[118] Ambrose *Hexaemeron* 1.1.1-2, p. 3, 10-13 (Schenkl). See J. Pépin, *Théologie cosmique et théologie chrétienne (Ambroise Exam. 1.1.1-4)* Paris 1964, 475-8, with comments in Richard Sorabji, *Time, Creation and the Continuum*, 283. The imprint in mud of a papyrus recently found in Afghanistan may contain another fragment of the *de Philosophia*, assigning to God a different causal role, that of making things participate in Forms: P. Hadot, 'Les textes littéraires grecs de la trésorie d'aï Khanoum', *Bulletin de correspondence hellénique* 111, 1987, 225-66.

[119] Arius Didymus is Stoicising, when he makes Aristotle's prime mover 'hold together' (*sunektikon*) the heavenly bodies, *Epitome* fr. phys. 9, in *Dox. Gr*. 450,16. The Alexandrian Neoplatonist Hierocles anticipates Ammonius' harmonisation of Plato and Aristotle, when he ascribes belief in divine creation not out of a substratum to both thinkers, ap. Photium *Bibliotheca* 171b33ff; 172a22ff (Bekker), crediting for the general thesis of harmony his teacher Plutarch of Athens (died 432 A.D.) and Ammonius Saccas, who taught Plotinus in the third century.

[120] Syrianus *in Metaph*. 117,25-118,11.

by Asclepius, goes further, and reports Syrianus as making Aristotle *tacitly* accept God as the efficient cause at least of eternal things:

So Aristotle appears not to be asserting an efficient cause for eternal things at all. But Syrianus says that the reason for his silence about the efficient cause is that he acknowledges God as efficient cause for *eternal* things, but acknowledges no determinate efficient cause for *generable* things, especially not for artefacts, since a house can be produced by different people.[121]

Ammonius' influence on his successors was decisive. His view of Aristotle's God as the world's efficient cause was not only published by Asclepius, but also endorsed by Philoponus, Simplicius, Olympiodorus, Elias, David and perhaps Stephanus.[122] The related thesis that Aristotle makes Ideas reside as creative *logoi* in the divine Intellect is found in Asclepius' edition of Ammonius, and is accepted, for example, by Philoponus, Simplicius and David.[123] But we can further trace how Ammonius' influence passed to the Islamic world. For one thing, Philoponus' endorsement of his interpretation would have been available in Arabic. But, more dramatically, Ammonius' interpretation of Aristotle's God is explicitly acknowledged by Farabi (c.873-950) in his work, the *Harmony of Plato and Aristotle*. Farabi says that Ammonius makes Aristotle's God a Creator. He endorses the interpretation, and further shares Ammonius' view that Aristotle makes Plato's ideal Forms exist as creative *logoi* in the divine mind. This is just one instance of the harmony of Plato and Aristotle which Farabi, like Ammonius, espouses. Tantalisingly, however, he says that Ammonius' arguments are too well known to need quoting. Despite this we see here the direct influence of Ammonius on the Islamic tradition.

Farabi's throw-away reference to Ammonius is not the only thing to tease us. There is an Arabic ms. of a work described as Ammonius', *On the Opinions of the Philosophers*, whose opening, on Greek theories of

[121] Asclepius *in Metaph*. 450,20-5.

[122] Philoponus *in Phys*. 189,10-26; 240,18-19; *in GC* 136,33-137,3 (cf. 286,7); *in DA* 37,18-31; *in An. Post*. 242,26-243,25; Simplicius *in Phys*. 256,16-25; 1360,25; 28-31; 1362,8; 16; 32; *in Cael*. 87,3-11; 143,9-144,4; 271,5-21; Olympiodorus *in Phaedonem*, lecture 13, §2, lines 37-40 Westerink (=76N); Elias *in Cat*. 120,16-17; David *in Isag*. 113,15-16; 114,34-115,3; 115,13-14; 24-32; Philoponus(?)/Stephanus(?) *in DA* 3 571,1-5.

[123] Asclepius *in Metaph*., from the voice of Ammonius, 44,32-7; 68,17-27; 75,27-8; 167,14-34; 183,14-16; 233,38-40; 363,1-5; 393,34-394,2; 441,27-31; 442,1-2. Simplicius *in Cael*. 87,3-11; Philoponus *in DA* 37,18-31; *in Phys*. 240,18-19; *in An. Post*. 242,26-243,25; David *in Isag*. 113,15-16; 114,34-115,3; 115,13-14; 24-32. Unlike Asclepius, Philoponus can be treated as offering an independent endorsement. For although his *in GC, in DA* and *in An. Post*. are also said to be 'from the voice of Ammonius', the description is explicitly qualified in the mss. to say that Philoponus is adding observations of his own. Moreover, the passage from *in An. Post*. was a late addition, according to a persuasive argument by Koenraad Verrycken, and was introduced long after Philoponus had passed out of Ammonius' direct sphere of influence (Koenraad Verrycken, *God en Wereld in de Wijsbegeerte van Ioannes Philoponus*, Ph.D diss. Louvain 1985, 509-23, summarised in Richard Sorabji, ed., *The Transformation of Aristotle*, London 1989).

creation, has been translated into English.[124] Unfortunately, it does not contain the arguments of our Ammonius.

Finally, Ibn al-Nadim's *Fihrist* (987) also records that Ammonius wrote on 'Aristotle's teachings concerning the Maker',[125] but this source is not necessarily independent of Farabi.

Farabi of course does not depend solely on Ammonius. In the chapter from which the following extracts are taken he cites *inter alia* the spurious *Theology of Aristotle*. It is probably from there that he gets the idea, which is not in Ammonius, that Aristotle's world is not eternal. The following passages have been rendered by Zimmermann from Farabi:

[Aristotle in the Theology: ...] Having established these premises, he ascends to the corporeal and incorporeal parts of the world and roundly declares that they have all come into being because the Creator has originated them, and that it is He who is the efficient cause, the true One, and the originator of everything. [That doctrine of Aristotle's is] in accordance with Plato's statements in his books on theology, such as the *Timaeus* and the *Republic*, and elsewhere.

Again, in those metaphysical treatises of his which bear the names of letters, Aristotle ascends from [certain general premises to] the Creator in book Lambda, in order then to return to the demonstration of those premises.

As far as we know, his achievement in that is unprecedented; neither has it been equalled by anyone since. How, then, can a thinker with such credentials be suspected of holding that there is no Maker and that the world is eternal?

There is a monograph by Ammonius in which he sets out the arguments of our two philosophers [viz. Plato and Aristotle] for the existence of a Maker [(*ṣāni* – translating, presumably, *dêmiourgos*)]. They are [or the book is] too well known for us to have to repeat [literally 'present'] them here.[126]

Since God is alive and brings into being this world and all that is in it, there must be in His mind, as part of His essence, forms of what He wants to bring into being. Again, since His essence endures without change or alteration, what is at His level of being must be likewise enduring for ever without change. If things to be had no form or trace in the essence of Him, the living one that wants to bring them into being, what would there have been for Him to bring into being? What paradigm after which to fashion what He was making and originating?[127]

---

[124] Ms. Aya Sofiya (Istanbul) no. 2450, opening translated by S.M. Stern, in A. Altmann and S.M. Stern, *Isaac Israeli: a Neoplatonic philosopher of the early tenth century*, Scripta Judaica 1, London 1958, 70-1.

[125] Ibn al-Nadim, *Fihrist*, ed. Tajaddod, 314,3f.

[126] Farabi, *Harmony of Plato and Aristotle*, ed. (with German translation) Dieterici, p. 24, line 13 – p. 25, line 1, abstract by Fritz Zimmermann.

[127] Farabi. op. cit., ed. Dieterici, p. 28, line 22, p.29, line 5, abstract by Fritz Zimmermann.

Can Ammonius' influence be traced any further? In the passage examined earlier, we saw Avicenna, who regarded Farabi as the best of the commentators, taking the same view of Aristotle's God as efficient cause of the world, although he did not share Ammonius' opinion that Aristotle gave *arguments* for God's creative role. Maimonides (1135-1204), who had read both Farabi and Avicenna, repeats that Aristotle's God is the efficient cause of the world's existence,[128] and he was available in Latin to the thirteenth century. So an indirect influence of Ammonius on this period seems positively likely. Thirteenth-century scholars disagreed, but such major figures as Bacon, Bonaventure and Thomas Aquinas thought that in various diverse senses Aristotle's God was a Creator.[129]

It would be droll if what started in Ammonius as an attempt to harmonise Aristotle's God with Plato's Creator finished by helping (only helping, of course) to make Aristotle's God safe for thirteenth-century Christianity.[130] Almost as surprising was our earlier finding that what started in the Christian Philoponus as an attempt to argue for the *beginning* of the universe finished up in Aquinas as an argument *against* its *ending*.

Dynamics

But in concentrating on the implications for Creation, I have neglected the implications for motion and dynamics. Are there any such implications? I believe there are. First, Proclus' use of Aristotle's 'infinite power' argument to establish a Creator-God is relevant to impetus theory. For Proclus provides the clearest example I know before Philoponus of an impetus or impressed force. I would not disagree with the scholarly findings cited in Chapter 14, which show

---

[128] Maimonides *Guide for the Perplexed*, part 2, ch. 21 (translations by M. Friedlander and S. Pines). Maimonides' Aristotle believes, but does not think he can prove, the creation to be beginningless, part 2, ch. 15.

[129] (i) For Thomas Aquinas' view that, in Aristotle's opinion, God was the efficient cause of the beginningless and endless existence of the world, see above. Thomas further recognised that despite the lack of a beginning that made God in a sense a Creator: *de aeternitate mundi* §§6, 7 and 8 (translated by Cyril Vollert, in Vollert, Kendzierski, Byrne, op. cit.); *Commentary on Lombard's Sentences* 2, d. 1, q. 1, a. 2, solutio; *Summa Theologiae* 1, q. 46, a. 1 and 2; *in Phys.* 8, lectio 2, 974. (ii) Bonaventure's view was that Aristotle's God did not create matter, but created the world, in the sense of imposing form on matter, either from some beginning point of time, or, as he later recognised, beginninglessly (references as above). (iii) Bacon simply believed that Aristotle's God gave the world a beginning: *de viciis contractis in studio theologie*, in *Opera hactenus inedita*, vol. 1, pp. 10-11; *Questions on Aristotle's Physics* in *Opera hactenus inedita*, vol. 13, pp. 148, lines 36-7; 171,14-37; 223,24-9; 376,4-6; 387,21-388,18; 390,1-4.

[130] The thirteenth-century condemnations of Aristotle, including that of 1277 mentioned above, were not to take permanent effect.

that in certain *other* contexts Proclus' conception differs from that of an impetus. But in the context of the infinite power argument, as shown *inter alia* by the passage translated above in this chapter, God is said to implant a power (*dunamis*) in repeated dollops into the cosmos.[131] In some further passages, Proclus even uses the same word, *endidonai*, as Philoponus uses after him for the implanting, although it is not a *power* (*dunamis*) which God is said to implant in the cosmos in these last passages, but simply immortality by way of repair and the bond which holds the cosmos together.[132] At least the former passages show that Proclus does anticipate the idea of an impressed power. Of course, the significance of this anticipation is reduced by the fact that Proclus confines the idea to the special case of the motion and existence of the *heavens*, and does not extend it to the most interesting case of projectiles. But since Philoponus discusses one of the relevant texts at length,[133] the question must be asked whether Philoponus could have been inspired by Proclus.

The answer is No. For one thing, Philoponus would have encountered *endidonai dunamin* as the expression for implanting a force in authors much earlier than Proclus. It is used by Alexander and Themistius in connexion with the Aristotelian theory, which Philoponus rejects, of a force being implanted not in a javelin, but in the pockets of air behind it.[134] Moreover, when Philoponus introduced the idea of impressed force in his *Physics* commentary, it did not occur to him to apply it, like Proclus, to the case of the *heavens*. He was concerned instead with the case of projectiles and of the fire belt. The first edition of the *Physics* commentary can be dated to 517, and it was well after its second edition in the *de Opificio Mundi* of 546-9, or even 557-60, that Philoponus first thought of using the idea of an impressed force in something closer to Proclus' manner, to explain the motion of *the heavens*. Even then, he could not in his own person have accepted Proclus' 'infinite power' argument as a reason for locating in the heavens an impressed force, for Proclus' argument depends upon his belief in the eternity of the cosmos, which Philoponus does not share.

The infinite power arguments carry a second implication for dynamics: if the rotation of the spheres is *frictionless*, why should Aristotle think that an infinite force was *required* to keep it going for ever? It might be objected that the question presupposes a concept of inertia, the idea that a moving body will continue moving, unless a force like friction acts on it to the contrary. And that idea was no more than fleetingly glimpsed in antiquity.[135] But there are analogous ideas in Aristotle.

[131] Proclus *in Tim.* 1.267,16-268,6; and *Examination of Aristotle's Objections to Plato's Timaeus*, as quoted by Philoponus *contra Proclum*. There is no anticipation in ps-Aristotle *de Mundo* 398b20, where God is said to transmit the power (*didôsi dunamin*) which moves the inner celestial spheres from the outermost one. For the simile of *endosis* is not that of implanting a force, but of striking a key note.

[132] Proclus *in Tim.* 1.260,14-15; 3.220,1-3. Same word at Philoponus *in Phys.* 384,33; 641,15; 642,5; 11.

[133] Philoponus *contra Proclum* 237,15-242,22; 297,21-304,9; 626,1-628,17.

[134] Alexander ap. Simplicium *in Phys.* 1347,4; Themistius *in Phys.* 235,4; 7.

[135] S. Pines is right not to find it in Galen's rejection of Plato's theorem that whatever

In his attack on motion in a vacuum, as we saw in Chapter 9, Aristotle produced two related objections. If there were motion in a vacuum, would not the speed be infinite, in the absence of resistance,[136] and where would the motion ever stop, if no stopping place in a vacuum differs from any other?[137] Avicenna endorses the second objection in his own person, and argues in detail that projectile motion in a vacuum would not stop.[138] Is there not, then, we may ask, a worry about the rotation of the celestial spheres? Although Aristotle surrounds them not with a vacuum, but (Chapter 8 above) with nothing at all, the question retains its force: why should they stop at one position or time rather than another? And this in turn creates a clash with the demand for infinite power. For on the one hand, no infinite power is needed to spin a finite body, and once it is spinning without differentiation in its surroundings, why, if Aristotle's argument is acceptable, should it stop? Conversely, if we take seriously the demand for infinite power, we presuppose that a finite power would be *exhausted*. But then we can *answer* Aristotle's question why something should stop after three miles, rather than four miles, of travel in a vacuum.

An analogous difficulty was noticed by Philoponus about Aristotle's other suggestion, that speed in a vacuum would be infinite, not because of the lack of differentiation, but because of the lack of resistance. Rotation, he pointed out, and in particular the rotation of the heavens, encounters no resistance, but is none the less finite in speed. The relevant discussion has already been partly translated in Chapter 14, but I shall translate here some extra lines:

> For if in general the reason why motion takes time were the physical [medium] that is cut through in the course of the motion, and for this reason things that moved through a vacuum would have to move without taking time because of there being nothing for them to cut through, this ought to happen all the more in the case of the fastest of all motions, I mean the [celestial] rotation. For what rotates does not cut through any physical [medium] either. But in fact this [timeless motion] does not happen: all rotation takes time, even without there being anything cut through in the motion.[139]

---

is moved, is moved *by* something ('Omne quod movetur necesse est ab alio moveri', *Isis* 52, 1961, 21-54), for Galen allows that a falling rock has an internal source of motion, even if he denies that it can be said to be moved *by* anything. Elizabeth Asmis has suggested to me that, in early atomist theory, atoms may continue moving sideways indefinitely by inertial motion, after a collision, instead of being propelled downwards by weight. On Aristotle *Phys.* 4.8, 215a19-20, see below.

[136] Aristotle *Phys.* 4.8, 215b22-216a4
[137] ibid. 4.8, 215a19-20.
[138] Avicenna, *Shifā* 1, p. 154f, translated into French and discussed by S. Pines, Études sur Awḥad al-Zamān Abu' l-Barakât al-Baghdâdî', *Revue des Études Juives* n.s. 3, 1938, repr. in his *Collected Works* I, Leiden 1979, 52-6.
[139] Philoponus *in Phys.* 690,34-691,5.

The point is repeated by Ibn Bajja or Avempace (died 1138), as noticed in Chapter 9.[140]

Averroes and Thomas Aquinas both try to rescue Aristotle, Averroes by arguing that it is the heavenly spheres themselves, and not any surrounding medium, which provide the resistance to motion.[141] Thomas suggests that Aristotle's connexion of lack of resistance with infinite speed is a merely dialectical move against opponents who postulate vacuum as the *sole* cause of motion,[142] but I have to say that I see no sign that Aristotle's opponents are as extreme as that.

One way or another, then, it looks as if Aristotle's dynamical theory must give. Or does he have a way out of the difficulties? As regards the lack of differentiation, someone might plead that we will find differences, if we look *beneath* the heavens, rather than beyond them. At some stage in the celestial rotation, the most prominent heavenly body, the sun, will be directly over the highest mountain. So it cannot be said that there are no differences between different possible stopping points for the celestial rotation. But it may still be wondered if these differences would provide a *good* reason for the celestial rotation to stop at one point rather than another. As regards the lack of resistance to celestial motion, an attempt might be made to find sources of friction by pointing out that the celestial spheres draw round with them (*sumperiagein*) the lower belts of fire and air, creating heat by inducing friction there (*parektribesthai*).[143] But whether they suffer friction in return is not said.

There is another reason for wondering why infinite power should be needed for everlasting celestial rotation.[144] In an earlier treatment of celestial motion in *On the Heavens*, we find no integral and undisputed reference to the role of God. Instead, Aristotle combines two ideas, that the heavens are made of an indestructible fifth element which can undergo no change but motion, and that circular motion is natural to that element. If it can undergo no other change, what reason could there be for its natural motion to cease? It cannot tire, or grow bored, or perish, for these would be changes. Nor does circular motion have a terminal destination. It looks, then, as if in his earlier work Aristotle allowed for a motion that was eternal not because of any infinite power,

---

[140] Avempace is reported by Averroes, *Long Commentary on Physics* 4, text 71, Latin in Juntine edition 1562, repr. Frankfurt 1912, p. 160r, b, E-F, translated in E.A. Moody, 'Galileo and Avempace, the dynamics of the leaning tower experiment', *Journal of the History of Ideas* 12, 1951, 163-93; 375-422. Thomas Aquinas gives the argument in his *Commentary on the Physics* 4, lectio 12, 534, translated by R.J. Blackwell, R.J. Spath, W.E. Thirlkel, London 1963.

[141] Averroes, *Long Commentary on Physics* 4, text 71; Thomas Aquinas, *Commentary on the Physics* 4, lectio 12, 535.

[142] Thomas Aquinas, *Commentary on the Physics* 4, lectio 12, 536. On the other hand, Thomas does not discount as dialectical Aristotle's 'why stop here?' question, lectio 11, 526.

[143] Aristotle *Meteor*. 1.3, 341a17-32; 1.7, 344a11-13; *Cael*. 2.7, 289a20-35.

[144] I owe the point to John Cleary.

but because of immunity to further change in something to which circular motion was natural.

There is a final corollary. For the question whether finite power would be *exhausted* has implications for the definition of impetus. It has been suggested that impetus is by definition something that gets exhausted, or that at least this is the pre-Galilean concept.[145] Philoponus does indeed treat impetus as exhaustible. This is his answer to Aristotle's question why motion in a vacuum would stop here rather than there: the force impressed in a projectile would be exhausted (*exasthenêsêi*), as Aristotle himself admits in regard to the *different* force which he postulates in the pockets of air behind a projectile.[146] However, it seems dangerous to elevate exhaustibility into a defining characteristic of all impetus theories, or all before Galileo's, since Avicenna maintains in the passage just discussed that only resistance exhausts the force in a projectile, and that in a vacuum the force would not be exhausted.[147] Similarly Buridan insists[148] that the impetus by which God sets the heavens moving at the time of Creation will never run out.[149]

I would conclude by drawing attention to the part played by the Neoplatonist commentators on Aristotle, not only in dynamics, but also on the question of Creation. The Aristotelian philosophy handed down to the middle ages was not just that of Aristotle, but a far richer compilation transformed by the preoccupations of the intervening commentators.

[145] For the bolder claim, see Michael Wolff, 'Philoponus and the rise of preclassical dynamics', in Richard Sorabji, ed., *Philoponus and the Rejection of Aristotelian Science*, at 84-5. For his reply to the doubt about Avicenna, see p. 85, n. 2. Galileo is treated as an innovator by Anneliese Maier, 'Galilei und die scholastische Impetustheorie', *in Saggi su Galileo Galilei*, Florence 1967 = her *Ausgehendes Mittelalter* 2, Rome 1967, 465-90, at 486-90.

[146] Philoponus *in Phys.* 644,16-22.

[147] Avicenna *Shifâ* 1, p. 154f.

[148] Buridan *in Phys.* 8.12, fol. 120, translated into French and discussed by P. Duhem, *Études sur Léonard de Vinci*, 3, Paris 1913, 40-2; also discussed by G. Sarton, *Introduction to the History of Science 3, part 1*, Baltimore 1947, 543; M. de Gandillac, *Le mouvement doctrinale du XIe au XIVe siècles*, Paris 1951, 459; C. Touati, *La pensée philosophique et théologie de Gersonide*, Paris 1973, 312-13; Marshall Claggett, *The Science of Mechanics in the Middle Ages*, Madison, Wisconsin 1959, 510-15. Inexhaustible forces are also postulated in *natural* motion by Avempace, according to S. Pines ('La dynamique d'Ibn Bâjja' in *Mélanges Alexandre Koyré* I, Paris 1964, 462, repr. in his *Collected Works*, vol. 2, 460) and by Gersonides, according to S. Feldman ('The end of the universe in medieval Jewish Philosophy', 63).

[149] I am grateful for help to David Barlow, John Dillon, Jill Kraye, Ian Mueller, Tom Settle, Robert Sharples, Andrew Smith and above all Fritz Zimmermann.

Principal philosophers discussed

(See General Index for others)

Early Greek thinkers

Alcmaeon, late sixth century B.C.?
Anaxagoras, c.500-c.427 B.C.
Democritus, the atomist, fl. c.435 B.C.
Archytas, Pythagorean, first half of fourth century B.C.
Pythagorean ideas on space and time were well developed by the fourth
century B.C.

Plato

Plato, c.427-348 B.C.

Aristotle and associates

Aristotle, 384-322 B.C.
Eudemus of Rhodes, his pupil, fourth century B.C.
Theophrastus, Aristotle's successor as head from 322 to c.287 B.C.
Alexander of Aphrodisias, greatest proponent of Aristotelianism, fl. c. A.D. 205.
Themistius, orator and sympathetic but independent commentator on
Aristotle, fl. late 340s-384/5 A.D.

Revival of Atomism

Epicurus, 341-270 B.C.

Stoics

Zeno of Citium, founded Stoa in Athens c.300 B.C.
Cleanthes, second head of Stoa 262-c.232 B.C.
Chrysippus, the most famous Stoic, c.280-c.206 B.C., third head of Stoa.
Posidonius, c.135-c.55 B.C., head of Stoic school in Rhodes.
Cleomedes, relayed Posidonius' arguments between A.D. 0 and 150.

Middle Platonists
(all dates are A.D. from here onwards)

Plutarch of Chaeroneia, c.46-120.
Altinus, his lectures attended by Galen between 149 and 157.

Christians

Gregory of Nyssa, c.311-394 or later.
Ambrose, 339-397.

Neoplatonists

Plotinus, c.205-260.
Porphyry, 232-309, Plotinus' pupil.
Iamblichus, c.250-c.325, Porphyry's pupil.
Dexippus, fourth century, Iamblichus' pupil.
Syrianus, died c.437.
Proclus, c.411-485, Syrianus' pupil.
Ammonius, 435/45-517/26, Proclus' pupil.
Damascius, head of Athenian school at closure, 529, Ammonius' pupil.
Philoponus, c.490-570s, Ammonius' pupil.
Simplicius, wrote after 529, pupil of Ammonius and Damascius.

Christians in the Islamic world

Job of Edessa, fl. c.817.
Yaḥyâ Ibn 'Adî, 892-973.

Islam

Ḍirâr, died 815 at latest.
Theology of Aristotle, compiled in circle of Kindi, early ninth century.
Farabi, c.873-950.
Avicenna, c.980-1038.
Ghazâlî, 1058-1111.
Avempace=Ibn Bâjja, died 1138.
Averroes, c.1126-c.1198.

Jewish philosophers

Saadia, died 942.
Maimonides, 1135-1204.
Gersonides, 1288-1344.
Crescas, 1340-1410.
Abravanel, 1437-1509.

The Latin West

Grosseteste, 1168-1253.
Roger Bacon, 1214/20-1292.
Bonaventure, c.1217-1274.
Thomas Aquinas, c.1224-1274.
Bishop Tempier, condemnation issued 1277.
Buridan, c.1295-1356.

Renaissance and Modern thinkers

Gianfrancesco Pico della Mirandola, 1469-1533.
Patrizi, 1529-1597.
Galileo, 1564-1642.
Descartes, 1596-1650.
Locke, 1632-1704.
Newton, 1642-1727.
Leibniz, 1646-1716.
Berkeley, 1685-1753.
Joseph Priestley, 1733-1804.

Select bibliography

Matter

General

C. Baeumker, *Das Problem der Materie in der griechischen Philosophie*, Münster 1890.

E. McMullin, ed., *The Concept of Matter*, Notre Dame, Indiana 1963.

S. Toulmin and J. Goodfield, *The Architecture of Matter*, London 1962.

Aristotle's theory of matter

H. Happ, *Hyle*, Berlin 1971, is the most extensive study, but offers a controversial interpretation. Especially relevant to the present book is:

I. Mueller, 'Aristotle on geometrical objects', *Archiv für Geschichte der Philosophie* 52, 1970, 156-71, repr. in J. Barnes, M. Schofield, R. Sorabji, eds, *Articles on Aristotle 3*, London and Ithaca N.Y. 1979.

On Aristotle on chemical mixture, see below.

Did Aristotle believe in prime matter? (The first three say 'no'):

Hugh R. King, 'Aristotle without prima materia', *Journal of the History of Ideas* 17, 1956, 370-89.

Willie Charlton, *Aristotle's Physics, Books I and II*, Oxford 1970, 129-45.

Willie Charlton, 'Prime matter: a rejoinder', *Phronesis* 28, 1983, 197-211.

F. Solmsen, 'Aristotle and prime matter', *Journal of the History of Ideas* 19, 1958, 243-52.

Alan Lacey, 'The Eleatics and Aristotle on some problems of change', *Journal of the History of Ideas* 26, 1965, 451-68.

H.M. Robinson, 'Prime matter in Aristotle', *Phronesis* 19, 1974, 168-88.

R.M. Dancy, 'Aristotle's second thoughts on substance', *Philosophical Review* 87, 372-413.

C.J.F. Williams, *Aristotle's De Generatione et Corruptione*, Oxford 1982, 211-19.

Atomist theories

Richard Sorabji, *Time, Creation and the Continuum*, London and Ithaca N.Y. 1983, chs 22-5 and literature cited there.

Interpretations of Plato's Timaeus on matter

Friedrich Solmsen, *Aristotle's System of the Physical World*, Ithaca N.Y. 1960, ch. 6.

F.M. Cornford, *Plato's Cosmology*, London 1937.

A.E. Taylor, *A Commentary on Plato's Timaeus*, Oxford 1928.

Harold Cherniss, *Aristotle's Criticism of Plato and the Academy*, Baltimore 1944.

J.C.M. van Winden, *Calcidius on Matter*, Leiden 1959.
Willie Charlton, *Aristotle's Physics, Books I and II*, Oxford 1970, 141-5.

Matter in Middle Platonism and Neoplatonism

J.C.M. van Winden, *Calcidius on Matter*, Leiden 1959.
F.R. Jevons, 'Dequantitation in Plotinus' cosmology', *Phronesis* 9, 1964, 64-71.
A.H. Armstrong, 'The origin of the non-materiality of body in Plotinus and the Cappadocians', *Studia Patristica* 5, Berlin 1962, repr. in his *Plotinian and Christian Studies*, London 1979.
Michael Wolff, *Fallgesetz und Massebegriff*, Berlin 1971 (on Philoponus).
C. Wildberg, *Philoponus' Criticism of Ether*, Ph.D. Cambridge 1984.
N. Tsouyopoulos, 'Die Entstehung physikalischer Terminologie aus der Neo-platonischen Metaphysik', *Archiv für Begriffsgeschichte* 13, 1969, 7-33 (on Simplicius).

Geometrical space in Late Neoplatonism

I. Mueller, 'Aristotle's doctrine of abstraction in some Aristotelian commen-tators and Neoplatonists', forthcoming in Richard Sorabji, ed., *The Trans-formation of Aristotle*.

Medieval treatments of matter

E. Sylla, 'Autonomous and handmaiden science: St. Thomas Aquinas and William of Ockham on the physics of the Eucharist', in J.E. Murdoch and E. Sylla, eds, *The Cultural Context of Medieval Learning, Boston Studies in the Philosophy of Science* 26, 1975, 349-96.
E. Sylla, 'Godfrey of Fontaines on motion with respect to quantity of the Eucharist', in A. Maierù and A. Paravicini Bagliani, eds, *Studi sul XIV secole in memoria di Anneliese Maier*, Rome 1981.
James A. Weisheipl, 'The place of John Dumbleton in the Merton School', *Isis* 50, 1959, 439-54 (esp. on Ockham).
James A. Weisheipl, 'The concept of matter in fourteenth-century science', in E. McMullin, ed., *The Concept of Matter in Greek and Medieval Philosophy*, Notre Dame, Indiana 1963, 1965[2], 147-69 (esp. on Ockham).
E. Grant, 'The principle of the impenetrability of bodies in the history of concepts of separate space from the middle ages to the seventeenth century', *Isis* 69, 1978, 551-71, repr. in his *Studies in Medieval Science and Natural Philosophy*, London 1981.
H.A. Wolfson, *Crescas' Critique of Aristotle*, Cambridge, Mass. 1929.

Newton against Descartes on matter

J.E. McGuire, 'Space infinity and divisibility: Newton on the creation of matter', in Z. Bechler, ed., *Contemporary Newtonian Research*, Dordrecht 1982, 145-90.

The history of field theory

P.M. Heimann and J.E. McGuire, 'Newtonian forces and Lockean powers: concepts of matter in eighteenth-century thought', *Historical Studies in the*

Physical Sciences 3, 1971, 233-306.
Milic Čapek, *The Philosophical Impact of Contemporary Physics*, Princeton 1981.
Louis de Broglie, *Nouvelles Perspectives en Microphysique*, translated by A.J. Pomerans as *New Perspectives in Physics*, New York 1962.
J.A. Wheeler, *Geometrodynamics*, New York 1962.
Charles W. Misner, 'Some topics for philosophical enquiry concerning the theories of Mathematical Geometrodynamics and of Physical Geometrodynamics', in K.F. Schafner and R.S. Cohen, eds, *Proceedings of the Biennial Meeting of the Philosophy of Science Association 1972*, Dordrecht 1974.
Lawrence Sklar, *Space, Time and Space-Time*, Berkeley and Los Angeles 1974.
Michael Redhead, 'Quantum field theory for philosophers', *Proc. of the Biennial Meeting of the Philosophy of Science Association 1982*, Dordrecht 1984.

Philosophical versions of field theory

Rudolf Carnap, *Logische Syntax der Sprache*, Vienna 1934, translated by Amethe Smeaton as *Logical Syntax of Language*, 1937.
Bertrand Russell, *Human Knowledge, its Scope and Limits*, London 1948 (contains a reply).
W V Quine, 'Whither physical objects?', *Boston Studies in the Philosophy of Science* 39, 1976, 497-504.
P.F. Strawson, *Individuals*, London 1959, ch. 6, part 2 and ch. 7.

Bodies as bundles of properties

A.C. Lloyd, 'Neoplatonic logic and Aristotelian logic', *Phronesis* 1, 1955-6, at 158-9.
A.C. Lloyd, *Form and Universal in Aristotle*, Liverpool 1981, 67-8 (Lloyd's account is the fullest and covers Platonist interpretations of Plato and Aristotle).
A.H. Armstrong, 'The origin of the non-materiality of body in Plotinus and the Cappadocians', *Studia Patristica* 5, Berlin 1962, repr. in his *Plotinian and Christian Studies*, London 1979.
Richard Sorabji, *Time, Creation and the Continuum*, London and Ithaca N.Y. 1983, ch. 18 (on Gregory of Nyssa).
Dermot Moran, *The Philosophy of John Scottus Eriugena*, forthcoming.

The main modern proponents are Stout and Russell:

Bertrand Russell, *An Inquiry into Meaning and Truth*, London 1940, ch. 6.
Bertrand Russell, *Human Knowledge, its Scope and Limits*, N.Y. 1948, part 2, ch. 3; part 4, ch. 8.
Bertrand Russell, *My Philosophical Development*, ch. 9.
G.F. Stout, 'The nature of universals and propositions', *Proceedings of the British Academy* 10, 1921, repr. in his *Studies in Philosophy and Psychology*, London 1936, 384-403.

For further discussion, see the articles reprinted in:

Michael J. Loux, *Universals and Particulars*, Notre Dame, Indiana 1968.

Anaxagoras on chemical combination

F.M. Cornford, 'Anaxagoras' theory of matter', *Classical Quarterly* 24, 1930, 14-30 and 83-95.
Gregory Vlastos, 'The physical theory of Anaxagoras', *Philosophical Review* 59, 1950, 31-57.
Colin Strang, 'The physical theory of Anaxagoras', *Archiv für Geschichte der Philosophie* 45, 1963, 101-18.

All the above are reprinted in:

R.E. Allen and David J. Furley, eds, *Studies in Presocratic Philosophy* vol. 2, London 1975.
George Kerferd, 'Anaxagoras and the concept of matter before Aristotle', *Bulletin of the John Rylands Library* 52, 1969, 129-43, repr. in Alexander P.D. Mourelatos, ed., *The Pre-Socratics*, Garden City N.Y. 1974.
Malcolm Schofield, *An Essay on Anaxagoras*, Cambridge 1980.
Jonathan Barnes, *The Presocratic Philosophers*, London 1972, vol. 2.

Aristotle on chemical combination

H.H. Joachim, 'Aristotle's conception of chemical combination', *Journal of Philology* 29, 1904, 72-86.
Richard Sharvy, 'Aristotle on mixtures', *Journal of Philosophy* 80, 1983, 441-8.
David J. Furley, 'The mechanics of *Meteorologica* IV: a prolegomenon to biology', in P. Moraux and J. Wiesner, eds, *Zweifelhaftes im Corpus Aristotelicum*, Berlin 1983, 73-93, repr. in his *Cosmic Problems*, Cambridge 1989.
James G. Lennox, reply to Sorabji, 'The Greek origins of the idea of chemical combination', in John J. Cleary, ed., *Proceedings of the Boston Area Colloquium in Ancient Philosophy* 4, 1989.

The Stoics on chemical combination

Robert B. Todd, *Alexander of Aphrodisias on Stoic Physics*, Leiden 1976 (includes text and translation of Alexander's critique).
H. Dörrie, *Porphyrios' 'Summikta Zetemata'*, Munich 1959, 24-35.
Michael J. White, 'Can unequal quantities of stuff be totally blended?', *History of Philosophy Quarterly* 3, 1986, 379-87.
Eric Lewis, 'Diogenes Laertius 7.151 and the Stoic theory of mixture', *Bulletin of the Institute of Classical Studies,* forthcoming, 1988.

Philoponus on mixture

Robert B. Todd, 'Some concepts in physical theory in John Philoponus' Aristotelian commentaries', *Archiv für Begriffsgeschichte* 1980, 151-70.

Stoics on cosmology, qualities and body

David Hahm, *The Origins of Stoic Cosmology*, Ohio State University 1977.
Jaap Mansfeld, 'Zeno of Citium', *Mnemosyne* 31, 1978, 134-78, at 158-67.
David Hahm, 'The Stoic theory of change', in Ronald H. Epp, ed., *Recovering the Stoics, The Southern Journal of Philosophy* 23, supplement, 1985, 39-56.
Margaret Reesor, 'The Stoic concept of quality', *American Journal of Philology* 75, 1954, 40-58.

M. Lapidge, '*Archai* and *stoicheia*: a problem in Stoic cosmology', *Phronesis* 18, 1973, 240-78.
M. Lapidge, 'Stoic cosmology', in John Rist, ed., *The Stoics*, Berkeley and Los Angeles 1978, 161-85.
Robert Todd, 'Monism and immanence in Stoic physics', in John Rist, ed., *The Stoics*, Berkeley and Los Angeles, 1978, 137-60.

The Stoic theory of categories

David Sedley, 'The Stoic criterion of identity', *Phronesis* 27, 1982, 255-75.

Christology and mystical union: appeals to theories of mixture

H.A. Wolfson, *The Philosophy of the Church Fathers*, Cambridge Mass. 1956, 364-403.
J. Pépin, 'Théories du mélange et théologie mystique', in *Miscellanea André Combes* (=*Divinitas* II 1967) 331-75, repr. in his *Ex Platonicorum Persona*, Amsterdam 1977.
T.F. Torrance, 'The relation of the incarnation to space in Nicene Theology', in Andrew Blane, ed., *The Ecumenical World of Orthodox Civilisation, Russia and Orthodoxy*, vol. 3, *Essays in Honor of Georges Florovsky*, The Hague, Paris 1973.

Stoic problems of mereology and fusion

David Sedley, 'The Stoic criterion of identity', *Phronesis* 27, 1982, 255-75.
Peter Geach, *Reference and Generality*, 3rd ed. only, Ithaca N.Y. 1980, sec. 110.
David Wiggins, 'On being in the same place at the same time', *Philosophical Review* 77, 1968, 90 5.
Harold W. Noonan, 'Wiggins on identity', *Mind* 1976, 559-75.
Judith Jarvis Thomson, 'Parthood and identity across time', *Journal of Philosophy* 80, 1983, 201-20.
Mark Johnston, 'Is there a problem about persistence?', *Proceedings of the Aristotelian Society* supp. vol. 61, 1987, 107-35.

Stoic knowledge of Aristotle

F.H. Sandbach, *Aristotle and the Stoics*, Cambridge Philological Society, supp. vol. 10, 1985.
Jaap Mansfeld, 'Zeno and Aristotle on mixture', *Mnemosyne* 36, 1983, 306-10.

Can some sub-atomic particles interpenetrate?

Alberto Cortes, 'Leibniz's principle of the identity of indiscernibles: a false principle', *Philosophy of Science* 43, 1976, 491-505.
R.L. Barnette, 'Does quantum mechanics disprove the principle of the identity of indiscernibles?' *Philosophy of Science* 45, 1978, 466-70.
Allen Ginsberg, 'Quantum theory and the identity of indiscernibles revisited', *Philosophy of Science* 48, 1981, 487-91.
Paul Teller, 'Quantum physics, the identity of indiscernibles and some unanswered questions', *Philosophy of Science* 50, 1983, 309-19.
Bas van Fraassen, 'Probabilities and the problem of individuation', in S.A.

Luckenbach, ed., *Probabilities, Problems and Paradoxes*, Encino, California 1972.

R.N. Hanson, 'The dematerialisation of matter', in E. McMullin, ed., *The Concept of Matter*, Notre Dame, Indiana 1963, 549-61.

Can any bodies interpenetrate?

David Wiggins, 'On being in the same place at the same time', *Philosophical Review* 77, 1968, 90-5.

David Sanford, 'Locke, Leibniz and Wiggins on being in the same place at the same time', *Philosophical Review* 79, 1970, 75-82.

David Sanford, 'The perception of shape', in S. Shoemaker and C. Ginet, eds, *Knowledge and Mind*, Oxford 1983.

Harold W. Noonan, 'Can one thing become two?', *Philosophical Studies* 33, 1978, 203-27.

Is vacuum impenetrable?

Brad Inwood, 'The origin of Epicurus' concept of void', *Classical Philology* 76, 1981, 273-85.

David Sedley, 'Two conceptions of vacuum', *Phronesis* 27, 1982, 175-93.

Edward Grant, 'The principle of the impenetrability of bodies in the history of concepts of separate space from the middle ages to the seventeenth century', *Isis* 69, 1978, 551-71.

Edward Grant, *Much Ado About Nothing*, Cambridge 1981, 31-8.

Moving holes and point-like bodies

David Sanford, 'Volume and solidity', *Australasian Journal of Philosophy* 45, 1967, 329-40.

David Lewis and Stephanie Lewis, 'Holes', *Australasian Journal of Philopophy* 48, 1970, 206-12, repr. in David Lewis' *Philosophical Papers* 1.

David Sedley, 'Two conception of vacuum', *Phronesis* 27, 1982, 175-93.

Space, place and vacuum

General

Edward Grant, *Much Ado About Nothing, Theories of Space and Vacuum from the Middle Ages to the Scientific Revolution*, Cambridge 1981.

Pierre Duhem, *Le Système du Monde, Histoire des doctrines cosmologiques de Platon à Copernic*, 10 vols, Paris 1913-1959 (see esp. vol. 1, chs 4 and 5 for Antiquity).

Alexandre Koyré, *From the Closed World to the Infinite Universe*, Baltimore 1957.

Max Jammer, *Concepts of Space*, Cambridge, Mass. 1954, 2nd ed. 1969.

The Presocratics

F.M. Cornford, 'The invention of space', in *Essays in Honour of Gilbert Murray*, London 1936.

Plato

For Plato's *Timaeus* and the interpretation of his space as matter, see above.
On the *Timaeus* more generally, see.

A.E. Taylor, *A Commentary on Plato's Timaeus*, Oxford 1928.
F.M. Cornford, *Plato's Cosmology*, London 1937.
D.T. Runia, *Philo of Alexandria and the Timaeus of Plato*, Leiden 1986.

Aristotle

W.D. Ross, *Aristotle's Physics, A revised text with introduction and commentary*, Oxford 1936.
Friedrich Solmsen, *Aristotle's System of the Physical World*, Ithaca N.Y. 1960.

Aristotle's definition of place, and criticism of it

Edward Hussey, *Aristotle's Physics, Books III and IV*, Oxford 1983.
G.E.L. Owen, *Logic, Science and Dialectic: collected papers in Greek philosophy*, London 1986, v. index s.v. 'Location'.
Henry Mendell, 'Topoi on topos: the development of Aristotle's concept of place', *Phronesis 32*, 1987, 206-31
Myles Burnyeat, 'The sceptic in his place and time', in Richard Rorty, J.B. Schneewind, Quentin Skinner, eds, *Philosophy in History*, Cambridge 1984.
Edward Grant, 'The medieval doctrine of place: some fundamental problems and solutions', in A. Maierù and A. Paravicini Bagliani, eds, *Studi sul XIV secolo in memoria di Anneliese Maier*, Rome 1981, 57-79.
David Furley, 'Summary of Philoponus' corollaries on place and void', in Richard Sorabji, ed., *Philoponus and the Rejection of Aristotelian Science*, London and Ithaca, N.Y. 1987.
Philippe Hoffmann, 'Simplicius: Corollarium de loco', in *L'Astronomie dans l'Antiquité Classique, Actes du colloque tenu à l'Université de Toulouse-le-Mirail, 1977*, Paris 1979, 143-63.

Change of place and change of relation

Concetta Luna, 'La relation chez Simplicius', in I. Hadot, ed., *Simplicius – sa vie, son oeuvre, sa survie*, Peripatoi vol. 15, Berlin 1987.

Aristotle on natural places and Theophrastus' criticism

Friedrich Solmsen, *Aristotle's System of the Physical World*, Ithaca N.Y. 1960, ch. 12.
S. Sambursky, *The Physical World of Late Antiquity*, London 1962, ch. 1.
Marlein van Raalte, 'The idea of the cosmos as an organic whole in Theophrastus' Metaphysics', in W. Fortenbaugh and R.W. Sharples, eds, *Theophrastus as a Natural Scientist*, Rutgers Studies in Classical Humanities 3, New Brunswick and London 1987 with contrasting assessments by Glenn Most and John Ellis.

Is place inert or active?

S. Sambursky, *The Concept of Place in Late Neoplatonism*, Jerusalem 1982 (introduction and texts with translation).

Aristotle on organic place

G.E.R. Lloyd, 'Right and left in Greek Philosophy', *Journal of Hellenic Studies* 82, 1962, 56-66.
James G. Lennox, 'Theophrastus on the limits of teleology', in W. Fortenbaugh, ed., *Theophrastus of Eresus, On his Life and Work*, Rutgers University Studies in Classical Humanities 2, New Brunswick and Oxford 1985.

Plato, Aristotle and Epicurus on up and down

Friedrich Solmsen, *Aristotle's System of the Physical World*, Ithaca, N.Y. 1960, ch. 13.
David Konstan, 'Epicurus on "up" and "down" ', *Phronesis* 17, 1972, 269-78.

Epicurus and earlier atomists on space

David Sedley, 'Two conceptions of vacuum', *Phronesis* 27, 1982, 175-93.
Brad Inwood, 'The origin of Epicurus' concept of void', *Classical Philology* 76, 1981, 273-85.
David Furley, 'Aristotle and the atomists on infinity', in I. Düring, ed., *Naturforschung bei Aristoteles und Theophrast*, Heidelberg 1969, 85-96.

Cleomedes and Stoic views of space

Robert B. Todd, 'Cleomedes and the Stoic concept of void', *Apeiron* 16, 1982, 129-36.
R. Goulet, *Cléomède: théorie élémentaire*, Paris 1980 (French translation of Cleomedes, with commentary).
Robert B. Todd, 'A note on Francesco Patrizi's use of Cleomedes', *Annals of Science* 39, 1982, 311-14.

Alexander of Aphrodisias on place

Robert B. Todd, 'Alexander of Aphrodisias and the case for the infinite universe (*Quaestiones* III 12)', *Eranos* 82, 1984, 185-93.
Robert B. Todd, 'Infinite body and infinite void: Epicurean physics and Peripatetic polemic', *Liverpool Classical Monthly* 7.6, 1982, 82-4.

Late Neoplatonists on place

S. Sambursky, *The Physical World of Late Antiquity*, London 1962.
S. Sambursky, *The Concept of Place in Late Neoplatonism*, Jerusalem 1982.
Pierre Duhem, *Le Système du Monde*, 10 vols, Paris 1913-1959, vol. 1, ch. 5.
Philippe Hoffmann, 'Simplicius: Corollariun de loco', in *L'Astronomie dans L'Antiquité Classique, Actes du Colloque tenu à l'Université de Toulouse-le-Mirail, 1977*, Paris 1979, 143-63.
Philippe Hoffmann, 'Les catégories *pou* et *pote* chez Aristote et Simplicius', in Pierre Aubenque, ed., *Concepts et Catégories dans la Pensée Antique*, Paris 1980, 217-45.

Gérard Verbeke, 'Ort und Raum nach Aristoteles und Simplikios, eine philosophische Topologie', in J. Irmscher and R. Müller, eds, *Aristoteles als Wissenschaftstheoretiker*, Berlin 1983, 113-22.

Wolfgang Wieland, 'Zur Raumtheorie des Johannes Philoponus', in *Festschrift J. Klein*, Göttingen 1967, 114-35.

Thirteenth- and fourteenth-century criticism of Aristotle on place

See above.

Pico on space

Charles Schmitt, *Gianfrancesco Pico della Mirandola (1469-1533) and his Critique of Aristotle*, The Hague 1967, ch. 5.

Patrizi on space

Benjamin Brickman, 'Francesco Patrizi, *On Physical Space*', *Journal of the History of Ideas* 4, 1943, 224-45 (a translation of Patrizi).

John Henry, 'Francesco Patrizi da Cherso's concept of space and its later influence', *Annals of Science* 36, 1979, 549-73.

Edward Grant, *Much Ado About Nothing*, Cambridge 1981, ch. 8.

Robert B. Todd, 'A note on Francesco Patrizi's use of Cleomedes', *Annals of Science* 39, 1982, 311-14.

Leibniz and Newton on space

H.G. Alexander, ed., *The Leibniz-Clarke Correspondence*, Manchester 1956 (with Clarke's English translation of Leibniz, originally written 1715-16).

J.E. McGuire, 'Existence, actuality and necessity: Newton on space and time', *Annals of Science* 35, 1978, 463-509.

J.E. McGuire, 'Body and void and Newton's *de Mundi Systemate*: some new sources', *Archive for the History of Exact Sciences* 3, 1966.

Infinite and extracosmic space

R. Mondolfo, *L'infinito nel pensiero dell' antichità classica*, Florence 1956, enlarged from a differently entitled edition of 1934.

Alexandre Koyré, *From the Closed World to the Infinite Universe*, Baltimore 1957.

Edward Grant, *Much Ado About Nothing*, Cambridge 1981, part II.

Edward Grant, 'The condemnation of 1277, God's absolute power and physical thought in the later Middle Ages', *Viator* 10, 1979, 211-44, repr. in his *Studies in Medieval Science and Natural Philosophy*, London 1981.

F.M. Cornford, 'The invention of space', in *Essays in Honour of Gilbert Murray*, London 1936.

David J. Furley, 'The Greek theory of the infinite universe', *Journal of the History of Ideas* 42, 1981, 571-85, repr. in his *Cosmic Problems*, Cambridge 1989.

David J. Furley, 'Aristotle and the atomists on infinity', in I. Düring, ed., *Naturforschung bei Aristoteles und Theophrast*, Heidelberg 1969, 85-96, repr. in his *Cosmic Problems*, Cambridge 1989.

Robert B. Todd, 'Cleomedes and the Stoic concept of void', *Apeiron* 16, 1982, 129-36.
Robert B. Todd, 'Alexander of Aphrodisias and the case for the infinite universe (*Quaestiones* III 12)', *Eranos* 82, 1984, 185-93.
Robert B. Todd, 'Infinite body and infinite void: Epicurean physics and Peripatetic polemic', *Liverpool Classical Monthly* 7.6, 1982, 82-4.
A.H. Armstrong, 'Plotinus' doctrine of the infinite and its significance for Christian thought', *Downside Review* 73, 1955, 47-58.

Closed space

Stephen F. Barker, 'Geometry', in Paul Edwards, ed., *The Encyclopaedia of Philosophy*, New York 1967.

Modern philosophical treatments of space

Bas van Fraassen, *An Introduction to the Philosophy of Time and Space*, New York 1970.
Adolf Grunbaum, *Philosophical Problems of Space and Time*, 2nd ed., Dordrecht 1973.
J.R. Lucas, *A Treatise on Time and Space*, London 1973.
Lawrence Sklar, *Space, Time and Space-Time*, Berkeley, Los Angeles 1974.
R.G. Swinburne, *Space and Time*, 2nd ed., London 1981.

Vacuum

Edward Grant, *Much Ado About Nothing*, Cambridge 1981.
Pierre Duhem, *Le Système du Monde*, 10 vols, Paris 1913-1959, vol 1, chs 4-5.
David Sedley, 'Two conceptions of vacuum', *Phronesis* 27, 1982, 175-93.
Brad Inwood, 'The origin of Epicurus' concept of void', *Classical Philology* 76, 1981, 273-85.
Robert B. Todd, 'Cleomedes and the Stoic concept of void', *Apeiron* 16, 1982, 129-36.
David J. Furley, 'Strato's Theory of the void', in J. Wiesner, ed., *Aristoteles Werk und Wirkung*, Paul Moraux gewidmet, vol. 1, Berlin 1985, 595-609, repr. in his *Cosmic Problems*, Cambridge 1989.
David Sedley, 'Philoponus' conception of space', in Richard Sorabji, ed., *Philoponus and the Rejection of Aristotelian Science*, London and Ithaca N.Y. 1987.
Charles Schmitt, *Gianfransesco Pico della Mirandola (1469-1533) and his Critique of Aristotle*, The Hague 1967, ch. 5.
Charles Schmitt, 'Experimental evidence for and against a void: the sixteenth-century arguments', *Isis* 58, 1967, 352-66.

Dynamics

Motion in a vacuum

David J. Furley, *Two Studies in the Greek Atomists*, Princeton 1967, ch. 8 (on Epicurus' response to Aristotle).
David J. Furley, 'Aristotle and the atomists on motion in a void', in Peter K. Machamer and Robert G. Turnbull, eds, *Motion and Time, Space and*

Matter, Ohio State University 1976, 83-100, repr. in his *Cosmic Problems*, Cambridge 1989.

David J. Furley, *The Greek Coomologioto*, Cambridge, vol. 1, 1987, vol. 2 forthcoming.

Pierre Duhem, *Le Système du Monde*, 10 vols, Paris 1913-1959 (see vol. 1, ch. 6 on Antiquity).

Michael Wolff, *Fallgesetz und Massebegriff*, Berlin 1971 (on Philoponus).

Michael Wolff, 'Philoponus and the rise of preclassical dynamics', in Richard Sorabji, ed., *Philoponus and the Rejection of Aristotelian Science*, London and Ithaca N.Y. 1987.

S. Pines, 'Études sur Awḥad al-Zamân Abu' l-Barakât al-Baghdâdî', *Revue des Études Juives*, n.s. 3, 1938, repr. in his *Collected Works*, vol. 1, Jerusalem and Leiden 1979 (Avicenna is discussed at 52-6, same pagination in both places).

S. Pines, 'La dynamique d'Ibn Bajjâ', in *Mélanges Alexandre Koyré* I, Paris 1964, repr. in his *Collected Works*, vol. 2, Jerusalem and Leiden 1986.

E.A. Moody, 'Galileo and Avempace: the dynamics of the leaning tower experiment', *Journal of the History of Ideas* 12, 1951, 163-93; 375-422, abbreviated in P.P. Wiener and A. Noland, eds, *Roots of Scientific Thought*, New York 1957, 176-206.

Edward Grant, 'Aristotle, Philoponus, Avempace and Galileo's Pisan dynamics', *Centaurus* 11, 1965, 79-85, repr. in his *Studies in Medieval Science and Natural Philosophy*, London 1981.

H.A. Wolfson, *Crescas' Critique of Aristotle*, Cambridge, Mass. 1929.

Anneliese Maier, 'Ergebnisse der spätscholastiken Naturphilosophie', *Scholastik* 35, 1960, 161-87, translated into English in Steven D. Sargent, ed., *On the Threshold of Exact Science*, Philadelphia 1982.

Marshall Claggett, *The Science of Mechanics in the Middle Ages*, Madison, Wisconsin 1959, ch. 7.

Edward Grant, *Much Ado About Nothing*, Cambridge 1981, ch. 3.

James A. Weisheipl, 'Motion in a void: Aquinas and Averroes', in A. Maurer, ed., *St. Thomas Aquinas 1274-1974: commemorative studies*, Toronto 1974, vol. 1, 467-88, repr. in his *Nature and Motion in the Middle Ages*, Washington D.C. 1985.

Charles Schmitt, *Gianfrancesco Pico della Mirandola (1469-1533) and his Critique of Aristotle*, The Hague 1967, ch. 5.

Impetus theory

E. Wohlwill, 'Über die Entdeckung des Beharrungsgesetzes', *Zeitschrift für Volkerpsychologie* 14, 1883, 365-410; 15, 1884, 70-135; 337-87 (identifies the theory).

E. Wohlwill, 'Ein Vorgänger Galileis im 6 Jahrhundert', *Verhandlungen der Gesellschaft deutscher Naturforscher und Ärzte* 77, part 2, 1905, repr. in *Physikalische Zeitschrift* 7, 1906, 23-32 (identifies Philoponus).

Pierre Duhem, *Études sur Léonard de Vinci*, Paris 1906-1913 (coins the name 'Impetus').

Pierre Duhem, *Le Système du Monde*, 10 vols, Paris 1913-1959 (see vol. 1, ch. 6 for Philoponus).

S. Pines, 'Omne quod movetur necesse est ab alio moveri', *Isis* 52, 1961, 21-54 (on Alexander and Galen).

S. Pines, 'Saint Augustin et la théorie de l'impetus', *Archives d'Histoire*

Doctrinale et Littéraire du Moyen Âge 44, 1969, 7-21, repr. in his *Collected Works*, vol. 2, Jerusalem and Leiden 1986.

Michael Wolff, *Geschichte der Impetustheorie*, Frankfurt 1978.

Michael Wolff, 'Philoponus and the rise of preclassical dynamics', in Richard Sorabji, ed., *Philoponus and the Rejection of Aristotelian Science*, London and Ithaca N.Y. 1987.

S. Sambursky, *The Physical World of Late Antiquity*, London 1962, ch. 3.

J. Christensen de Groot, 'Philoponus on *De Anima* 2.5, *Physics* 3.3, and the propagation of light', *Phronesis* 28, 177-96.

G.A. Lucchetta, *Una fisica sensa matematica: Democrito, Aristotele, Filopono*, Trento 1978.

G.A. Lucchetta, 'Ipotesi per l'applicazione dell' "impetus" ai cieli in Giovanni Filopono', *Atti e Memorie dell' Accademia patavina di scienze, lettere ed arti* 87, 1974-5, 339-52.

G.A. Lucchetta, 'Dinamica dell' *impetus* e Aristotelismo Veneto', in *Aristotelismo Veneto e Scienza Moderna, Saggi e Testi* 18, Padua 1983.

S. Pines, 'Un précurseur Bagdadien de la théorie de l'impetus', *Isis* 44, 1953, 246-51, repr. in his *Collected Works*, vol. 2, Jerusalem and Leiden 1986.

S. Pines, 'Études sur Awḥad al-Zamân Abu' l-Barakât al-Baghdâdî', *Revue des Études Juives*, n.s. 3 and 4, 1938, repr. in his *Collected Works*, vol. 1, Jerusalem and Leiden 1979.

S. Pines, 'Les précurseurs musulmans de la théorie de l'impetus', *Archeion* 21, 1938, 298-306, repr. in his *Collected Works*, vol. 2, Jerusalem and Leiden 1986.

Fritz Zimmermann, 'Philoponus' impetus theory in the Arabic tradition', in Richard Sorabji, ed., *Philoponus and the Rejection of Aristotelian Science*, London and Ithaca N.Y. 1987.

Anneliese Maier, *Zwei Grundprobleme der scholastischen Naturphilosophie*[2], Rome 1951.

Marshall Claggett, *The Science of Mechanics in the Middle Ages*, Madison, Wisconsin 1959, ch. 8.

William A. Wallace, 'Galileo and scholastic theories of impetus', in A. Maierù and A. Paravicini Bagliani, eds, *Studi sul XIV secolo in memoria di Anneliese Maier*, Rome 1981.

Inertia

S. Pines, 'Omne quod movetur necesse est ab alio moveri', *Isis* 52, 1961, 21-54, repr. in his *Collected Works*, vol. 2, Jerusalem and Leiden 1986.

Richard S. Westfall, 'Circular motion in seventeenth-century mechanics', *Isis* 63, 1972, 184-9.

J.E. McGuire, comment on I.B. Cohen, in R. Palter, ed., *The Annus Mirabilis of Sir Isaac Newton 1666-1966*, Cambridge, Mass. 1970, 186-91.

Circular versus rectilinear motion

David J. Furley, 'The Greek theory of the infinite universe', *Journal of the History of Ideas* 42, 1981, 571-85, repr. in his *Cosmic Problems*, Cambridge 1989.

Richard S. Westfall, 'Circular motion in seventeenth-century mechanics', *Isis* 63, 1972, 184-9.

Natural motion in Aristotle

Friedrich Solmsen, *Aristotle's System of the Physical World*, Ithaca N.Y. 1960.
Sarah Waterlow, *Nature, Change and Agency in Aristotle's Physics*, Oxford 1982.
David J. Furley, 'Self-movers', in G.E.R. Lloyd and G.E.L. Owen, eds, *Aristotle on Mind and the Senses, Proceedings of the 7th Symposium Aristotelicum*, Cambridge 1978, repr. in his *Cosmic Problems*, Cambridge 1989.

Natural motion in Alexander and Philoponus

S. Pines, 'Omne quod movetur necesse est ab alio moveri', *Isis* 52, 1961, 21-54 (on Alexander), repr. in his *Collected Works*, vol. 2, Jerusalem and Leiden 1986.
J E. McGuire, 'Philoponus on *Physics* II 1: *phusis, dunamis* and the motion of simple bodies', *Ancient Philosophy* 5, 1985, 241-67.
Michael Wolff, *Fallgesetz und Massebegriff*, Berlin 1971, ch. 3.
Michael Wolff, 'Philoponus and the rise of preclassical dynamics', in Richard Sorabji, ed., *Philoponus and the Rejection of Aristotelian Science*, London and Ithaca N.Y. 1987.

Stoic dynamics

S. Sambursky, *Physics of the Stoics*, London 1959.
David Hahm, *The Origins of Stoic Cosmology*, Ohio State University 1977.
Michael Wolff, 'Hipparchus and the Stoic theory of motion', in J. Barnes and M. Mignucci, eds, *The Bounds of Being*, Naples, forthcoming.

Aristotle on celestial motion

H. von Arnim, *Die Entstehung der Gotteslehre des Aristoteles*, Vienna 1931.
W.K.C. Guthrie, 'The development of Aristotle's Theology I', *Classical Quarterly* 27, 1933, 162-71.
W.K.C. Guthrie, introduction to Loeb edition of Aristotle *On the Heavens*, London and Cambridge, Mass. 1960.
H. Cherniss, *Aristotle's Criticism of Plato and the Academy*, Baltimore 1944, Appendix 19, 581-602.
Friedrich Solmsen, *Aristotle's System of the Physical World*, Ithaca N.Y. 1960.
Sarah Waterlow, *Nature, Change and Agency*, Oxford 1982, ch. 5.

Attacks on Aristotle on celestial motion: Xenarchus and Philoponus

S. Sambursky, *The Physical World of Late Antiquity*, London 1962, ch. 5.
C. Wildberg, *John Philoponus' Criticism of Aristotle's theory of Ether*, Ph.D. diss., Cambridge 1984.
C. Wildberg, *Philoponus, Against Aristotle on the Eternity of the World*, fragments assembled and translated into English, vol. 1 of Richard Sorabji, ed., *The Ancient Commentators on Aristotle*, London and Ithaca N.Y. 1987.
P. Moraux, *Der Aristotelismus*, vol. 1, Berlin 1973 (on Xenarchus).

Supernatural motion and creation

E. Evrard, 'Les convictions réligieuses de Jean Philopon et la date de son commentaire aux Météorologiques', *Bulletin de l'academie royale de Belgique, classe des lettres* 5, 1953, 299-357.

Lindsay Judson, 'God or nature? Philoponus on generability and perishability', in Richard Sorabji, ed., *Philoponus and the Rejection of Aristotelian Science*, London and Ithaca N.Y. 1987.

Robert W. Sharples, 'Alexander of Aphrodisias: problems about possibility 2', *Bulletin of the Institute of Classical Studies* 30, 1983, 99-103.

Robert W. Sharples, 'The unmoved mover and the motion of the heavens in Alexander of Aphrodisias', *Apeiron* 17, 1983, 62-6.

On the concept of the supernatural see:

Henri de Lubac, *Surnaturel*, Paris 1946.

Klaus Kremer, *Der Metaphysikbegriff in den Aristoteles-Kommentaren der Ammonius-Schule*, Münster 1960.

Infinite power argument for a Creator

Carlos Steel, 'Proclus et Aristote sur la causalité efficiente de l'Intellect divin', in J. Pépin and H.D. Saffrey, eds, *Proclus – lecteur et interprète des Anciens*, Actes du Colloque Proclus C.N.R.S., Paris 1987.

Carlos Steel, '"Omnis corporis potentia est finita." L'interprétation d'un principe aristotélicien: de Proclus à S. Thomas', in Jan P. Beckmann, Ludger Honnefelder, Gangolf Schrimpf, Georg Wieland, eds, *Philosophie im Mittelalter*, Hamburg 1987.

Lindsay Judson, 'God or nature? Philoponus on generability and perishability', in Richard Sorabji, ed., *Philoponus and the Rejection of Aristotelian Science*, London and Ithaca N.Y. 1987.

S. Pines, 'An Arabic summary of a lost work of John Philoponus', *Israel Oriental Studies* 2, 1972, 320-52, repr. in his *Collected Works*, vol. 2, Jerusalem and Leiden 1986.

H.A. Wolfson, 'The Kalam arguments for creation in Saadia, Averroes, Maimonides and St. Thomas', in *Saadia Anniversary Volume, American Academy of Jewish Research, Texts and Studies* 2, New York 1943, 197-245.

H.A. Wolfson, *The Philosophy of the Kalam*, Cambridge, Mass. 1976, 373-82.

H.A. Davidson, 'John Philoponus as a source of medieval Islamic and Jewish proofs of creation', *Journal of the American Oriental Society* 89, 1969, 357-91.

H.A. Davidson, 'The principle that a finite body can contain only finite power', in S. Stein and R. Loewe, eds, *Studies in Jewish Religious and Intellectual History Presented to Alexander Altmann*, Alabama 1979, 75-92.

Michael Wolff, *Fallgesetz und Massebegriff*, Berlin 1971, 94-9.

L.G. Westerink, *The Greek Commentaries on Plato's Phaedo*, vol. 1, Olympiodorus, Amsterdam 1973, note to 13.2, 76, 37-40 on use of verb *epinaein* for the influx of infinite power.

S. Pines, 'A tenth-century philosophical correspondence', *Proceedings of the American Academy for Jewish Research* 24, 1955, 114-15.

Joel L. Kraemer, 'A lost passage from Philoponus' *contra Aristotelem* in Arabic translation', *Journal of the American Oriental Society* 85, 1965, 318-27.

C. Genequand, *Ibn Rushd's Metaphysics*, Leiden 1984.

H.A. Wolfson, *Crescas' Critique of Aristotle*, Cambridge, Mass. 1929.

Seymour Feldman, 'The end of the universe in medieval Jewish philosophy', *Association for Jewish Studies Review*, 1986.

C. Touati, *La pensée philosophique et théologique de Gersonide*, Paris 1973, 308-15.

Aristotle's God as Creator

J. Pépin, *Théologie cosmique et théologie chrétienne (Ambroise Exam. 1.1.1-4)*, Paris 1964, 475-8 (on St. Ambrose).

Koenraad Verrycken, 'Ammonius' Theology', in Richard Sorabji, ed., *The Transformation of Aristotle*, forthcoming.

Fritz Zimmermann, 'The origins of the so-called *Theology of Aristotle*', in Jill Kraye, W.F. Ryan, C.B. Schmitt, eds, *Pseudo-Aristotle in the Middle Ages*, London 1986, 110-240.

R. Jolivet, 'Aristote et Saint Thomas ou la notion de création', in his *Essai sur les rapports entre la pensée grecque et la pensée chrétienne*, Paris 1955.

Luca Bianchi, *L'errore di Aristotele: la polemica contro l'eternità del mondo nel XIII secolo*, Florence 1984.

I. Hadot, *Le problème du néoplatonisme Alexandrin: Hieroclès et Simplicius*, Paris 1978, chs 4 and 5 (on Hierocles).

P. Moraux, *Der Aristotelismus*, Berlin 1973, 277-305 (on Arius Didymus).

Platonic Ideas as creative causes in God's mind

Julia Annas, 'Aristotle on inefficient causes', *Philosophical Quarterly* 32, 1982, 311-26.

John Dillon, *The Middle Platonists*, London 1977.

The harmony of Plato and Aristotle

Pierre Hadot, 'L'Harmonie des philosophies de Plotin et d'Aristote selon Porphyre dans le commentaire de Dexippe sur les *Catégories*', in *Plotino e il Neoplatonismo in Oriente e in Occidente*, Accademia Nazionale dei Lincei, Rome 1974.

Richard Sorabji, ed., *The Transformation of Aristotle*, forthcoming, will include this article in translation along with other relevant articles.

Circular time

A.A. Long, 'The Stoics on world-conflagration and everlasting recurrence', in R. Epp, ed., *Recovering the Stoics, Southern Journal of Philosophy*, vol. 23 supplement, 1985, 13-37.

Adolf Grunbaum, *Philosophical Problems of Space and Time*[2], Dordrecht 1973.

Bas van Fraassen, *An Introduction to the Philosophy of Time and Space*, New York 1970.

Lawrence Sklar, *Space, Time and Space-Time*, Berkeley, Los Angeles 1974, 303-17.

W.H. Newton-Smith, *The Structure of Time*, London 1980, 65-8.

W.V. Quine, 'Comments on Newton-Smith', *Analysis* 39, 1979, 66-7.

Susan Weir, *An Inquiry into the Possibility and Implications of a Closed Temporal Topology*, Ph.D. diss., Bristol 1985.

Time travel

David Lewis, 'The paradoxes of time travel', *American Philosophical Quarterly* 13, 1976, 145-52.

Murray Macbeath, 'Who was Dr. Who's father?', *Synthese* 51, 1982, 397-430.

Paul Horwich, *Asymmetries in Time*, Boston Mass. 1987, ch. 6, 'Time travel'.

Kurt Gödel, 'A remark about the relationship between Relativity Theory and Idealistic Philosophy', in P.A. Schilpp, ed., *Albert Einstein, Philosopher-Scientist*, New York 1951, 555-62.

David Malament, '"Time Travel" in the Gödel universe', *Proceedings of the Biennial Meeting of the Philosophy of Science Association* 1984, vol. 2, 1985.

Stephen Hawking, *A Brief History of Time*, 1988.

Is backwards causation possible?

Yes:

Michael Dummett, 'Bringing about the past', *Philosophical Review* 73, 1964, 338-59.

Michael Dummett, 'Can an effect precede its cause?', *Proceedings of the Aristotelian Society*, supp. vol. 28, 1954, 27-44.

Replies:

A. Flew, 'Can an effect precede its cause?', *Proceedings of the Aristotelian Society*, supp. vol. 28, 1954, 27-44.

Max Black, 'Why cannot an effect precede its cause?', *Analysis* 16, 1956, 49-58.

A. Flew, 'Effects before their causes? Addenda and corrigenda', *Analysis* 16, 1956, 104-10.

A. Flew, 'Causal disorder again', *Analysis* 17, 1957, 81-6.

Michael Scriven, 'Randomness and the causal order', *Analysis* 17, 1957, 5-9.

D.F. Pears, 'The priority of causes', *Analysis* 17, 1957, 54-63.

William Dray, 'Taylor and Chisholm on making things to have happened', *Analysis* 20, 1959-60, 79-82:

Samuel Gorovitz, 'Leaving the past alone', *Philosophical Review* 73, 1964, 360-71.

R.G. Swinburne, 'Affecting the past', *Philosophical Quarterly* 16, 1966, 341-7.

R.G. Swinburne, *Space and Time*, ch. 8, London 1968, 2nd ed. 1981.

Richard M. Gale, 'Why a cause cannot be later than its effect', *Review of Metaphysics* 19, 1966, 209-34.

Richard M. Gale, *The Language of Time*, London 1968, ch. 7.

D.H. Mellor, *Real Time*, Cambridge 1981, ch. 10.

Appraisal, with reference to further literature:

J.L. Mackie, *The Cement of the Universe*, Oxford 1974, ch. 7.

The identity of indiscernibles and 'thisness'

R.M. Adams, 'Primitive thisness and primitive identity', *Journal of Philosophy* 76, 1979, 5-26.

The direction of time

John Earman, 'An attempt to add a little direction to "the problem of the direction of time"', *Philosophy of Science* 41, 1974, 15-17.

Problems of immortality and personhood

Bernard Williams, 'The Makropoulos case: reflections on the tedium of immortality', in his *Problems of the Self*, Cambridge 1973.
Derek Parfit, *Reasons and Persons*, Oxford 1984.
Mark Johnston, 'Human beings', *Journal of Philosophy* 84, 1987, 59-83, offers a view on much of the recent literature.

General index

Abravanel

  preserving motion of heavens preserves them in existence, 251, 266

  associates infinite power argument for God as sustainer and idea of God overriding nature with Avicenna, 260-1

  argues from finite power to finite duration of world, 272

abstraction

  objects of mathematics reached by abstraction, 7-8, 15-17, 77

  prime matter reached by abstraction, 5-8

  mechanics abstracts less than mathematics, 17

Abu Bishr Matta

  Nestorian Christian of Baghdad, translator of Aristotle, Alexander, Themistius, 261

  Aristotle's God only a mover, existence of heavens necessary in itself independently of God, 261-3, 266

Adams, Robert M., 180, 304

Alamanno, Cosmo, 147

Albertus Magnus: harmonisation of Aristotle with Christianity, 274

Albinus (?), author of *Didaskalikos*: individual body as a bundle of qualities and prime matter, 33, 49

Alcmaeon

  friend of Pythagoreans, 182

  immortality and circular time, 182-3

Alexander of Aphrodisias

  greatest exponent of Aristotelianism, 273

  his *Quaestiones* reflect seminars in which Aristotelian doctrine presupposed, 136

  misinterprets Stoics, 86, 88, 89, 93-8

  attacks Stoics on mixture, 88, 89, 93, 94-5, 95-8, 98-104

  attacks Stoic theology, 93-8

  attacks Stoic dynamics, 154

  soul vehicles would have to interpenetrate with our flesh, 107

  takes Aristotle to deny that geometrical solids can be in the same place as physical bodies, 77

  theory of mixture: original powers of four elements do not persist in mixture, 69

  answers Archytas' stretching argument for infinite space, 126-8

  refutes 'limited by' argument for infinite universe, 136-8

  argues for finite universe through definitions of 'whole' and 'limit', 139-40

  Plato's 'space' a metaphor, 32, 36

  definite and indefinite extension, 10

  rotation is not a change of place, 193, 194

  genuine change in respect of relatives?, 198

  the world is imperishable by its nature, not thanks to God, 255, 259, 264

  if the world had a beginning, as Plato thought, it would be perishable by nature and God could not override this, 258

  God is the efficient, as well as final, cause of motion, not a cause in any way of the world's existence, 264

  the fifth element is a substratum, but not matter, 42

  the heavens can't have the same prime matter as the four elements, nor different prime matter, 14-15

  objects of mathematics are mind-dependent, 16

  projectiles propelled by force implanted in air, 144, 282

  elemental motion due to inner nature (no mention of obstacle-remover), 223

  'generator' merely shows inner nature is responsible, 224

  A. does not anticipate impetus theory, 235, 236-7

ps.-Alexander in Averroes

  credited with Philoponus' infinite power arguments, 260-1, 264-6

  Thomas Aquinas takes at face value, 246, 269, 270

ps.-Alexander, comm. *in Metaph.*

  prime matter outside the categories, 14

  takes Aristotle to deny geometrical solids can be in the same place as physical bodies, 77

ps.-Alexander in the *Theology of Aristotle*: creation of world with a beginning, 274

Allaire, Edwin B., 9, 57

Index locorum

To *Necessity, Cause and Blame* (NCB), *Time, Creation and the Continuum* (TCC) and *Matter, Space and Motion* (MSM). Compiled by John Ellis, Harry Ide and Eric Lewis.

Commentaries on Aristotle are cited, unless otherwise stated, from the Prussian Academy edition, *Commentaria in Aristotelem Graeca*. Where the same work has been referred to in different ways in the three books, NCB, TCC and MSM, the index varies correspondingly in its form of reference, remaining close to that actually given on the page cited. Abbreviations:

DG = H.Diels, *Doxographici Graeci*.
DK = H.Diels-W.Kranz, *Die Fragmente der Vorsokratiker*.
PG = Migne, *Patrologia Graeca*.
SVF = von Arnim, *Stoicorum Veterum Fragmenta*.